Electrification of Smart Cities

Electrification of Smart Cities

Editors

Chun Sing Lai
Kim-Fung Tsang
Yinhai Wang

MDPI • Basel • Beijing • Wuhan • Barcelona • Belgrade • Manchester • Tokyo • Cluj • Tianjin

Editors

Chun Sing Lai
Department of Electronic and
Electrical Engineering
Brunel University London
London
United Kingdom

Kim-Fung Tsang
Department of Electrical
Engineering
City University of Hong Kong
Hong Kong

Yinhai Wang
Faculty of Civil and
Environmental Engineering
University of Washington
Washington DC
United States

Editorial Office
MDPI
St. Alban-Anlage 66
4052 Basel, Switzerland

This is a reprint of articles from the Special Issue published online in the open access journal *Applied Sciences* (ISSN 2076-3417) (available at: www.mdpi.com/journal/applsci/special_issues/electrification_smart_cities).

For citation purposes, cite each article independently as indicated on the article page online and as indicated below:

LastName, A.A.; LastName, B.B.; LastName, C.C. Article Title. *Journal Name* **Year**, *Volume Number*, Page Range.

ISBN 978-3-0365-7305-2 (Hbk)
ISBN 978-3-0365-7304-5 (PDF)

© 2023 by the authors. Articles in this book are Open Access and distributed under the Creative Commons Attribution (CC BY) license, which allows users to download, copy and build upon published articles, as long as the author and publisher are properly credited, which ensures maximum dissemination and a wider impact of our publications.

The book as a whole is distributed by MDPI under the terms and conditions of the Creative Commons license CC BY-NC-ND.

Contents

About the Editors vii

Chun Sing Lai, Kim-Fung Tsang and Yinhai Wang
Electrification of Smart Cities
Reprinted from: *Appl. Sci.* **2023**, *13*, 4499, doi:10.3390/app13074499 1

Mohammed AlShamsi, Mostafa Al-Emran and Khaled Shaalan
A Systematic Review on Blockchain Adoption
Reprinted from: *Appl. Sci.* **2022**, *12*, 4245, doi:10.3390/app12094245 7

Yujie Yuan, Xiushan Jiang and Chun Sing Lai
A Perfect Decomposition Model for Analyzing Transportation Energy Consumption in China
Reprinted from: *Appl. Sci.* **2023**, *13*, 4179, doi:10.3390/app13074179 25

Maarit Vehviläinen, Rita Lavikka, Seppo Rantala, Marko Paakkinen, Janne Laurila and Terttu Vainio
Setting Up and Operating Electric City Buses in Harsh Winter Conditions
Reprinted from: *Appl. Sci.* **2022**, *12*, 2762, doi:10.3390/app12062762 39

Mohammed Mahedi Hasan, Nikos Avramis, Mikaela Ranta, Mohamed El Baghdadi and Omar Hegazy
Parameter Optimization and Tuning Methodology for a Scalable E-Bus Fleet Simulation Framework: Verification Using Real-World Data from Case Studies
Reprinted from: *Appl. Sci.* **2023**, *13*, 940, doi:10.3390/app13020940 59

Guodong Guo and Yanfeng Gong
Research on a Day-Ahead Grouping Coordinated Preheating Method for Large-Scale Electrified Heat Systems Based on a Demand Response Model
Reprinted from: *Appl. Sci.* **2022**, *12*, 10758, doi:10.3390/app122110758 87

Nikos Andriopoulos, Konstantinos Plakas, Christos Mountzouris, John Gialelis, Alexios Birbas and Stylianos Karatzas et al.
Local Energy Market-Consumer Digital Twin Coordination for Optimal Energy Price Discovery under Thermal Comfort Constraints
Reprinted from: *Appl. Sci.* **2023**, *13*, 1798, doi:10.3390/app13031798 105

Dennis Agnew, Nader Aljohani, Reynold Mathieu, Sharon Boamah, Keerthiraj Nagaraj and Janise McNair et al.
Implementation Aspects of Smart Grids Cyber-Security Cross-Layered Framework for Critical Infrastructure Operation
Reprinted from: *Appl. Sci.* **2022**, *12*, 6868, doi:10.3390/app12146868 127

Yiqiang Duan, Haoliang Yuan, Chun Sing Lai and Loi Lei Lai
Fusing Local and Global Information for One-Step Multi-View Subspace Clustering
Reprinted from: *Appl. Sci.* **2022**, *12*, 5094, doi:10.3390/app12105094 147

Yang Wei, Kim Fung Tsang, Chung Kit Wu, Hao Wang and Yucheng Liu
A Multi-Leak Identification Scheme Using Multi-Classification for Water Distribution Infrastructure
Reprinted from: *Appl. Sci.* **2022**, *12*, 2128, doi:10.3390/app12042128 167

Paula Bendiek, Ahmad Taha, Qammer H. Abbasi and Basel Barakat
Solar Irradiance Forecasting Using a Data-Driven Algorithm and Contextual Optimisation
Reprinted from: *Appl. Sci.* **2021**, *12*, 134, doi:10.3390/app12010134 181

Shengchao Jian, Xiangang Peng, Haoliang Yuan, Chun Sing Lai and Loi Lei Lai
Transmission Line Fault-Cause Identification Based on Hierarchical Multiview Feature Selection
Reprinted from: *Appl. Sci.* **2021**, *11*, 7804, doi:10.3390/app11177804 . . . . **201**

Hirotaka Takano, Naohiro Yoshida, Hiroshi Asano, Aya Hagishima and Nguyen Duc Tuyen
Calculation Method for Electricity Price and Rebate Level in Demand Response Programs
Reprinted from: *Appl. Sci.* **2021**, *11*, 6871, doi:10.3390/app11156871 . . . . **217**

Zhen Xu, Ping Yang, Zhuoli Zhao, Chun Sing Lai, Loi Lei Lai and Xiaodong Wang
Fault Diagnosis Approach of Main Drive Chain in Wind Turbine Based on Data Fusion
Reprinted from: *Appl. Sci.* **2021**, *11*, 5804, doi:10.3390/app11135804 . . . . **233**

Chun Sing Lai, Mengxuan Yan, Xuecong Li, Loi Lei Lai and Yang Xu
Coordinated Operation of Electricity and Natural Gas Networks with Consideration of Congestion and Demand Response
Reprinted from: *Appl. Sci.* **2021**, *11*, 4987, doi:10.3390/app11114987 . . . . **251**

About the Editors

Chun Sing Lai

Dr. Chun Sing Lai received his B.Eng. (First Class Hons.) in electrical and electronic engineering from Brunel University London, London, UK, in 2013, and his D.Phil. degree in engineering science from the University of Oxford, Oxford, UK, in 2019. He is currently a Lecturer with the Department of Electronic and Electrical Engineering and Course Director of MSc Electric Vehicle Systems at Brunel University London. His current research interests are in power system optimization and electric vehicle systems. Dr. Lai was a Technical Program Co-Chair for the 2022 IEEE International Smart Cities Conference. He is the Vice-Chair of the IEEE Smart Cities Publications Committee. He is an Associate Editor for *IEEE Transactions on Systems, Man, and Cybernetics*: Systems, and IET Energy Conversion and Economics. He is the Working Group Chair for the IEEE P2814 and P3166 Standards; an Associate Vice President of Systems Science and Engineering of IEEE Systems, Man, and Cybernetics Society (IEEE/SMCS) and Co-Chair of the IEEE SMC Intelligent Power and Energy Systems Technical Committee. He is a also recipient of the 2022 Meritorious Service Award from the IEEE SMC Society for "meritorious and significant service to IEEE SMC Society technical activities and standards development". Additionally, he is an IET Member, IEEE Senior Member, Chartered Engineer, and Fellow of the Higher Education Academy.

Kim-Fung Tsang

Dr. Kim-Fung Tsang received an Associateship degree in electrical engineering from The Hong Kong Polytechnic University, Hong Kong, in 1983, and M.Eng. (by research) and Ph.D. degrees in electrical engineering from the University of Wales College of Cardiff (formerly known as the University of Wales Institute of Science and Technology), Cardiff, U.K., in 1987 and 1995, respectively.

In 1988, he joined the City University of Hong Kong, where he is currently an Associate Professor in the Department of Electronic Engineering. He has authored or coauthored about 200 technical papers and 4 books/chapters.

Dr. Tsang is a Fellow of The Hong Kong Institution of Engineers, a Chartered Engineer and Member of IET, an Associate Editor and Guest Editor of the *IEEE Transactions on Industrial Informatics*, an Associate Editor for the *IEEE Industrial Electronics Magazine*, an Associate Editor for the *IEEE ITeN*, and an Editor of the *KSII Transactions on Internet and Information Systems*. He is also an IEEE Senior Member. Since 2022, Dr. Tsang has been the Editor-in-Chief of the *IEEE Transactions on Consumer Electronics*.

Yinhai Wang

Prof. Yinhai Wang received his Ph.D. degree in transportation engineering from the University of Tokyo in 1998; his master's degree in computer science and engineering from the University of Washington (UW); and his master's degree in construction management and bachelor's degree in civil engineering from Tsinghua University, China. He is currently a Professor in the Civil and Environmental Engineering (CEE) Department and the Electrical and Computer Engineering (ECE) Department, UW, where he is also the founder of the Smart Transportation Applications and Research Laboratory (STAR Laboratory). He also serves as the Director for Pacific Northwest Transportation Consortium (PacTrans), USDOT University Transportation Center for Federal Region 10. He has conducted extensive research in traffic sensing, transportation big data management and analytics, large-scale transportation system analysis, transportation data science, traffic operations, and decision support. He is also an IEEE Senior Member.

Editorial

Electrification of Smart Cities

Chun Sing Lai [1,*], Kim-Fung Tsang [2] and Yinhai Wang [3]

[1] Brunel Interdisciplinary Power Systems Research Centre, Department of Electronic and Electrical Engineering, Brunel University London, London UB8 3PH, UK
[2] Department of Electrical Engineering, City University of Hong Kong, Hong Kong 999077, China; ee330015@cityu.edu.hk
[3] Department of Civil and Environmental Engineering, University of Washington, More Hall, Seattle, WA 98195, USA
* Correspondence: chunsing.lai@brunel.ac.uk

1. Introduction

Electrification plays a critical role in decarbonizing energy consumption for various sectors, including transportation, heating, and cooling. Several essential infrastructures are incorporated in smart cities, including smart grids and transportation networks. These infrastructures are complementary solutions to the development of novel services, offering enhanced energy efficiency and energy security.

The purpose of this Special Issue is to collect high-quality papers that address issues related to cutting-edge technologies employed by smart cities undergoing electrification. Some of the topics of interest for this Special Issue include:

- The electrification of building environments and transportation systems;
- The role of smart grids in smart cities and their impacts;
- The influence of ICT and IoT infrastructures incorporating big data on smart cities' electrification;
- Market, services, and business models for smart cities' electrification;
- Standards for smart cities' electrification and their implementation;
- The integration of advanced smart grid technology in smart cities, including in terms of energy storage, demand-side management, and distributed energy resources.

2. Short Summary of the Papers

The variation of energy consumption in transportation and the main influencing factors of decomposition contribute to reducing transportation energy consumption and realizing the sustainable development of the transportation industry. Yuan, Jiang, and Lai [1] proposed an improved decomposition model according to the factors governing the direction of change based on existing index decomposition methods. The influencing factors of transportation energy consumption are quantitatively decomposed according to a transportation energy consumption decomposition model. The contributions of transportation turnover, transportation structure, and transportation energy consumption intensity changes to the variation of transportation energy consumption are quantitatively calculated. The results demonstrate that there is great energy conservation potential in the adjustment of transportation structures and that transportation energy intensity is the main factor of energy conservation.

Hasan et al. [2] presented the optimization and tuning of a simulation framework to improve its simulation accuracy while evaluating the energy utilization of electric buses under various mission scenarios. The simulation framework was developed using a low-fidelity (Lo-Fi) model of a forward-facing electric bus's (e-bus) powertrain to achieve the fast simulation speeds necessary for real-time fleet simulations. The measurement data required verification that the proper tuning of the simulation framework was provided by the bus's original equipment manufacturers (OEMs), and these data were obtained from various demonstrations of 12 m and 18 m buses in the cities of Barcelona, Gothenburg, and Osnabruck.

Recently, the increasing winter load peak has applied great pressure on power grids. The demand response on the load side helps alleviate the expansion of the power grid and promote the consumption of renewable energy. However, the response of large-scale electric heat loads to the same electricity price curve leads to new load peaks and regulation failure. Guo and Gong [3] proposed a grouping-coordinated preheating framework based on a demand response model that realizes the integration of information between the central controller and each regulation group. A thermal parameter model of a room and a performance map of an inverter air conditioner/heat pump are integrated into the demand response model. In this framework, a coordination mechanism is adopted to avoid regulation failure, an edge computing structure is applied to consider the users' preferences and plans, and a grouped and parallel computing structure is proposed to improve computing efficiency.

Communication networks in power systems constitute a major component of the smart grid paradigm. These networks enable the automation of power grid operation and self-recovery in negative contingencies. However, this dependency on communication networks attracts cyber threats. An adversary can launch an attack on a communication network, which, in turn, can influence a power grid's operation. Such attacks may constitute the injection of false data into system measurements, the flooding of communication channels with unnecessary data, or the interception of messages. The use of machine learning to process data gathered from communication networks and the power grid is a promising solution for detecting cyber threats. Agnew et al. [4] presented a co-simulation of cyber-security for a cross-layer strategy. The advantage of such a framework is the augmentation of valuable data, which enhances the detection and identification of anomalies in the operation of a power grid. The framework is implemented on the IEEE 118-bus system. The system is constructed in Mininet to simulate a communication network and obtain data for analysis. A distributed three-controller software-defined networking (SDN) framework is proposed that utilizes an Open Network Operating System (ONOS) cluster. According to the findings of our recommended program, it outperforms a single SDN controller framework by a factor of more than ten times the throughput.

Multi-view subspace clustering has drawn significant attention in the pattern recognition and machine learning research communities. However, most of the existing multi-view subspace clustering methods are still limited in two aspects. (1) The subspace representation yielded by the self-expression reconstruction model ignores the local structural information of the data. (2) The development of subspace representation and clustering are used as two individual procedures, thereby failing to account for their interactions. To address these problems, Duan et al. [5] proposed a novel multi-view subspace-clustering method fusing local and global information for one-step multi-view clustering.

The city of Tampere in Finland aims to be carbon-neutral by 2030 and seeks to determine how the electrification of public transport would help achieve this climate goal. Thus far, research has covered topics related to electric buses, ranging from battery technologies to lifecycle assessment and cost analysis. However, less is known about electric city buses' performance in cold climatic zones. Vehviläinen et al. [6] collected and analyzed weather and electric-city-bus-related data to ascertain the effects of temperature and weather conditions on the electric buses' efficiency. Data were collected from four battery-electric buses and one hybrid bus as a reference. The buses were fast-charged at a market and slow-charged at a depot. The test route ran downtown.

Water distribution infrastructure (WDI) has been well established and significantly improves living quality. Nonetheless, aging WDI poses a challenging worldwide problem entailing the wasting of natural resources, leading to direct and indirect economic losses. The total losses due to leaks are valued at USD 7 billion per year. However, Wei et al. [7] developed a multi-classification multi-leak identification (MC-MLI) scheme to combat this problem. In this MC-MLI scheme, a novel adaptive kernel (AK) program is developed to adapt to different WDI scenarios. The AK program improves overall identification capacity by customizing a weighting vector and transforming into the extracted feature vector. Afterwards, a multi-classification (MC) scheme is designed to facilitate efficient adaptation to potentially hostile inhomogeneous WDI scenarios. The MC scheme comprises multiple classifiers for customizing the network to different pipelines.

Solar forecasting plays a crucial role in the renewable energy transition. Major challenges related to load balancing and grid stability emerge when a high percentage of energy is provided by renewables. These can be tackled by new energy management strategies guided by power forecasts. Bendiek et al. [8] presented a data-driven and contextual optimization-forecasting (DCF) algorithm for solar irradiance that was comprehensively validated using short- and long-term predictions in three US cities: Denver, Boston, and Seattle. Moreover, step-by-step implementation guidelines with which to follow and reproduce the results were proposed.

Fault-cause identification plays a significant role in transmission line maintenance and fault disposal. With the increasing types of monitoring data, i.e., micrometeorology and geographic information, multiview learning can be used to realize the fusion of information for better fault-cause identification. To reduce the amount of redundant information in different types of monitoring data, Jian et al. [9] proposed a hierarchical multiview feature selection (HMVFS) method to address the challenge of combining waveform and contextual fault features. To enhance the discriminant ability of the model, an ε-dragging technique is introduced to enlarge the boundary between different classes. To effectively select the useful feature subset, two regularization terms, namely, l2,1-norm and Frobenius norm penalties, are adopted to conduct hierarchical feature selection for multiview data.

Demand response programs (DRs) can be implemented with fewer investment costs than those incurred in power plants or facilities and enable us to control power demand. Therefore, they are widely regarded as an efficient option for power supply–demand-balancing operations. On the other hand, DRs bring new difficulties regarding the evaluation of consumers' cooperation and the setting of electricity prices or rebate levels while reflecting their results. Takano et al. [10] presented a theoretical approach that calculates electricity prices and rebate levels in DRs based on the framework of social welfare maximization. In the authors' proposal, the DR-originated changes in the utility functions of power suppliers and consumers are used to set a guide for DR requests. Moreover, optimal electricity prices and rebate levels are defined from the standpoint of minimal burden in DRs. Through numerical simulations and a discussion of their results, the validity of the authors' proposal is verified.

The construction and operation of wind turbines have become important aspects of the development of smart cities. However, a fault in the main drive chain often causes wind turbine outages, thereby seriously impacting the normal operation of wind turbines in smart cities. To overcome the shortcomings of the commonly used main drive chain fault diagnosis method, which only uses a single data source, Xu et al. [11] proposed a fault feature extraction and fault diagnosis approach based on data source fusion. By fusing two data sources, that is, the supervisory control and data acquisition (SCADA) real-time monitoring system data and the main drive chain vibration monitoring data, the fault features of the main drive chain are jointly extracted, and an intelligent fault diagnosis model for the main drive chain in the wind turbine based on data fusion is established.

Lai et al. [12] presented a new coordinated operation (CO) framework for electricity and natural gas networks that considers network congestion and demand response. A credit rank (CR) indicator for coupling units was introduced, and gas consumption constraint information of natural-gas-fired units (NGFUs) was provided. A natural gas network operator (GNO) delivers this information to an electricity network operator (ENO). A major advantage of this operation framework is that frequent information interaction between GNO and ENO is unnecessary. The entire framework contains two participants and three optimization problems, namely, GNO optimization sub-problem-A, GNO optimization sub-problem-B, and ENO optimization sub-problem.

Blockchain technologies have received considerable attention from academia and industry due to their distinctive characteristics, such as data integrity, security, decentralization, and reliability. However, their adoption rate is still low, which is one of the primary reasons behind conducting studies related to users' satisfaction and adoption. Determining the factors impacting the use and adoption of blockchain technologies can efficiently address their adoption challenges. Alshamsi, Al-Emran, and Shaalan [13] performed a systematic review of blockchain technologies to offer a thorough understanding of what impacts their adoption and discuss the main challenges and opportunities across various sectors. Of the 902 studies collected, 30 empirical studies met the eligibility criteria and were thoroughly analyzed. The results confirmed that the technology acceptance model (TAM) and technology–organization–environment (TOE) model were the most common models for studying blockchain adoption.

The upward trend of adopting Distributed Energy Resources (DER) reshapes the energy landscape and supports the transition toward a sustainable, carbon-free electricity system. The integration of the Internet of Things (IoT) in Demand Response (DR) enables the transformation of energy flexibility, stemming from electricity consumers/prosumers, into a valuable DER asset, thus placing consumers/prosumers at the center of the electricity market. Andriopoulos et al. [14] showed how Local Energy Markets (LEM) act as a catalyst by providing a digital platform where the prosumers' energy needs and offerings can be efficiently settled locally while minimizing grid interaction. This paper unveils how IoT technology, which enables the control and coordination of numerous devices, further unleashes the flexibility potential of the distribution grid, offered as an energy service to both the LEM participants and the external grid.

Author Contributions: Writing—original draft preparation, C.S.L.; writing—review and editing, K.-F.T. and Y.W. All authors have read and agreed to the published version of the manuscript.

Conflicts of Interest: The authors declare no conflict of interest.

References

1. Yuan, Y.; Jiang, X.; Lai, C.S. A Perfect Decomposition Model for Analyzing Transportation Energy Consumption in China. *Appl. Sci.* **2023**, *13*, 4179. [CrossRef]
2. Hasan, M.M.; Avramis, N.; Ranta, M.; Baghadadi, M.E.; Hegazy, O. Parameter Optimization and Tuning Methodology for a Scalable E-Bus Fleet Simulation Framework: Verification Using Real-World Data from Case Studies. *Appl. Sci.* **2023**, *13*, 940. [CrossRef]
3. Guo, G.; Gong, Y. Research on a Day-Ahead Grouping Coordinated Preheating Method for Large-Scale Electrified Heat Systems Based on a Demand Response Model. *Appl. Sci.* **2023**, *12*, 10758. [CrossRef]
4. Agnew, D.; Aljohani, N.; Mathieu, R.; Boamah, S.; Nagaraj, K.; McNair, J.; Bretas, A. Implementation Aspects of Smart Grids Cyber-Security Cross-Layered Framework for Critical Infrastructure Operation. *Appl. Sci.* **2022**, *12*, 6868. [CrossRef]
5. Duan, Y.; Yuan, H.; Lai, C.S.; Lai, L.L. Fusing Local and Global Information for One-Step Multi-View Subspace Clustering. *Appl. Sci.* **2022**, *12*, 5094. [CrossRef]
6. Vehviläinen, M.; Rita, L.; Rantala, S.; Paakkinen, M.; Laurilla, J.; Terttu, V. Setting Up and Operating Electric City Buses in Harsh Winter Conditions. *Appl. Sci.* **2022**, *12*, 2762. [CrossRef]
7. Wei, Y.; Tsang, K.F.; Wu, C.K.; Wang, H.; Liu, Y. A Multi-Leak Identification Scheme Using Multi-Classification for Water Distribution Infrastructure. *Appl. Sci.* **2022**, *12*, 2128. [CrossRef]
8. Bendiek, P.; Taha, A.; Abbasi, Q.H.; Barakat, B. Solar Irradiance Forecasting Using a Data-Driven Algorithm and Contextual Optimisation. *Appl. Sci.* **2022**, *12*, 134. [CrossRef]
9. Jian, S.; Peng, X.; Yuan, H.; Lai, C.S.; Lai, L.L. Transmission Line Fault-Cause Identification Based on Hierarchical Multiview Feature Selection. *Appl. Sci.* **2021**, *11*, 7804. [CrossRef]
10. Takano, H.; Yoshida, N.; Asano, H.; Hagishima, A.; Tuyen, N.D. Calculation Method for Electricity Price and Rebate Level in Demand Response Programs. *Appl. Sci.* **2021**, *11*, 6871. [CrossRef]
11. Xu, Z.; Yang, P.; Zhao, Z.; Lai, C.S.; Lai, L.L.; Wang, X. Fault Diagnosis Approach of Main Drive Chain in Wind Turbine Based on Data Fusion. *Appl. Sci.* **2021**, *11*, 5804. [CrossRef]
12. Lai, C.S.; Yan, M.; Li, X.; Lai, L.L.; Xu, Y. Coordinated Operation of Electricity and Natural Gas Networks with Consideration of Congestion and Demand Response. *Appl. Sci.* **2021**, *11*, 4987. [CrossRef]

13. AlShamsi, M.; Al-Emran, M.; Shaalan, K. A Systematic Review on Blockchain Adoption. *Appl. Sci.* **2022**, *12*, 4245. [CrossRef]
14. Andriopoulos, N.; Plakas, K.; Mountzouris, C.; Gialelis, J.; Birbas, A.; Karatzas, S.; Papalexopoulos, A. Local Energy Market-Consumer Digital Twin Coordination for Optimal Energy Price Discovery under Thermal Comfort Constraints. *Appl. Sci.* **2023**, *13*, 1798. [CrossRef]

Disclaimer/Publisher's Note: The statements, opinions and data contained in all publications are solely those of the individual author(s) and contributor(s) and not of MDPI and/or the editor(s). MDPI and/or the editor(s) disclaim responsibility for any injury to people or property resulting from any ideas, methods, instructions or products referred to in the content.

Review

A Systematic Review on Blockchain Adoption

Mohammed AlShamsi [1,2], Mostafa Al-Emran [1,3,*] and Khaled Shaalan [1]

[1] Faculty of Engineering & IT, The British University in Dubai, Dubai P.O. Box 345015, United Arab Emirates; 20199983@student.buid.ac.ae (M.A.); khaled.shaalan@buid.ac.ae (K.S.)
[2] Technical Support Section, Emirates Health Authority, Dubai P.O. Box 4545, United Arab Emirates
[3] Department of Computer Techniques Engineering, Dijlah University College, Baghdad 00964, Iraq
* Correspondence: mostafa.alemran@buid.ac.ae

Abstract: Blockchain technologies have received considerable attention from academia and industry due to their distinctive characteristics, such as data integrity, security, decentralization, and reliability. However, their adoption rate is still scarce, which is one of the primary reasons behind conducting studies related to users' satisfaction and adoption. Determining what impacts the use and adoption of Blockchain technologies can efficiently address their adoption challenges. Hence, this systematic review aimed to review studies published on Blockchain technologies to offer a thorough understanding of what impacts their adoption and discuss the main challenges and opportunities across various sectors. From 902 studies collected, 30 empirical studies met the eligibility criteria and were thoroughly analyzed. The results confirmed that the technology acceptance model (TAM) and technology–organization–environment (TOE) were the most common models for studying Blockchain adoption. Apart from the core variables of these two models, the results indicated that trust, perceived cost, social influence, and facilitating conditions were the significant determinants influencing several Blockchain applications. The results also revealed that supply chain management is the main domain in which Blockchain applications were adopted. Further, the results indicated inadequate exposure to studying the actual use of Blockchain technologies and their continued use. It is also essential to report that existing studies have examined the adoption of Blockchain technologies from the lens of the organizational level, with little attention paid to the individual level. This review is believed to improve our understanding by revealing the full potential of Blockchain adoption and opening the door for further research opportunities.

Keywords: Blockchain; technology adoption theories; technology adoption models; systematic review

Citation: AlShamsi, M.; Al-Emran, M.; Shaalan, K. A Systematic Review on Blockchain Adoption. *Appl. Sci.* **2022**, *12*, 4245. https://doi.org/10.3390/app12094245

Academic Editors: Chun Sing Lai, Yinhai Wang and Kim-Fung Tsang

Received: 13 March 2022
Accepted: 18 April 2022
Published: 22 April 2022

Publisher's Note: MDPI stays neutral with regard to jurisdictional claims in published maps and institutional affiliations.

Copyright: © 2022 by the authors. Licensee MDPI, Basel, Switzerland. This article is an open access article distributed under the terms and conditions of the Creative Commons Attribution (CC BY) license (https://creativecommons.org/licenses/by/4.0/).

1. Introduction

The creator of Bitcoin, Satoshi Nakamoto, proficiently described the Blockchain technology as a dispersed "peer-to-peer linked-structure" that could be used to resolve the apprehensions of maintaining the transaction order along with dodging double-spending issues [1]. With Bitcoin, transactions are commanded by grouping them into constrained-size structures called blocks, and a similar timestamp is shared between all blocks. In a Blockchain, miners such as network nodes connect the blocks preferably in chronological order, with each block containing a hash of the previous one [2]. Hence, the structure of a Blockchain often succeeds in holding an auditable and robust registry for all the related transactions. Any technology has its negative and positive sides. Negatively, Blockchain technologies have some disadvantages [3]. For example, Blockchains are harder to scale because of their consensus approach. Processing can be slow on Blockchains when many users exist on the network. Some solutions require high energy consumption. It's challenging to integrate Blockchains with several systems, specifically the legacy ones.

Positively, Blockchain technologies have brought many opportunities for various sectors. For instance, the banking sector can benefit from Blockchain to drive customer transactions under similar Blockchain standards. Blockchain allows for the transparent auditing of transactions. It is the case that different organizations have invested in this technology for a variety of reasons, such as reducing transaction costs and making architectures more transparent, safe, and quick. The significance of the Blockchain is demonstrated by the number of crypto-currencies, which have surpassed 1900 and are still growing [4]. This growth pace often creates interoperability problems due to the assortment of cryptocurrency-related applications [5,6]. Such a landscape is quickly evolving, as Blockchain is being applied in other fields outside of cryptocurrencies, where smart contracts play a primary role. Smart contracts are described as "a computerized transaction protocol that executes the terms of a contract" [7]. These contracts enable an individual to transform contractual clauses into embeddable code [8], consequently limiting the external risks and participation. Thus, a smart contract is an agreement between two parties where the agreed terms and conditions are automatically imposed even if the parties do not trust each other. As such, smart contracts in the context of Blockchains are scripts that run in a decentralized manner and are saved in the Blockchain without relying on third parties [9].

In healthcare, Blockchains can be used to reduce the communication and computational burden in data management [10]. This can be achieved through a secure transaction for a group of networks. Large amounts of healthcare data can also be managed using smart contract systems [11]. Additionally, linking medical devices to a Blockchain platform can connect patients, doctors, and providers to better understand who is complying with treatments and their consequences. In education, Blockchain applications can be used for certificate/degree verification, students' assessments, credit transfer, data management, and admission purposes. Academic credential verification is essential for employers and other authorities to affirm the validity of an academic degree. Blockchain technologies allow students to access their official certificates under the protection and control of their universities [12].

The significance of Blockchain technology is increasing [13]. IBM added that around 33% of C-suite executives itemized that Blockchain is being discovered by them or were involved actively in the past projects [14]. The research and development community at large is already aware of the potential of upcoming technologies, along with discovering many different applications across a wide array of industries [9]. The development of Blockchain applications can be classified into three generations: (i) Blockchain 1.0 is used for cryptocurrency transactions, (ii) Blockchain 2.0 is used for financial applications, and (iii) Blockchain 3.0 is used for other industrial applications, such as government, health, science, and Internet of Things (IoT) [13].

Blockchain assists in tracing and verifying multistep transactions that require traceability and verification. Blockchains minimize compliance costs and accelerate data transfer processing. Perhaps the worthiest value of adopting Blockchains is the enhanced security provided to users while making transactions. This feature builds confidence between consumers and industry partners, protects privacy, and increases transparency in tracing transactions. Despite the tremendous opportunities of Blockchain technologies, their adoption across many domains is still in short supply [15]. Their low adoption rates stem from inadequate knowledge regarding the factors affecting their use [16]. To draw a holistic view of the factors affecting Blockchain adoption, we need to understand the theories/models through which these factors are derived. Understanding those factors through the lenses of these theories/models would help scholars and practitioners prepare future policies and procedures for effectively employing Blockchain technologies across various sectors. By inspecting the existing reviews on Blockchain, it has been observed that there is inadequate knowledge about the main research methods used in Blockchain adoption and the primary domains involving Blockchain applications. To understand these issues, this systematic review aimed to provide a holistic view of Blockchain adoption through the lenses of technology adoption theories and models, and to identify the main research gaps that would guide future research. Therefore, this review study poses the following research questions:

Q1: What are the main research methods and domains in the selected studies?

Q2: What are the main theories/models used for studying the use of Blockchain?
Q3: What are the most frequent external factors affecting the use of Blockchain?
Q4: What is the primary purpose of the reviewed studies?
Q5: Who are the target participants in the selected studies?

2. Related Work

Blockchain technology is a distributed ledger introduced for cryptocurrency in 2008 by Satoshi Nakamoto. In October 2008, Bitcoin was released [17]. In the second generation, smart contracts were introduced for assets and trust agreements. It was initiated by Ethereum, one of the most renowned Blockchain-based software platforms. The next wave of Blockchain technologies will focus on scaling and addressing transaction processing times and bottlenecking problems. There are three categories of Blockchains, including public, proprietary, and permissioned. As a public Blockchain, anyone can join, leave, contribute, read, and audit the Blockchain network, as it is decentralized, self-governed, and authority-free. Bitcoin represents an instance of public Blockchain. The private Blockchain, in contrast, is a closed network with a verified and authentic invitation that is only available for trusted and selected parties. This means only the Blockchain owner has the authority to edit, delete, or override entries on the Blockchain. The last type is referred to as permissioned Blockchain, which permits anyone to join after their identity is verified. Each individual is given specific permissions on the network to perform specific processes. In the supply chain, suppliers, for instance, could manage a permissioned Blockchain for their business partners and customers with different access rights. On the other hand, customers can only be allowed to read product documents, whereas wholesalers and suppliers have access to edit information about the goods and delivery.

Due to its intrinsic characteristics in maintaining transaction transparency across various entities, Blockchain has received much attention from different industries. The primary example is the use of cryptocurrencies in finance [18,19]. Further, pharmaceutical, transportation, origin-to-consumer, legal, and regulatory areas are other non-financial domains that have witnessed prompt adoption and use of Blockchain applications [20]. Moreover, other applications have emerged regarding the use of Blockchain in the healthcare industry [21,22], the chemical industry [23], and big data [24]. Blockchain technology is viewed as a significant component of the fourth industrial revolution that has facilitated changing the structure of the global economy and enhancing the opportunities for innovation, development, and improved quality of life. Furthermore, a Blockchain-based digital government often streamlines processes, protects data, and reduces abuse and fraud while instantaneously boosting accountability and trust. Governments, businesses, and individuals share resources through a distributed ledger protected by cryptography. By eliminating single points of failure, it protects governments and citizen data.

A synthesis of the previously published reviews was performed to understand the current state of the art of Blockchain technologies. Table 1 shows the earlier review studies conducted on Blockchain technology. This subject has recently gained extensive international interest and attention. It can be noticed that Blockchain technology has been studied across several disciplines, including energy, healthcare, agriculture, education, logistics, and supply chain management. Some reviews have examined the underlying Blockchain technology, such as cryptography, peer-to-peer networking, distributed storage, consensus algorithms, and smart contracts [25–27]. Other reviews were interested in highlighting the laws and regulations governing this technology [28]. Some of the reviews focused on the educational applications built using Blockchain technologies, their benefits, and the obstacles to implementation [29]. Another review identified organizational theories and discussed their application in adopting Blockchain technologies in logistics and supply chain management [30]. It can be observed that the existing reviews have neglected to review the factors affecting Blockchain adoption from the perspective of technology adoption theories/models. In addition, there is insufficient knowledge about the main research methods used in Blockchain adoption and the primary domains involving Blockchain applications. Therefore, this systematic review aimed to provide a comprehensive review of Blockchain adoption by examining the main research methods, domains, technology acceptance models/theories, influential factors, research objectives, and target participants.

Table 1. Previous review studies on Blockchain technologies.

Source	Review Type	Number of Reviewed Studies	Domain	Aim
[31]	Systematic review	65 studies	Healthcare	To review the use of Blockchain technology in healthcare.
[25]	Systematic review	140 studies	Energy	To review and examine the basic ideas that drive Blockchain technology, such as systems design and distributed consensus methods. It also concentrated on Blockchain solutions for the energy industry and enlightened the state-of-the-art issues by extensively analyzing the literature and existing business cases.
[32]	Systematic review	33 studies	Healthcare	To demonstrate the potential use of Blockchain technologies, their obstacles, and future research directions in healthcare.
[33]	Systematic review	27 studies	Supply chain management	To explain the most common Blockchain applications in supply chain management (SCM). It also covered the critical disruptions and problems resulting from Blockchain adoption in SCM, and how the future of Blockchains in SCM holds.
[29]	Systematic review	31 studies	Education	To review the applications of Blockchain in education and provide an insight into the main benefits and obstacles of implementation.
[28]	Systematic review	29 studies	Supply chain	To evaluate how Blockchain technologies would affect supply chain practices and policies in the future.
[26]	Systematic review	61 studies	Healthcare	To review the prototypes, frameworks, and implementations of Blockchain in healthcare.
[34]	Systematic review	10 studies	Agriculture	To review current research subjects, significant contributions, and benefits of using Blockchain technologies in agriculture.
[30]	Systematic review	22 studies	Logistics and supply chain management	To identify the most relevant organizational theories used in Blockchain literature in the context of logistics and supply chain management (LSCM). It also examined the content of those organizational theories to formulate relevant research questions for investigating the adoption of Blockchain technologies in LSCM.
[35]	Review	-	General (not specific to a particular domain)	To examine the Blockchain and its related essential features, concerns (IoT, security, and data management), and industrial applications. It also provides potential difficulties and future directions.
[36]	Systematic review	42 studies	Healthcare	To figure out how Blockchain technologies can be used in the healthcare domain.
[37]	Systematic review	35 studies	Governance	To provide scholars and practitioners with directions on using Blockchain applications in governance research.
[38]	Systematic review	32 studies	General (not specific to a particular domain)	To offer the most recent state of research on the potential combination of AI and Blockchain technologies and discuss the possible advantages of such a combination.
[39]	Review	-	Supply chain and logistics	To explain and describe the idea of Blockchain and its use in logistics and supply chains.
[40]	Systematic review	35 studies	IoT	To evaluate academic solutions and approaches of integrating Blockchain with IoT.
[41]	Review	-	General (not specific to a particular domain)	To discuss the fundamentals of Blockchain technologies and their technical details.
This study	Systematic review	30 studies	General (not specific to a particular domain)	To provide a thorough review of Blockchain adoption by examining the main research methods, domains, technology acceptance models/theories, influential factors, research objectives, and target participants.

3. Materials and Methods

This study applied the systematic review approach to review the existing studies on Blockchain adoption. This approach uncovers sources relevant to a research topic and provides a rich synthesis of the subject under examination. This research follows the systematic review guiding principles introduced by Kitchenham and Charters [42] and other related systematic reviews [43–45]. The following subsections detail the phases followed during the review process.

3.1. Inclusion and Exclusion Criteria

Table 2 lists the inclusion and exclusion criteria for the publications that were critically evaluated in this review.

Table 2. Inclusion and exclusion criteria.

Inclusion Criteria	Exclusion Criteria
Should be published between 2010 and 2021.	Studies involving Blockchain but without a theoretical model.
Should involve a theoretical model for evaluating Blockchain.	Studies involving a theoretical model but without a Blockchain.
Should measure the adoption, acceptance, or continued use of Blockchain.	Studies written in languages other than English.
Should be written in English language.	

3.2. Data Sources and Search Strategies

In this systematic review, the surveyed articles were collected from a wide range of online databases, including Emerald, IEEE, ScienceDirect, Springer, MDPI, and Google Scholar. The search for these studies was undertaken in April 2021. The keywords used in the search include ((("Blockchain") AND ("adoption" OR "acceptance" OR "use" OR "intention to use" OR "continued use" OR "continuous intention")). Choosing the keywords is essential since it determines which articles are to be retrieved [46]. Using the above search strategies, the search results retrieved 902 articles. Of those, 218 were marked as duplicates, so we removed them from the analysis. Thus, the overall number of the remaining articles becomes 684. We have applied the inclusion and exclusion criteria for each of these studies. Accordingly, 30 studies met these criteria and were kept for the final analysis. The search and refinement stages were carried out using the "Preferred Reporting Items for Systematic Reviews and Meta-Analyses (PRISMA)" [47]. Figure 1 shows the PRISMA flow diagram.

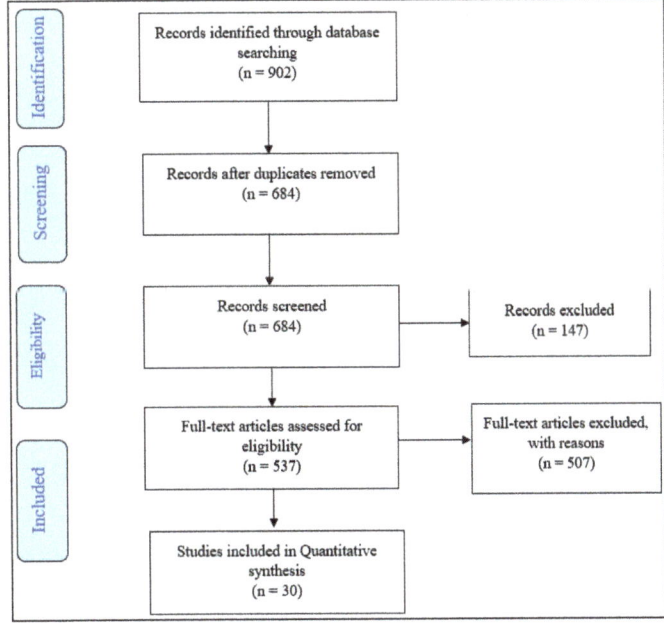

Figure 1. PRISMA flowchart.

3.3. Quality Assessment

Along with the inclusion and exclusion criteria, quality assessment is another crucial factor to consider [48]. A nine-criteria checklist was adopted from Alqudah et al. [49] and Al-Emran et al. [48] as a quality assessment and used to provide a method for evaluating the quality of the research papers that were kept for the final analysis ($n = 30$). Table 3 illustrates the quality assessment checklist. The primary purpose of the checklist was not to criticize any scholar's work, and the checklist was adapted from those suggested by Kitchenham and Charters [42]. Each question in the checklist was scored according to the three-point scale, an answer of "Yes" being worth 1 point, an answer of "No" being worth 0 points, and an answer of "Partially" being worth 0.5 points. Therefore, each study could receive an accumulated score between 0 and 9. The higher the total scores a study attained, the higher the degree to which the study addressed the research questions. This was ensured by assessing each study against the nine quality assessment criteria. For each study, the first and second authors assigned scores to the nine quality assessment criteria independently to ensure accuracy. Differences in assigning scores between the two authors were resolved through discussion and further review of the disputed articles. Table 4 presents the quality assessment results of all 30 studies. It is evident that all studies passed the quality assessment, and they were eligible for final analysis.

Table 3. Quality assessment checklist.

#	Questions
1	Is the research aim specified clearly?
2	Did the study achieve its aim?
3	Are the variables considered by the study clearly indicated?
4	Is the context/discipline of the study clearly defined?
5	Are the data collection methods sufficiently detailed?
6	Are the measures' reliability and validity clearly described?
7	Are the statistical techniques used to analyze the data sufficiently described?
8	Do the findings add to the literature?
9	Does the study add to your knowledge or understanding?

Table 4. Quality assessment results.

Study	Q1	Q2	Q3	Q4	Q5	Q6	Q7	Q8	Q9	Total	Percentage
S1	1	1	1	0.5	1	1	1	1	1	8.5	94.44%
S2	1	1	1	1	1	1	1	1	1	9	100%
S3	1	0.5	1	1	1	1	1	0.5	0.5	7.5	83.33%
S4	1	1	1	1	1	1	1	1	1	9	100%
S5	1	0.5	1	1	1	1	1	0.5	0.5	7.5	83.33%
S6	1	1	1	1	1	1	1	1	1	9	100%
S7	1	1	1	1	0	1	0	1	0.5	6.5	72.22%
S8	1	0.5	1	0.5	1	0.5	1	0.5	0.5	6.5	72.22%
S9	1	1	1	1	1	1	1	1	1	9	100%
S10	1	1	1	1	1	1	1	1	0.5	8.5	94.44%
S11	1	1	0.5	0.5	0.5	0.5	0.5	1	0.5	6	66.66%
S12	1	1	1	1	1	1	1	1	1	9	100%
S13	1	1	1	1	1	1	1	1	1	9	100%
S14	1	1	1	1	0	0.5	0	1	0.5	6	66.66%
S15	1	1	1	1	1	1	1	1	1	9	100%
S16	1	1	1	1	1	1	0.5	1	1	8.5	94.44%
S17	1	1	1	1	1	1	1	1	1	9	100%
S18	1	1	1	1	1	1	1	1	1	9	100%
S19	1	1	1	1	1	0.5	0.5	1	1	8	88.88%
S20	1	1	1	1	1	1	1	1	1	9	100%
S21	1	0.5	1	1	1	0.5	1	0.5	1	7.5	83.33%
S22	1	1	1	1	1	1	1	1	1	9	100%
S23	1	1	1	1	0.5	1	0.5	1	1	8	88.88%
S24	1	1	1	1	1	1	1	1	1	9	100%
S25	1	1	1	1	0.5	1	1	1	1	8.5	94.44%
S26	1	1	1	1	1	1	1	1	1	9	100%
S27	1	1	1	1	1	1	0.5	1	0.5	8	88.88%
S28	1	1	1	1	0.5	1	1	1	1	8.5	94.44%
S29	1	1	1	1	1	1	0.5	1	1	8.5	94.44%
S30	1	1	1	1	0.5	1	1	1	1	8.5	94.44%

3.4. Data Coding and Analysis

For the sake of answering the research questions of this review, we have coded the final list of the remaining articles (n = 30) based on several characteristics, including authors, publication year, methods, countries, factors, domains, theories/models, research aims, and participants.

4. Results

Drawing upon the 30 research studies analyzed in this systematic review, we have reported the findings to answer the formulated research questions. Table A1 (Appendix A) provides a brief description of all the analyzed studies.

4.1. Main Research Methods

Figure 2 depicts the distribution of studies according to the research method used in data collection. It is evident that questionnaire surveys represent the primary research method used in 77% of the analyzed Blockchain adoption studies. However, only 10% of the Blockchain adoption studies relied on interviews in collecting their data.

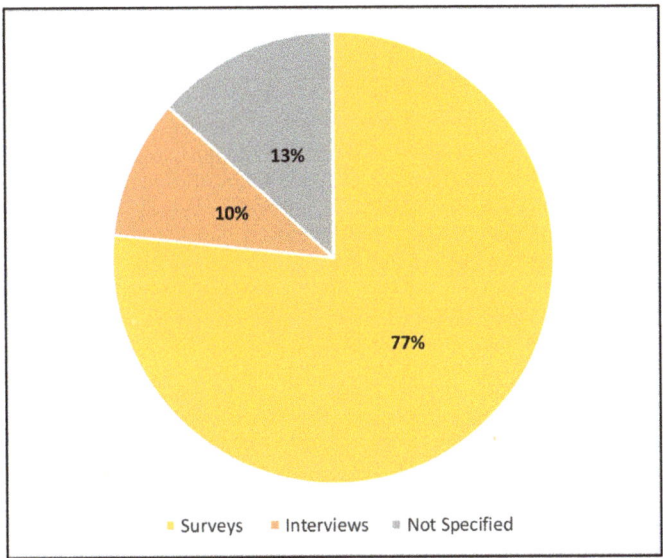

Figure 2. Distribution of studies by research methods.

4.2. Main Domains

Blockchain applications have been extensively used across many domains/sectors. The collected studies were analyzed according to these domains/sectors to provide an overview of the current status of Blockchain applications. Figure 3 depicts the main domains/sectors in which Blockchain applications were adopted. It can be seen that supply chain management dominates the list, with 12 studies. This is followed by education and agriculture, with three studies each. In the supply chain, organizations can automate physical assets and create a decentralized steady record of all transactions, making it possible to track assets from production to delivery or use by end-users. Other applications include maritime shipping [50], organizing decisions [51], and executing operations [52].

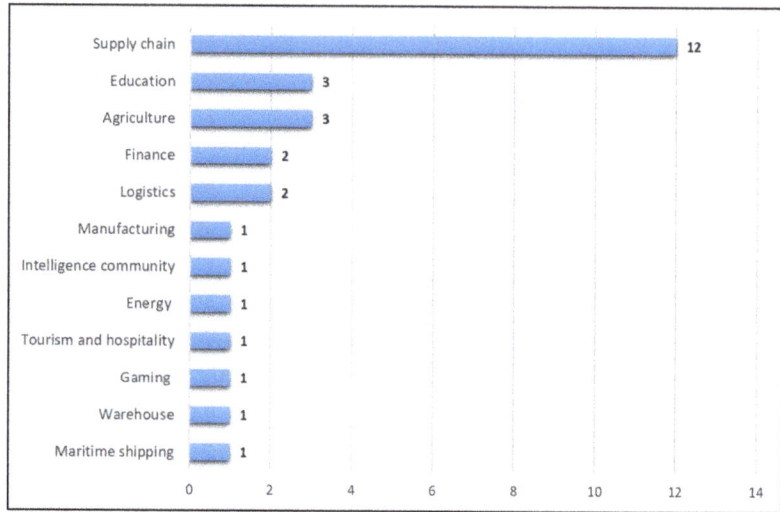

Figure 3. Main domains in Blockchain adoption.

4.3. Prevailing Theories/Models in Blockchain Adoption

As we aimed to examine the adoption of Blockchain technologies, the collected articles were analyzed from the perspective of technology adoption theories/models, as shown in Figure 4. It can be seen that the "technology acceptance model (TAM)" is the most common model in studying Blockchain adoption, with 14 studies. This is followed by the "technology-organization-environment (TOE)" ($n = 8$), "unified theory of acceptance and use of technology (UTAUT)" ($n = 7$), and "innovation diffusion theory (IDT)" ($n = 5$). The rest of the theories/models appeared only once in the examined studies (i.e., TRI2, ISS, TTF, TPB, TRA, and TAM3).

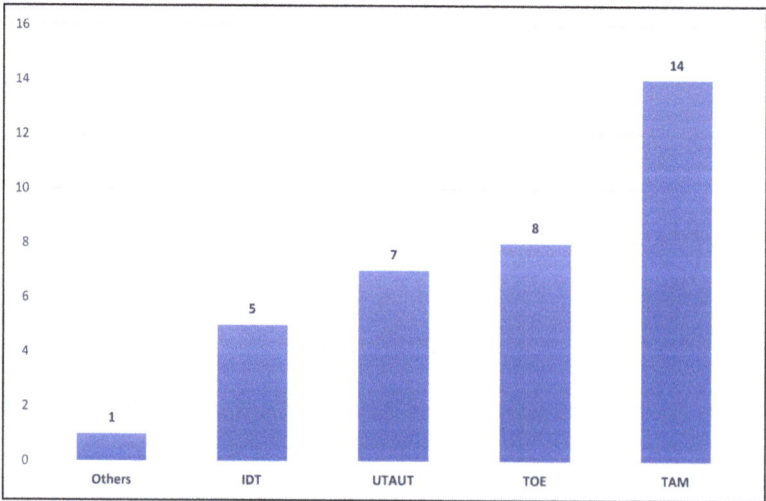

Figure 4. Distribution of studies by technology adoption models/theories.

4.4. Most Frequent External Factors Affecting the Use of Blockchain

The low adoption rates of many technologies, with no exceptions to Blockchain, stem from the inadequate knowledge regarding the factors affecting their use. Therefore, we have analyzed the collected studies to identify the most common external factors affecting the adoption of Blockchain technologies, as shown in Figure 5. Trust appeared to be the most common factor affecting the adoption of Blockchain technologies ($n = 17$). This is followed by the perceived cost and social influence with 11 studies each, then by facilitating conditions ($n = 10$), performance expectancy, effort expectancy, and information security, with seven studies each.

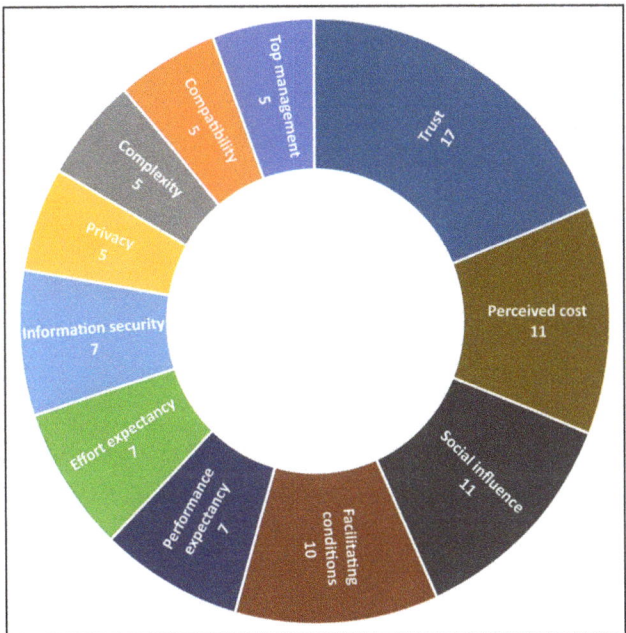

Figure 5. Most common external factors affecting Blockchain adoption.

Apart from the collected studies, we have also analyzed the existing literature on Blockchain to determine the barriers affecting its adoption. Security risk [53,54] and privacy risk [53,55,56] were among the main risks that negatively affect the use of Blockchain technologies. High energy costs [53,54,57] and investment costs [58–60] represent the main costs of using Blockchain technologies. Organizations also impose some barriers to using Blockchain technologies, such as organizational policies [61–63], organizational culture [55,61,64,65], lack of knowledge and management support [63,65,66], and lack of collaboration and coordination [63,64,67]. It is also imperative to mention that adopting Blockchain is hindered by some technological barriers, such as technological immaturity [53,54,65], reluctance to change [60,68,69], interoperability issues [60,63,65,70,71], and scalability issues [53,56,63]. Cultural differences [61,64,72] are also considered a barrier to Blockchain adoption. This is mainly because users rely on themselves when seeking advice related to using Blockchains in individualistic societies, while they rely on others in collectivistic cultures.

4.5. Primary Purpose of the Reviewed Studies

There are three different concepts within the technology adoption domain, including adoption, acceptance, and post-adoption/continuous intention. The adoption is usually measured by potential users who have not yet used the technology, whereas actual users measure the acceptance. The post-adoption/continuous intention measures the continued use of the technology after a sufficient period of users' experience. It is essential to understand the purpose of the analyzed studies concerning the previously mentioned concepts to understand where we stand on Blockchain adoption. It has been noticed that 80% of the analyzed studies concentrated on measuring Blockchain adoption, followed by 10% for both acceptance and continuous intention.

4.6. Target Participants in the Selected Studies

To understand who evaluated the use of Blockchain technologies, we have classified the analyzed studies in terms of participants, as depicted in Figure 6. Fifty percent of the analyzed studies relied on the top management to evaluate the use of Blockchain technologies. This was followed by experts and consultants (28%), academics (13%), and students (5%).

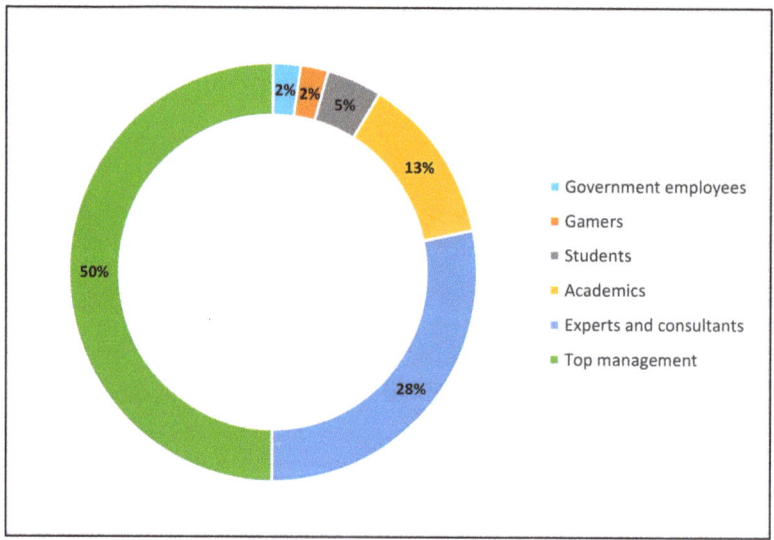

Figure 6. Distribution of studies by participants.

5. Discussion

It is imperative to understand what impacts the adoption of new technologies, such as Blockchain. Adopting any technology relies on the determinants affecting its users [73–75]. A forecasting report illustrates the positive evolution of the Blockchain market between 2017 and 2024 [76]. While this magnitude was USD 800 million in 2017, it is estimated to reach USD 20,550 million in 2024. Therefore, it is vital to gain more insights into what impacts the adoption of Blockchain technologies across many sectors to improve and sustain their usage. Hence, this systematic review was carried out to analyze the adoption of Blockchain technologies from the lenses of technology adoption theories/models.

The results showed that questionnaire surveys represent the primary research method used in 77% of the analyzed Blockchain adoption studies. These outcomes agree with some of the earlier systematic reviews in the technology adoption domain [77–80], which concluded that questionnaire surveys were the most common data collection method. In terms of the Blockchain, these results contradict what Frizzo-Barker et al. [81] reported, in which comparative studies were the primary method used in most of the analyzed articles. Drawing upon the findings of this systematic review, it is suggested that further research would consider the mixed-research approach by involving interviews or focus groups besides using questionnaire surveys. This is because qualitative approaches provide more insights into the cause–effect relationships among the factors affecting Blockchain adoption.

Regarding the main domains/sectors through which the analyzed studies were carried out, supply chain management dominates the list with 12 studies, followed by education and agriculture, with three studies each. While our findings agree with [82], who found that supply chain management is the dominating sector concerning studies related to Blockchain adoption, it contradicts what is reported by [81], who suggested that banking and finance was the primary sector. Regardless of the differences between this review and the previously conducted reviews, understanding Blockchain adoption across many domains is still in short supply due to the limited number of applications.

Concerning the prevailing technology adoption theories/models, the results pointed out that the TAM is the most common model in studying Blockchain adoption, with 14 studies. This is followed by the TOE ($n = 8$), UTAUT ($n = 7$), and IDT ($n = 5$). In the same vein, Taherdoost [82] found that TAM and TOE were the most dominating models in studying Blockchain adoption. TAM is still a valid model to evaluate large-scale emerging technologies [83,84], where Blockchain is not an exception. These results indicate that studies focusing on the individual level have mainly relied on TAM, while those examining the organizational level have relied on TOE. Further research might consider other adoption models that have yet to be used in the existing literature, such as UTAUT2, ECM, PMT, etc.

For the influential factors affecting Blockchain adoption, trust was seen to be the most common factor affecting the adoption of Blockchain technologies ($n = 17$). This was followed by the perceived cost and social influence, with 11 studies each, as well as facilitating conditions ($n = 10$), performance expectancy, effort expectancy, and information security, with seven studies each. Studies on the individual level have examined trust, social influence, performance expectancy, effort expectancy, and information security. In contrast, those focusing on the organizational level studied mainly the organizational perspective's factors in delivering Blockchain-based services, such as trust, perceived cost, and facilitating conditions. Still, there is abundant room for other factors to be investigated from the perspective of other technology adoption theories/models and Blockchain-related specific characteristics. On the other side, we have also analyzed the existing literature on Blockchain to determine the barriers affecting its adoption. Being aware of these barriers and considering them when implementing Blockchains would improve their adoption rate. Security risk and privacy risk were among the main risks that negatively affect the use of Blockchain technologies. High energy and investment costs represent the main costs of using Blockchain technologies. Organizations also impose some barriers to using Blockchain technologies, such as organizational policies, organizational culture, lack of knowledge and management support, and lack of collaboration and coordination. It is also imperative to mention that adopting Blockchain is hindered by some technological barriers, such as technological immaturity, reluctance to change, interoperability issues, and scalability issues. Cultural differences are also considered a barrier to Blockchain adoption. This is mainly because users rely on themselves when seeking advice related to using Blockchains in individualistic societies, while they rely on others in collectivistic cultures.

To understand the primary purpose of the analyzed studies, the results showed that 80% of those studies concentrated on measuring Blockchain adoption, followed by 10% for both acceptance and continuous intention. These results clearly indicate that the majority of existing studies have examined the adoption stage of Blockchain technologies, the step that precedes the actual use of the technology. The results provided evidence that there is inadequate exposure to studying the actual use of Blockchain technologies and their continued use, which furnish a good space for further research.

The results reported that 50% of the analyzed studies relied on the top management to evaluate the use of Blockchain technologies, followed by experts and consultants (28%), academics (13%), and students (5%). This shows that most of the existing studies have examined the adoption of Blockchain technologies from the lens of the organizational level, with little attention paid to the individual level.

This systematic review differs from previous reviews in several ways. This review did not limit the data collection to a specific domain, while most of the earlier reviews did. Most of the earlier reviews concentrated on using Blockchain applications in healthcare [26,31,32,36] and supply chain and logistics [28,30,33,39]. While some of the previously conducted reviews were general in the domain, their aims and scope were entirely different from the current study. For instance, Lu [35] investigated the Blockchain and its essential features, concerns (IoT, security, and data management), and industrial applications. Karger [38] offered the most recent research on the potential combination of AI and Blockchain technologies and discussed the possible advantages of such a combination. Besides, Namasudra et al. [41] discussed the fundamentals of Blockchain technologies and their technical details. To make it distinct, this systematic review provided a thorough review of Blockchain adoption by examining the main research methods, domains, influential factors, research objectives, and target participants through the lenses of technology acceptance models/theories.

6. Conclusions and Future Work

Despite the immense opportunities of Blockchain technologies, their adoption across many domains is still in short supply [15]. This is one of the main reasons behind conducting studies related to users' satisfaction and adoption. Determining what impacts the use and adoption of Blockchain technologies can efficiently address their adoption challenges. Therefore, we have reviewed the Blockchain adoption studies from the perspective of technology adoption theories/models to identify the most influential factors, main research methods, domains/sectors, research objectives, and target participants. It is believed that this systematic review would be a valuable guide for scholars and practitioners seeking to understand the challenges and opportunities related to the adoption of Blockchain technologies across various sectors.

This review shed light on several gaps in research. First, the TAM and TOE were the most common models for understanding the factors affecting the use and adoption of Blockchain technologies. Little attention has been paid to the role of technical, social, and psychological elements in understanding the adoption of Blockchain applications. This gap requires further research by considering other adoption theories/models such as UTAUT2, ECM, PMT, etc. Second, although Blockchain adoption is still in short supply, the findings showed that supply chain management was the dominating sector among others in the examined studies. We found a dearth of empirical research in the other domains, which necessitates the need for future research to look at how Blockchain technologies are adopted. Third, trust, perceived cost, and social influence were the most common factors affecting the adoption of Blockchain technologies. The other factors were mainly adapted from the most common theories, such as TAM and TOE. By involving other theories/models, understanding what impacts the use and adoption of Blockchain technologies would be enlightened, specifically when the factors are related to Blockchain-specific characteristics.

Fourth, 77% of the analyzed Blockchain adoption studies relied on questionnaire surveys for data collection. Hence, it is suggested that further research would consider the mixed-research approach by involving interviews or focus groups, besides using questionnaire surveys. This is because qualitative methods can explain the interrelationships among the factors affecting the adoption of Blockchain. Fifth, unlike the previous systematic reviews that analyzed conceptual and empirical studies, it is imperative to mention that this review has concentrated only on empirical Blockchain studies. Since there is still a limited number of studies across the world, more empirical research is required to examine the users' maturity levels and capabilities of adopting Blockchain applications across many collectivistic and individualistic societies. The implications of Blockchain applications, with their negative or positive sides, in certain cultural environments would assist in developing these applications both socially and economically.

Sixth, the findings showed that 80% of the examined studies concentrated on measuring Blockchain adoption, with a limited number of studies focusing on the acceptance and continuous intention perspectives. There is insufficient knowledge of what impacts the actual use of Blockchain technologies and their continued use, which opens the door for further research trials. Seventh, 50% of the analyzed studies relied on the top management, experts, and consultants to evaluate the use of Blockchain technologies. This shows that most of the existing studies have examined the adoption of Blockchain technologies from the organizational level perspective, with little attention paid to the individual level.

This review is limited in two ways. First, we focused on specific online databases to collect articles, such as Emerald, IEEE, ScienceDirect, Springer, MDPI, and Google Scholar. However, these online databases do not represent the entire literature published on Blockchain adoption. Further reviews might thus extend this review by involving studies indexed in other databases, such as Scopus and Web of Science. Second, this systematic review involved analyzing only empirical quantitative studies. Considering qualitative studies in future reviews would add more insights into the observed results.

Author Contributions: Conceptualization, M.A., M.A.-E. and K.S.; methodology, M.A. and M.A.-E.; validation, M.A. and M.A.-E.; formal analysis, M.A.; investigation, M.A.; resources, M.A.; writing—original draft preparation, M.A.; writing—review and editing, M.A.-E. and K.S.; supervision, M.A.-E. and K.S.; project administration, M.A.-E. and K.S. All authors have read and agreed to the published version of the manuscript.

Funding: This research received no external funding.

Institutional Review Board Statement: Not applicable.

Informed Consent Statement: Not applicable.

Data Availability Statement: The data presented in this study are available on request from the authors.

Conflicts of Interest: The authors declare no conflict of interest.

Appendix A

Table A1. List of analyzed studies.

#	Source	Year	Method	Country	Domain	Theories/Models
S1	[85]	2020	Survey	Not specified	Finance	TAM
S2	[51]	2021	Survey	India	Supply chain management	TAM and TOE
S3	[52]	2020	Survey	Malaysia	Supply chain management	TOE
S4	[86]	2020	Survey	India	Supply chain management	TAM and IDT
S5	[87]	2020	Survey	India	Agriculture supply chain	ISM-DEMATEL
S6	[50]	2019	Survey and interviews	Taiwan	Maritime shipping	TAM
S7	[88]	2020	Not specified	Malaysia	Warehouse industry	UTAUT
S8	[89]	2020	Survey	Nigeria	Logistics	TOE
S9	[90]	2019	Survey	Brazil	Supply chain management	UTAUT
S10	[91]	2019	Survey	Indonesia	Gaming	TAM
S11	[92]	2017	Survey	Taiwan	Finance	IDT and TAM
S12	[93]	2019	Survey	USA	Academia	UTAUT
S13	[94]	2019	Survey	USA and India	Logistics and supply chain management	TAM and UTAUT
S14	[95]	2018	Not specified	Not specified	Supply chain management	UTAUT
S15	[96]	2021	Survey	Kenya	Finance	TAM and IDT
S16	[97]	2019	Survey	Canada	Research community	TAM

Table A1. Cont.

#	Source	Year	Method	Country	Domain	Theories/Models
S17	[98]	2019	Survey	Taiwan	Tourism and hospitality	TAM
S18	[99]	2020	Survey	Developed countries	Energy	TAM and DOI
S19	[100]	2021	Survey	Malaysia	Education	TAM and DOI
S20	[101]	2020	Survey	Malaysia	Intelligence community	TAM 3 and TRI 2
S21	[102]	2021	Semi-structured interviews	Middle East and North Africa	N/A	DOI and TOE
S22	[103]	2021	Survey	India	Agri-food supply chain	ISM and DEMATEL
S23	[61]	2021	Survey	Not specified	Supply chain management	TOE
S24	[104]	2021	Survey	Malaysia	Manufacturing	TOE
S25	[105]	2021	Survey	Australia	Supply chain management	UTAUT, TTF, and ISS
S26	[106]	2018	Survey	India	Supply chain management	TAM, TRI, and TPB
S27	[17]	2019	Interviews	Ireland	Mixed contexts	TOE
S28	[107]	2020	Not specified	Not specified	Finance	TAM
S29	[108]	2020	Survey	Brazil	Supply chain management	UTAUT
S30	[109]	2020	Case study	Indonesia	Agriculture	TOE and the theory of mindfulness of adoption

References

1. Nakamoto, S. RAIN: A Bio-Inspired Communication and Data Storage Infrastructure. *Artif. Life* **2008**, *23*, 552–557. [CrossRef]
2. Crosby, M.; Pattanayak, P.; Verma, S.; Kalyanaraman, V. Integration von Big Data-Komponenten in die Business Intelligence. *Controlling* **2016**, *27*, 222–228. [CrossRef]
3. Iredale, G. Top Disadvantages of Blockchain Technology. Available online: https://101blockchains.com/disadvantages-of-blockchain/ (accessed on 1 April 2022).
4. CoinMarketCap Cryptocurrency Prices, Charts And Market Capitalizations. Available online: https://coinmarketcap.com/ (accessed on 20 December 2021).
5. Haferkorn, M.; Diaz, J.M.Q. Seasonality and interconnectivity within cryptocurrencies—An analysis on the basis of bitcoin, litecoin and namecoin. In *Lecture Notes in Business Information Processing*; Springer: Cham, Switzerland, 2015; Volume 217, pp. 106–120. [CrossRef]
6. Tschorsch, F.; Scheuermann, B. Bitcoin and beyond: A technical survey on decentralized digital currencies. *IEEE Commun. Surv. Tutor.* **2016**, *18*, 2084–2123. [CrossRef]
7. Nick Szabo Smart Contracts. Available online: http://szabo.best.vwh.net/smart.contracts.html (accessed on 15 December 2021).
8. Szabo, N. Smart Contracts: Formalizing and Securing Relationships on Public Networks. *First Monday* **1997**, *2*, 9. [CrossRef]
9. Christidis, K.; Devetsikiotis, M. Blockchains and Smart Contracts for the Internet of Things. *IEEE Access* **2016**, *4*, 2292–2303. [CrossRef]
10. Ismail, L.; Materwala, H.; Zeadally, S. Lightweight Blockchain for Healthcare. *IEEE Access* **2019**, *7*, 149935–149951. [CrossRef]
11. Khatoon, A. A Blockchain-Based Smart Contract System for Healthcare Management. *Electronics* **2020**, *9*, 94. [CrossRef]
12. Vidal, F.R.; Soares, C.; Blockchain, A. Analysis of Blockchain Technology for Higher Education. In Proceedings of the 2019 International Conference on Cyber-Enabled Distributed Computing and Knowledge Discovery (CyberC), Guilin, China, 17–19 October 2019; pp. 28–33. [CrossRef]
13. Zhao, J.L.; Fan, S.; Yan, J. Overview of business innovations and research opportunities in blockchain and introduction to the special issue. *Financ. Innov.* **2016**, *2*, 28. [CrossRef]
14. IBM Blockchain for Business. *Enterprise Adoption Patterns Use Case Examples from Practice Into Hyperledger Fabric v1*; IBM Blockchain for Business: Frankfurt, Germany, 2017.
15. Agi, M.A.; Jha, A.K. Blockchain technology in the supply chain: An integrated theoretical perspective of organizational adoption. *Int. J. Prod. Econ.* **2022**, *247*, 108458. [CrossRef]
16. Lu, L.; Liang, C.; Gu, D.; Ma, Y.; Xie, Y.; Zhao, S. What advantages of blockchain affect its adoption in the elderly care industry? A study based on the technology–organisation–environment framework. *Technol. Soc.* **2021**, *67*, 101786. [CrossRef]
17. Clohessy, T.; Acton, T. Investigating the influence of organizational factors on blockchain adoption: An innovation theory perspective. *Ind. Manag. Data Syst.* **2019**, *119*, 1457–1491. [CrossRef]
18. Tapscott, T. How Blockchain is Changing Insurance. *Itnow* **2017**, *60*, 16–17. [CrossRef]
19. Fosso Wamba, S.; Kala Kamdjoug, J.R.; Epie Bawack, R.; Keogh, J.G. Bitcoin, Blockchain and Fintech: A systematic review and case studies in the supply chain. *Prod. Plan. Control* **2018**, *31*, 115–142. [CrossRef]

20. Perez, G. Blockchain: A Study Rooted in Reality. 2018. Available online: https://news.sap.com/2018/04/blockchain-a-study-rooted-in-reality/ (accessed on 20 December 2021).
21. Grover, P.; Kar, A.K.; Davies, G. "Technology enabled Health"—Insights from twitter analytics with a socio-technical perspective. *Int. J. Inf. Manag.* **2018**, *43*, 85–97. [CrossRef]
22. Sodhro, A.H.; Luo, Z.; Sangaiah, A.K.; Baik, S.W. Mobile edge computing based QoS optimization in medical healthcare applications. *Int. J. Inf. Manag.* **2019**, *45*, 308–318. [CrossRef]
23. Takhar, S.S.; Liyanage, K. Blockchain application in supply chain chemical substance reporting—A Delphi study. *Int. J. Internet Technol. Secur. Trans.* **2018**, *11*, 75–107. [CrossRef]
24. Chae, B. A General framework for studying the evolution of the digital innovation ecosystem: The case of big data. *Int. J. Inf. Manag.* **2019**, *45*, 83–94. [CrossRef]
25. Andoni, M.; Robu, V.; Flynn, D.; Abram, S.; Geach, D.; Jenkins, D.; McCallum, P.; Peacock, A. Blockchain technology in the energy sector: A systematic review of challenges and opportunities. *Renew. Sustain. Energy Rev.* **2019**, *100*, 143–174. [CrossRef]
26. Chukwu, E.; Garg, L. A systematic review of blockchain in healthcare: Frameworks, prototypes, and implementations. *IEEE Access* **2020**, *8*, 21196–21214. [CrossRef]
27. Li, J.; Greenwood, D.; Kassem, M. Blockchain in the built environment and construction industry: A systematic review, conceptual models and practical use cases. *Autom. Constr.* **2019**, *102*, 288–307. [CrossRef]
28. Wang, Y.; Han, J.H.; Beynon-Davies, P. Understanding blockchain technology for future supply chains: A systematic literature review and research agenda. *Supply Chain Manag.* **2018**, *24*, 62–84. [CrossRef]
29. Alammary, A.; Alhazmi, S.; Almasri, M.; Gillani, S. Blockchain-based applications in education: A systematic review. *Appl. Sci.* **2019**, *9*, 2400. [CrossRef]
30. Kummer, S.; Herold, D.M.; Dobrovnik, M.; Mikl, J.; Schäfer, N. A systematic review of blockchain literature in logistics and supply chain management: Identifying research questions and future directions. *Futur. Internet* **2020**, *12*, 60. [CrossRef]
31. Agbo, C.; Mahmoud, Q.; Eklund, J. Blockchain Technology in Healthcare: A Systematic Review. *Healthcare* **2019**, *7*, 56. [CrossRef]
32. Hölbl, M.; Kompara, M.; Kamišalić, A.; Zlatolas, L.N. A systematic review of the use of blockchain in healthcare. *Symmetry* **2018**, *10*, 470. [CrossRef]
33. Queiroz, M.M.; Ivanov, D.; Dolgui, A.; Fosso Wamba, S. Impacts of epidemic outbreaks on supply chains: mapping a research agenda amid the COVID-19 pandemic through a structured literature review. *Ann. Oper. Res.* **2020**, 1–38. [CrossRef]
34. Bermeo-Almeida, O.; Cardenas-Rodriguez, M.; Samaniego-Cobo, T.; Ferruzola-Gómez, E.; Cabezas-Cabezas, R.; Bazán-Vera, W. Blockchain in agriculture: A systematic literature review. *Commun. Comput. Inf. Sci.* **2018**, *883*, 44–56. [CrossRef]
35. Lu, Y. The blockchain: State-of-the-art and research challenges. *J. Ind. Inf. Integr.* **2019**, *15*, 80–90. [CrossRef]
36. Tandon, A.; Dhir, A.; Islam, N.; Mäntymäki, M. Blockchain in healthcare: A systematic literature review, synthesizing framework and future research agenda. *Comput. Ind.* **2020**, *122*, 103290. [CrossRef]
37. Liu, Y.; Lu, Q.; Zhu, L.; Paik, H.-Y.; Staples, M. A Systematic Literature Review on Blockchain Governance. *SSRN Electron. J.* **2021**. [CrossRef]
38. Karger, E. Combining blockchain and artificial intelligence—Literature review and state of the art. In Proceedings of the 41st International Conference on Information Systems, ICIS 2020, Making Digital Inclusive: Blending the Locak and the Global, Hyderabad, India, 13–16 December 2020.
39. Dujak, D.; Sajter, D. *Blockchain Applications in Supply Chain*; Springer International Publishing: Berlin/Heidelberg, Germany, 2021; ISBN 9783319916682.
40. Lo, S.K.; Liu, Y.; Chia, S.Y.; Xu, X.; Lu, Q.; Zhu, L.; Ning, H. Analysis of Blockchain Solutions for IoT: A Systematic Literature Review. *IEEE Access* **2019**, *7*, 58822–58835. [CrossRef]
41. Namasudra, S.; Deka, G.C.; Johri, P.; Hosseinpour, M.; Gandomi, A.H. The Revolution of Blockchain: State-of-the-Art and Research Challenges. *Arch. Comput. Methods Eng.* **2021**, *28*, 1497–1515. [CrossRef]
42. Kitchenham, B.; Charters, S. *Guidelines for Performing Systematic Literature Reviews in Software Engineering*; Keele University: Keele, UK, 2007; Volume 4, pp. 5356–5373.
43. Al-Qaysi, N.; Mohamad-Nordin, N.; Al-Emran, M. Employing the technology acceptance model in social media: A systematic review. *Educ. Inf. Technol.* **2020**, *25*, 4961–5002. [CrossRef]
44. Al-Saedi, K.; Al-Emran, M.; Abusham, E.; El-Rahman, S.A. Mobile Payment Adoption: A Systematic Review of the UTAUT Model. In Proceedings of the 2019 International Conference on Fourth Industrial Revolution, ICFIR 2019, Manama, Bahrain, 19–21 February 2019.
45. Alsharida, R.A.; Hammood, M.M.; Al-Emran, M. Mobile Learning Adoption: A Systematic Review of the Technology Acceptance Model from 2017 to 2020. *Int. J. Emerg. Technol. Learn.* **2021**, *16*, 147–162. [CrossRef]
46. Costa, V.; Monteiro, S. Key knowledge management processes for innovation: A systematic literature review. *VINE J. Inf. Knowl. Manag. Syst.* **2016**, *46*, 386–410.
47. Moher, D.; Liberati, A.; Tetzlaff, J.; Altman, D.G.; Altman, D.; Antes, G.; Atkins, D.; Barbour, V.; Barrowman, N.; Berlin, J.A.; et al. Preferred reporting items for systematic reviews and meta-analyses: The PRISMA statement. *PLoS Med.* **2009**, *6*, e1000097. [CrossRef]
48. Al-Emran, M.; Mezhuyev, V.; Kamaludin, A.; Shaalan, K. The impact of knowledge management processes on information systems: A systematic review. *Int. J. Inf. Manag.* **2018**, *43*, 173–187. [CrossRef]

49. Alqudah, A.A.; Al-Emran, M.; Shaalan, K. Technology Acceptance in Healthcare: A Systematic Review. *Appl. Sci.* **2021**, *11*, 10537. [CrossRef]
50. Yang, C. Maritime shipping digitalization: Blockchain-based technology applications, future improvements, and intention to use. *Transp. Res. Part E* **2019**, *131*, 108–117. [CrossRef]
51. Kamble, S.S.; Gunasekaran, A.; Kumar, V.; Belhadi, A.; Foropon, C. A machine learning based approach for predicting blockchain adoption in supply Chain. *Technol. Forecast. Soc. Chang.* **2021**, *163*, 120465. [CrossRef]
52. Wong, L.W.; Leong, L.Y.; Hew, J.J.; Tan, G.W.H.; Ooi, K.B. Time to seize the digital evolution: Adoption of blockchain in operations and supply chain management among Malaysian SMEs. *Int. J. Inf. Manag.* **2020**, *52*, 101997. [CrossRef]
53. Esmaeilian, B.; Sarkis, J.; Lewis, K.; Behdad, S. Blockchain for the future of sustainable supply chain management in Industry 4.0. *Resour. Conserv. Recycl.* **2020**, *163*, 105064. [CrossRef]
54. Böckel, A.; Nuzum, A.K.; Weissbrod, I. Blockchain for the Circular Economy: Analysis of the Research-Practice Gap. *Sustain. Prod. Consum.* **2021**, *25*, 525–539. [CrossRef]
55. Hatzivasilis, G.; Ioannidis, S.; Fysarakis, K.; Spanoudakis, G.; Papadakis, N. The green blockchains of circular economy. *Electron.* **2021**, *10*, 2008. [CrossRef]
56. Kouhizadeh, M.; Sarkis, J.; Zhu, Q. At the nexus of blockchain technology, the circular economy, and product deletion. *Appl. Sci.* **2019**, *9*, 1712. [CrossRef]
57. Ajwani-Ramchandani, R.; Figueira, S.; Torres de Oliveira, R.; Jha, S.; Ramchandani, A.; Schuricht, L. Towards a circular economy for packaging waste by using new technologies: The case of large multinationals in emerging economies. *J. Clean. Prod.* **2021**, *281*, 125139. [CrossRef]
58. Demestichas, K.; Daskalakis, E. Information and communication technology solutions for the circular economy. *Sustainability* **2020**, *12*, 7272. [CrossRef]
59. Upadhyay, A.; Mukhuty, S.; Kumar, V.; Kazancoglu, Y. Blockchain technology and the circular economy: Implications for sustainability and social responsibility. *J. Clean. Prod.* **2021**, *293*, 126130. [CrossRef]
60. Yildizbasi, A. Blockchain and renewable energy: Integration challenges in circular economy era. *Renew. Energy* **2021**, *176*, 183–197. [CrossRef]
61. Kouhizadeh, M.; Saberi, S.; Sarkis, J. Blockchain technology and the sustainable supply chain: Theoretically exploring adoption barriers. *Int. J. Prod. Econ.* **2021**, *231*, 107831. [CrossRef]
62. Hew, J.J.; Wong, L.W.; Tan, G.W.H.; Ooi, K.B.; Lin, B. The blockchain-based Halal traceability systems: A hype or reality? *Supply Chain Manag.* **2020**, *25*, 863–879. [CrossRef]
63. Erol, I.; Peker, I.; Ar, I.M.; Turan, İ.; Searcy, C. Towards a circular economy: Investigating the critical success factors for a blockchain-based solar photovoltaic energy ecosystem in Turkey. *Energy Sustain. Dev.* **2021**, *65*, 130–143. [CrossRef]
64. Ada, N.; Kazancoglu, Y.; Sezer, M.D.; Ede-Senturk, C.; Ozer, I.; Ram, M. Analyzing barriers of circular food supply chains and proposing industry 4.0 solutions. *Sustainability* **2021**, *13*, 6812. [CrossRef]
65. Bekrar, A.; El Cadi, A.A.; Todosijevic, R.; Sarkis, J. Digitalizing the closing-of-the-loop for supply chains: A transportation and blockchain perspective. *Sustainability* **2021**, *13*, 2895. [CrossRef]
66. Gopalakrishnan, P.K.; Hall, J.; Behdad, S. Cost analysis and optimization of Blockchain-based solid waste management traceability system. *Waste Manag.* **2021**, *120*, 594–607. [CrossRef]
67. Çetin, S.; De Wolf, C.; Bocken, N. Circular digital built environment: An emerging framework. *Sustainability* **2021**, *13*, 6348. [CrossRef]
68. Kouhizadeh, M.; Zhu, Q.; Sarkis, J. Blockchain and the circular economy: Potential tensions and critical reflections from practice. *Prod. Plan. Control* **2020**, *31*, 950–966. [CrossRef]
69. Shojaei, A.; Ketabi, R.; Razkenari, M.; Hakim, H.; Wang, J. Enabling a circular economy in the built environment sector through blockchain technology. *J. Clean. Prod.* **2021**, *294*, 126352. [CrossRef]
70. Bressanelli, G.; Pigosso, D.C.A.; Saccani, N.; Perona, M. Enablers, levers and benefits of Circular Economy in the Electrical and Electronic Equipment supply chain: A literature review. *J. Clean. Prod.* **2021**, *298*, 126819. [CrossRef]
71. Wang, B.; Luo, W.; Zhang, A.; Tian, Z.; Li, Z. Blockchain-enabled circular supply chain management: A system architecture for fast fashion. *Comput. Ind.* **2020**, *123*, 103324. [CrossRef]
72. Romero-Hernández, O.; Romero, S. Maximizing the value of waste: From waste management to the circular economy. *Thunderbird Int. Bus. Rev.* **2018**, *60*, 757–764. [CrossRef]
73. Al-Emran, M.; Mezhuyev, V.; Kamaludin, A. Students' Perceptions towards the Integration of Knowledge Management Processes in M-learning Systems: A Preliminary Study. *Int. J. Eng. Educ.* **2018**, *34*, 371–380.
74. Al-Emran, M.; Mezhuyev, V.; Kamaludin, A.; AlSinani, M. Development of M-learning Application based on Knowledge Management Processes. In Proceedings of the 2018 7th International conference on Software and Computer Applications (ICSCA 2018), Kuantan, Malaysia, 8–10 February 2018; pp. 248–253.
75. Al Shamsi, J.H.; Al-Emran, M.; Shaalan, K. Understanding key drivers affecting students' use of artificial intelligence-based voice assistants. *Educ. Inf. Technol.* **2022**, 1–21. [CrossRef]
76. Chernov, A.; Chernova, V. Global blockchain technology market analysis-current situations and forecast. In Proceedings of the Economic and Social Development: Book of Proceedings, Warsaw, Poland, 26–27 September 2018; pp. 143–152.

77. Wu, W.-H.; Wu, Y.-C.J.; Chen, C.-Y.; Kao, H.-Y.; Lin, C.-H.; Huang, S.-H. Review of trends from mobile learning studies: A meta-analysis. *Comput. Educ.* **2012**, *59*, 817–827. [CrossRef]
78. Crompton, H.; Burke, D.; Gregory, K.H. The use of mobile learning in PK-12 education: A systematic review. *Comput. Educ.* **2017**, *110*, 51–63. [CrossRef]
79. Al-Saedi, K.; Al-Emran, M. A Systematic Review of Mobile Payment Studies from the Lens of the UTAUT Model. In *Recent Advances in Technology Acceptance Models and Theories*; Springer: Cham, Switzerland, 2021; Volume 335, pp. 79–106.
80. Al-Maroof, R.A.; Al-Emran, M. Research Trends in Flipped Classroom: A Systematic Review. In *Recent Advances in Intelligent Systems and Smart Applications*; Springer: Berlin/Heidelberg, Germany, 2021; pp. 253–275.
81. Frizzo-Barker, J.; Chow-White, P.A.; Adams, P.R.; Mentanko, J.; Ha, D.; Green, S. Blockchain as a disruptive technology for business: A systematic review. *Int. J. Inf. Manag.* **2020**, *51*, 102029. [CrossRef]
82. Taherdoost, H. A Critical Review of Blockchain Acceptance Models—Blockchain Technology Adoption Frameworks and Applications. *Computers* **2022**, *11*, 24. [CrossRef]
83. Al-Emran, M.; Granić, A. Is It Still Valid or Outdated? A Bibliometric Analysis of the Technology Acceptance Model and Its Applications From 2010 to 2020. In *Recent Advances in Technology Acceptance Models and Theories*; Springer: Cham, Germany, 2021.
84. Al-Emran, M.; Al-Maroof, R.; Al-Sharafi, M.A.; Arpaci, I. What impacts learning with wearables? An integrated theoretical model. *Interact. Learn. Environ.* **2020**. [CrossRef]
85. Albayati, H.; Kyoung, J.; Jeung, J. Technology in Society Accepting financial transactions using blockchain technology and cryptocurrency: A customer perspective approach. *Technol. Soc.* **2020**, *62*, 101320. [CrossRef]
86. Karamchandani, A.; Srivastava, S.K.; Srivastava, R.K. Perception-based model for analyzing the impact of enterprise blockchain adoption on SCM in the Indian service industry. *Int. J. Inf. Manag.* **2020**, *52*, 102019. [CrossRef]
87. Yadav, V.S.; Singh, A.R.; Raut, R.D.; Govindarajan, U.H. Blockchain technology adoption barriers in the Indian agricultural supply chain: An integrated approach. *Resour. Conserv. Recycl.* **2020**, *161*, 104877. [CrossRef]
88. Wahab, S.N.; Loo, Y.M.; Say, C.S. Antecedents of Blockchain Technology Application among Malaysian Warehouse Industry. *Int. J. Logist. Syst. Manag.* **2020**, *1*, 1. [CrossRef]
89. Orji, I.J.; Kusi-Sarpong, S.; Huang, S.; Vazquez-Brust, D. Evaluating the factors that influence blockchain adoption in the freight logistics industry. *Transp. Res. Part E Logist. Transp. Rev.* **2020**, *141*, 102025. [CrossRef]
90. Wamba, S.F.; Queiroz, M.M. The role of social influence in blockchain adoption: The Brazilian supply chain case. *IFAC-PapersOnLine* **2019**, *52*, 1715–1720. [CrossRef]
91. Pantouw, R.T.; Aruan, D.T.H. Influence of Game Design and Playability Toward Continuance Intention Using TAM Framework. *IPTEK J. Proc. Ser.* **2019**, 307. [CrossRef]
92. Lou, A.T.F.; Li, E.Y. Integrating innovation diffusion theory and the technology acceptance model: The adoption of blockchain technology from business managers' perspective. In Proceedings of the International Conference on Electronic Business, Dubai, United Arab Emirates, 4–8 December 2017; pp. 299–302.
93. Lee, C.C.; Kriscenski, J.C.; Lim, H.S. An empirical study of behavioral intention to use blockchain technology. *J. Int. Bus. Discip.* **2019**, *14*, 1–21.
94. Queiroz, M.M.; Fosso Wamba, S. Blockchain adoption challenges in supply chain: An empirical investigation of the main drivers in India and the USA. *Int. J. Inf. Manag.* **2019**, *46*, 70–82. [CrossRef]
95. Francisco, K.; Swanson, D. The Supply Chain Has No Clothes: Technology Adoption of Blockchain for Supply Chain Transparency. *Logistics* **2018**, *2*, 2. [CrossRef]
96. Aketch, S.; Mwambia, F.; Baimwera, B. Effects of Blockchain Technology on Performance of Financial Markets in Kenya. *Int. J. Financ. Account.* **2021**, *6*, 1–15. [CrossRef]
97. Shrestha, A.K.; Vassileva, J. User Acceptance of Usable Blockchain-Based Research Data Sharing System: An Extended TAM-Based Study. In Proceedings of the 2019 First IEEE International Conference on Trust, Privacy and Security in Intelligent Systems and Applications (TPS-ISA), Los Angeles, CA, USA, 12–14 December 2019.
98. Nuryyev, G.; Wang, Y.P.; Achyldurdyyeva, J.; Jaw, B.S.; Yeh, Y.S.; Lin, H.T.; Wu, L.F. Blockchain technology adoption behavior and sustainability of the business in tourism and hospitality SMEs: An empirical study. *Sustainability* **2019**, *12*, 1256. [CrossRef]
99. Ullah, N.; Alnumay, W.S.; Al-Rahmi, W.M.; Alzahrani, A.I.; Al-Samarraie, H. Modeling cost saving and innovativeness for blockchain technology adoption by energy management. *Energies* **2020**, *13*, 4783. [CrossRef]
100. Ullah, N.; Al-Rahmi, W.M.; Alzahrani, A.I.; Alfarraj, O.; Alblehai, F.M. Blockchain technology adoption in smart learning environments. *Sustainability* **2021**, *13*, 1801. [CrossRef]
101. Muhamad, W.N.W.; Razali, N.A.M.; Wook, M.; Ishak, K.K.; Zainudin, N.M.; Hasbullah, N.A.; Ramli, S. Evaluation of Blockchain-based Data Sharing Acceptance among Intelligence Community. *Int. J. Adv. Comput. Sci. Appl.* **2020**, *11*, 597–606. [CrossRef]
102. Toufaily, E.; Zalan, T.; Dhaou, S. Ben A framework of blockchain technology adoption: An investigation of challenges and expected value. *Inf. Manag.* **2021**, *58*, 103444. [CrossRef]
103. Saurabh, S.; Dey, K. Blockchain technology adoption, architecture, and sustainable agri-food supply chains. *J. Clean. Prod.* **2021**, *284*, 124731. [CrossRef]
104. Fernando, Y.; Rozuar, N.H.M.; Mergeresa, F. The blockchain-enabled technology and carbon performance: Insights from early adopters. *Technol. Soc.* **2021**, *64*, 101507. [CrossRef]

105. Alazab, M.; Alhyari, S.; Awajan, A.; Abdallah, A.B. Blockchain technology in supply chain management: An empirical study of the factors affecting user adoption/acceptance. *Clust. Comput.* **2021**, *24*, 83–101. [CrossRef]
106. Kamble, S.; Gunasekaran, A.; Arha, H. Understanding the Blockchain technology adoption in supply chains-Indian context. *Int. J. Prod. Res.* **2018**, *57*, 2009–2033. [CrossRef]
107. Singh, H.; Jain, G.; Munjal, A.; Rakesh, S. Blockchain technology in corporate governance: Disrupting chain reaction or not? *Corp. Gov.* **2020**, *20*, 67–86. [CrossRef]
108. Queiroz, M.M.; Fosso Wamba, S.; De Bourmont, M.; Telles, R. Blockchain adoption in operations and supply chain management: Empirical evidence from an emerging economy. *Int. J. Prod. Res.* **2021**, *59*, 6087–6103. [CrossRef]
109. Rijanto, A. Business financing and blockchain technology adoption in agroindustry. *J. Sci. Technol. Policy Manag.* **2020**, *12*, 215–235. [CrossRef]

Article

A Perfect Decomposition Model for Analyzing Transportation Energy Consumption in China

Yujie Yuan [1,2], Xiushan Jiang [1,*] and Chun Sing Lai [2,*]

1 School of Traffic and Transportation, Beijing Jiaotong University, Beijing 100044, China
2 Department of Electronic and Electrical Engineering, Brunel University London, London UB8 3PH, UK
* Correspondence: xshjiang@bjtu.edu.cn (X.J.); chunsing.lai@brunel.ac.uk (C.S.L.)

Abstract: Energy consumption in transportation industry is increasing. Transportation has become one of the fastest energy consumption industries. Transportation energy consumption variation and the main influencing factors of decomposition contribute to reduce transportation energy consumption and realize the sustainable development of transportation industry. This paper puts forwards an improved decomposition model according to the factors of change direction on the basis of the existing index decomposition methods. Transportation energy consumption influencing factors are quantitatively decomposed according to the transportation energy consumption decomposition model. The contribution of transportation turnover, transportation structure and transportation energy consumption intensity changes to transportation energy consumption variation is quantitatively calculated. Results show that there exists great energy-conservation potential about transportation structure adjustment, and transportation energy intensity is the main factor of energy conservation. The research achievements enrich the relevant theory of transportation energy consumption, and help to make the transportation energy development planning and carry out related policies.

Keywords: transportation; energy consumption; influencing factors; index decomposition approach

Citation: Yuan, Y.; Jiang, X.; Lai, C.S. A Perfect Decomposition Model for Analyzing Transportation Energy Consumption in China. *Appl. Sci.* 2023, 13, 4179. https://doi.org/10.3390/app13074179

Academic Editors: Wenming Yang and Luca Fiori

Received: 31 January 2023
Revised: 8 March 2023
Accepted: 21 March 2023
Published: 25 March 2023

Copyright: © 2023 by the authors. Licensee MDPI, Basel, Switzerland. This article is an open access article distributed under the terms and conditions of the Creative Commons Attribution (CC BY) license (https:// creativecommons.org/licenses/by/ 4.0/).

1. Introduction

Transportation has become one of the fastest growing energy consumption industries worldwide. Quantitative assessment of various factors affecting energy consumption is essential not only for a better understanding of past behaviors of transportation energy consumption, but also for estimating energy requirements of alternative industrialization strategies.

To study the related issues of energy consumption in the transportation system, we should first clarify the composition of the transportation system. The national transportation system divides into the domestic inter-city transportation system composed of railway, highway, waterway, civil aviation, and pipeline, and the urban transportation system formed by urban road and rail transportation. Therefore, waterway transportation does not include ocean transportation. According to the nature of transportation tasks, road transportation can divide into operational transportation completed by operating vehicles, and non-operational transportation completed by non-operating vehicles. According to the different railway transportation systems undertaken, railway transportation can be divided into passenger and freight transportation completed by the national and local railway transportation systems with or without a network, and urban passenger transportation completed by the urban rail transit system. The inter-city transportation system, composed of five modes of transportation, undertakes the road transportation of operational and non-operational vehicles, the same as the transportation tasks of railway, waterway, civil aviation, and pipeline, defined as complex transportation. The transportation system is composed of the domestic inter-city transportation system, and the intra-city transportation system is a total transportation system. To facilitate the analysis and calculation and the availability of data, the scope of the transportation system studied is the operational transportation system, which includes five modes of transportation (highway, railway, waterway, aviation, and pipeline).

Decomposition methodology is an effective method dealing with energy consumption-related issues analysis. Studies can be traced back to the early 1980s. Laspeyres index decomposition was first proposed to analyze the influence factors of industrial energy consumption [1]. Many researchers subsequently applied this method to decompose the energy consumption change [2–4]. However, an important factor, called residual known as the sum of all the interactions of the main effects, was usually ignored, which caused large estimation errors.

Boyd et al. is likely to be the first to analyze energy consumption problems using Divisia index decomposition [5–7]. The same decomposition methods were also directly applied in Howarth et al. [8,9] and Li [10], where the residual was not resolved. Though Sun, J. [11–13] proposed a complete decomposition model, which disposed the residual according to the principle of "common creation, equal distribution", the error will become bigger when the time span of analysis is enlarged.

Among Divisia index decomposition methods, two methods are widely applied: arithmetic mean Divisia index method (AMDI) [14] and logarithmic mean Divisia index decomposition method (LMDI) [15]. In the formulae of AMDI, logarithmic terms were introduced, which might lead to computational problems when zero values appear in the data set (i.e., denominator is zero). A framework for additive and multiplicative decomposition [16,17] was extended based on the two general parametric Divisia index methods, i.e., additive and multiplicative decompositions [14]. The LMDI was proposed with the continuous development of the Divisia Decomposition [18,19]. It is reasonable to replace arithmetic mean weight function by logarithmic mean weight function, because the latter can decompose the residual completely without generating unexplained residual. Ang B.W. [20–27] analyzed many index decomposition approaches and pointed out the advantages of LMDI including eliminating residual term and using time independence. Many researchers analyzed the problems of the LMDI method, which occurred when processing negative numbers and zero values [28–31]. A new decomposition method called the LMDII was introduced [21]. This approach could completely decompose the remaining items and deal with zeros appearing in the data set in the decomposition process. However, it is lack of the consideration of changes of intermediate demand, and it ignores the influence of the energy consumption or consumption structure changes. Another method introduced a 'mean rate-of-change index' (MRCI) [28] to give different weights for decomposed terms. This method provides more plausible and reasonable results, because it ensures residual-free decomposition even when data contain negative values, which cannot be handled by the LMDI method.

In addition, the Shapley decomposition, which calculates the influence of factors on the energy consumption variation according to the total contribution of various factors [24], makes it possible to present a correct and symmetric decomposition without residual [32]. Thus, the residual can be resolved completely.

In summary, every method has its own advantages and disadvantages. The Laspeyres and Divisia index decompositions are the most primitive methods. However, neglecting residual term is their common problem. The complete decomposition model is widely used to solve the residual, the index weight and the change of the positive and negative numbers are neglected. In LMDI decomposition, the residual term can be totally decomposed, while zero and negative values remain as a problem in data processing.

Analyzing energy consumption trends and strength is beneficial to solving the problem of energy distribution imbalance and then to improving energy efficiency [33–36]. The calculation results also have many errors according to different decomposition methods of the residual items. Therefore, it is necessary to seek a more scientific decomposition method to accurately analyze influence factors of energy consumption. The residual term is related to changes in both quantity and direction of influence factors, and is more likely decided by negatively changed factors. To this end, this paper proposes a perfect decomposition method that considers the changes of influence factors and the changing direction. The remainder of this paper is organized as follows: Section 2 proposes the methodology; Sections 3 and 4 describe an empirical case study and discuss the results; Conclusions are made at last.

2. Methodology
2.1. Decomposition Model Construction According to Factor Direction

A perfect decomposition model is proposed according to the different changing directions of factors. This principle can be extended from two factors and three factors to multiple factors.

2.1.1. Two-Factor Decomposition Model

We take a two-factor model as a sample example to describe this principle. Figure 1 illustrated the process of the factor changes in different directions, i.e., x_1 decreases by Δx_1, and then x_2 increases by Δx_2.

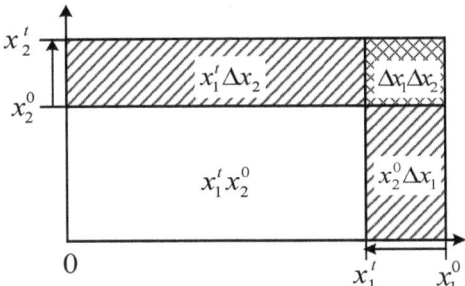

Figure 1. Index change when two factors change at different directions.

Assume that $v = x_1 x_2$, i.e., variable v is determined by factors x_1 and x_2, within the time period $[0, t]$, $x_1^t = x_1^0 + \Delta x_1$, $x_2^t = x_2^0 + \Delta x_2$, the change of variable Δv can be represented as:

$$\Delta v = v^t - v^0 = x_1^t x_2^t - x_1^0 x_2^0 = x_2^0 \Delta x_1 + x_1^0 \Delta x_2 + \Delta x_1 \Delta x_2 \quad (1)$$

where $x_2^0 \Delta x_1$ and $x_1^0 \Delta x_2$ represent the contributions of x_1 and x_2 to the total change of variable v, respectively; $\Delta x_1 \Delta x_2$ is the residual term. There are two situations need to be discussed.

The factors change at the same direction;

If x_1 increases by Δx_1, x_2 increases by Δx_2, accordingly. The complete decomposition of two factors is as follows:

$$x_{1-effect} = x_2^0 \Delta x_1 + \frac{1}{2} \Delta x_1 \Delta x_2 \quad (2)$$

$$x_{2-effect} = x_1^0 \Delta x_2 + \frac{1}{2} \Delta x_1 \Delta x_2 \quad (3)$$

The term $\Delta x_1 \Delta x_2$ is the residual term in the traditional decomposition method, which can be divided equally to the contributions of x_1 and x_2. Both the changes of x_1 and x_2, i.e., Δx_1 and Δx_2, determines the contributions. If one of Δx_1 and Δx_2 is zero, the other is also zero.

The factors change in different directions;

Figure 1 illustrated the process of the factor changes in different directions, i.e., x_1 decreases by Δx_1, and then x_2 increases by Δx_2.

It can be seen from Figure 1 that when x_2^0 increases by Δx_2, $x_1^0 \Delta x_2$ is the contribution of the change of x_2 to the total change of v that contains two parts, i.e., $x_1^t \Delta x_2$ and $\Delta x_1 \Delta x_2$. When x_1^0 decreases by Δx_1, $x_2^0 \Delta x_1$ is the contribution of the change of x_2 to the total change of v, which counteracts $\Delta x_1 \Delta x_2$ because x_2^0 increase by Δx_2, the contribution of x_1 and x_2 to the total change of variable v can be respectively calculated as follows:

$$x_{1-effect} = x_2^0 \Delta x_1 + \Delta x_1 \Delta x_2 \tag{4}$$

$$x_{2-effect} = x_1^0 \Delta x_2 \tag{5}$$

From Figure 1 and Equations (2)–(5), it can be summarized that for the two-factor model, if the two factors change at same direction, the residual term can be divide equally to the two factors; however, if the two factors change at different directions, the residual term belongs to the factor that changes negatively.

2.1.2. Three-Factor Decomposition Model

Assume that the variable $v = x_1 x_2 x_3$, where the variable v is determined by x_1, x_2 and x_3, within the time period $[0, t]$. The change of variable v, i.e., Δv, can be calculated as:

$$\Delta v = v^t - v^0 = \underbrace{x_1^t x_2^t x_3^t - x_1^0 x_2^0 x_3^0}_{\text{similar items}} = \underbrace{(x_1^0 + \Delta x_1)(x_2^0 + \Delta x_2)(x_3^0 + \Delta x_3)}_{\text{joint effect items}} - \underbrace{x_1^0 x_2^0 x_3^0}_{\text{residual item}} \tag{6}$$

$$= x_2^0 x_3^0 \Delta x_1 + x_1^0 x_3^0 \Delta x_2 + x_1^0 x_2^0 \Delta x_3 + x_3^0 \Delta x_1 \Delta x_2 + x_2^0 \Delta x_1 \Delta x_3 + x_1^0 \Delta x_2 \Delta x_3 + \Delta x_1 \Delta x_2 \Delta x_3$$

where Δv is composed of the following three parts: the first part is the contributions of the change of single factor x_1, x_2, or x_3 to the total change of v, which is the sum of $x_2^0 x_3^0 \Delta x_1$, $x_1^0 x_3^0 \Delta x_2$, and $x_1^0 x_2^0 \Delta x_3$; the second part $x_3^0 \Delta x_1 \Delta x_2$, $x_2^0 \Delta x_1 \Delta x_3$, and $x_1^0 \Delta x_2 \Delta x_3$ are the joint effects of the change of two factors; the third part $\Delta x_1 \Delta x_2 \Delta x_3$ is a residual item produced by the change of the three factors simultaneously.

The factors change at the same direction;

In the three-factor model, when all factors change at the same direction, there are two situations, i.e., all factors increase or decrease simultaneously. The common effect and contribution of the changes of the factors are the same, which can be equally assigned to each factor as follows:

$$x_{1-effect} = x_2^0 x_3^0 \Delta x_1 + \frac{1}{2} \Delta x_1 (x_3^0 \Delta x_2 + x_2^0 \Delta x_3) + \frac{1}{3} \Delta x_1 \Delta x_2 \Delta x_3 \tag{7}$$

$$x_{2-effect} = x_1^0 x_3^0 \Delta x_2 + \frac{1}{2} \Delta x_2 (x_3^0 \Delta x_1 + x_1^0 \Delta x_3) + \frac{1}{3} \Delta x_1 \Delta x_2 \Delta x_3 \tag{8}$$

$$x_{3-effect} = x_1^0 x_2^0 \Delta x_3 + \frac{1}{2} \Delta x_3 (x_2^0 \Delta x_1 + x_1^0 \Delta x_2) + \frac{1}{3} \Delta x_1 \Delta x_2 \Delta x_3 \tag{9}$$

The factors change at different directions;

When the three factors change at different directions, two cases are needed to discuss. The change of two factors is positive, and one factor is negative;

Assume that the change of x_3 is negative, i.e., x_3 decreases, while the changes of x_1 and x_2 are positive, i.e., both x_1 and x_2 increase. Figure 2 illustrates the changes of the three factors.

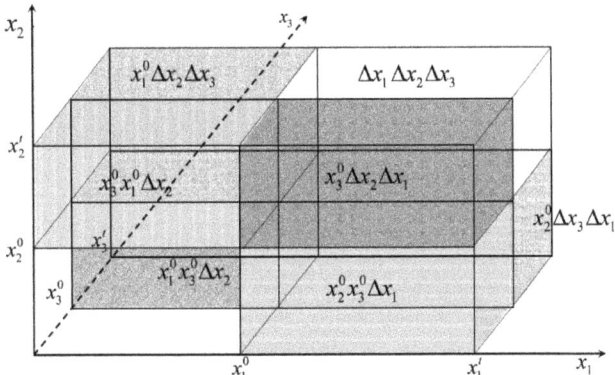

Figure 2. Index change when three factors change at different directions.

It can be seen from Figure 2 that, when x_2 increase by Δx_2, the total change of v increases by $x_1^0 x_3^0 \Delta x_2$, $(x_1^0 x_3^0 \Delta x_2 = x_3^t x_1^0 \Delta x_2 + x_1^0 \Delta x_2 \Delta x_3)$; when x_1 increases by Δx_1, the total change of v increases by $x_2^0 x_3^0 \Delta x_1 = x_3^t x_2^0 \Delta x_1 + x_2^0 \Delta x_1 \Delta x_3$; when x_3 decreases by Δx_3, the total change of v decreases by $x_2^0 x_1^0 \Delta x_3$, $x_2^0 \Delta x_1 \Delta x_3$, $x_1^0 \Delta x_2 \Delta x_3$, and $\Delta x_1 \Delta x_2 \Delta x_3$. At the same time, it also cancels out the total change of $x_2^0 \Delta x_1 \Delta x_3$, $x_1^0 \Delta x_2 \Delta x_3$, and $\Delta x_1 \Delta x_2 \Delta x_3$, because x_2 increases by Δx_2 and x_1 increases by Δx_1. Each factor increases or decreases to offset the other. The contribution of x_1, x_2 and x_3 to the total change of v is represented in the following formulas, respectively.

$$x_{1-effect} = \Delta x_1 x_2^0 x_3^0 + \frac{\Delta x_1 x_3^0 \Delta x_2}{2} \quad (10)$$

$$x_{2-effect} = \Delta x_2 x_1^0 x_3^0 + \frac{\Delta x_2 x_3^0 \Delta x_1}{2} \quad (11)$$

$$x_{3-effect} = x_1^0 x_2^0 \Delta x_3 + \Delta x_2 \Delta x_3 x_1^0 + \Delta x_1 \Delta x_3 x_2^0 + \Delta x_1 \Delta x_2 \Delta x_3 \quad (12)$$

It can be seen that, when the factors change at different directions, an amount of changes are negative, such as $x_2^0 x_1^0 \Delta x_3$, $x_2^0 \Delta x_1 \Delta x_3$, $x_1^0 \Delta x_2 \Delta x_3$, the residual $\Delta x_1 \Delta x_2 \Delta x_3$ belongs to the change of the negative factors.

The change of two factors is negative, and one factor is positive;

Assume that the change of x_1 and x_2 are negative, while the change of x_3 is positive. Then, the contribution of x_1, x_2 and x_3 can be respectively calculated as follows:

$$x_{1-effect} = \Delta x_1 x_2^0 x_3^0 + \Delta x_1 x_2^0 \Delta x_3 + \frac{\Delta x_1 \Delta x_2 x_3^0}{2} + \frac{\Delta x_1 \Delta x_2 \Delta x_3}{2} \quad (13)$$

$$x_{2-effect} = \Delta x_2 x_1^0 x_3^0 + \Delta x_2 x_1^0 \Delta x_3 + \frac{\Delta x_1 \Delta x_2 x_3^0}{2} + \frac{\Delta x_1 \Delta x_2 \Delta x_3}{2} \quad (14)$$

$$x_{3-effect} = x_1^0 x_2^0 \Delta x_3 \quad (15)$$

2.1.3. Multi-Factor Decomposition Model According to Factors Changing Direction

In general, if variable v is determined by n factors, denoted by $x_1, x_2, \ldots, x_n$, i.e., $v = x_1 x_2 \ldots x_n = \prod_{i=1}^{n} x_i$, we can analyze the changes as follows.

All factors change at the same direction;

According to the two-factor and three-factor decomposition processes, the common effect of the changes of the factors is the same, when the changing directions of the factors are the same. The influence of the interaction in Δv can be separated into related factors, and the contribution of each factor to the total change of v can be respectively calculated as follows:

$$x_{i-effect} = \frac{v^0}{x_i^0}\Delta x_i + \sum_{j \neq i}\frac{v^0}{2x_i^0 x_j^0}\Delta x_i \Delta x_j + \sum_{j \neq r \neq i}\frac{v^0}{3x_i^0 x_j^0 x_r^0}\Delta x_i \Delta x_j \Delta x_r + \ldots + \frac{1}{n}\Delta x_1 \Delta x_2 \ldots \Delta x_n \quad (16)$$

Factors change at different directions;

According to the decomposition process of two factors and three factors, whose changing directions are different, the perfect decomposed principles can be generalized. During the decomposition, it is considered that the positive and negative terms offset each other, $x_{i-effect}$ must contain Δx_i, for the negative variable x_i, all the variables x_i and increment Δx_i are needed to consider in the process of decomposition. For the positive variable x_i, only variables x_i are considered, and the increment Δx_i, which is more than 0, cannot contain the variables less than zero. Unified expression of multiple variables can be deduced.

When the changing directions are different, there are $n-1$ cases that need to be discussed, i.e., $\Delta x_k < 0$, where $k = 1, 2, 3 \ldots, n-1$.

- One factor changes less than 0, i.e., $k = 1$;

$$x_{1-effect} = \Delta x_1 \prod_{j=2}^{n} x_j^t = \Delta x_1 \prod_{j=2}^{n}\left(x_j^0 + \Delta x_j\right), \quad \text{if } i = 1 \quad (17)$$

$$x_{i-effect} = \frac{v^0}{x_i^0}\Delta x_i + \sum_{j \neq i \neq 1}\frac{v^0}{2x_i^0 x_j^0}\Delta x_i \Delta x_j + \sum_{j \neq r \neq i \neq 1}\frac{v^0}{3x_i^0 x_j^0 x_r^0}\Delta x_i \Delta x_j \Delta x_r + \ldots + \frac{x_1}{n-1}\Delta x_2 \ldots \Delta x_n, \quad \text{if } i > 1 \quad (18)$$

- More than one factor change less than 0, i.e., $k \geq 2$;

According to the size of i and k, two cases are needed to discuss.

$$x_{i-effect} = \frac{v^0}{x_i^0}\Delta x_i + \sum_{j \neq i > k}\frac{v^0}{x_i^0 x_j^0}\Delta x_i \Delta x_j + \ldots + \sum_{j \neq r \neq i > k}\frac{v^0}{x_i^0 x_j^0 x_r^0}\Delta x_i \Delta x_j \Delta x_r$$
$$+ \frac{v^0}{x_i x_{k+1} \ldots x_n}\Delta x_i \Delta x_{k+1} \ldots \Delta x_n + \sum_{j \neq i < k}\frac{x_1^0 x_2^0 \ldots x_k^0}{2x_i^0 x_j^0}\Delta x_i \Delta x_j P(X)$$
$$+ \sum_{j \neq r \neq i < k}\frac{x_1^0 x_2^0 \ldots x_k^0}{3x_i^0 x_j^0 x_r^0}\Delta x_i \Delta x_j \Delta x_r P(X) + \ldots + \frac{v^0}{k x_1^0 x_2^0 \ldots x_k^0}\Delta x_1 \Delta x_2 \ldots \Delta x_k$$
$$\text{if } i \leq k \quad (19)$$

where $P(X)$ is a mixed term that can be expressed as $P(X) = \sum p(x)$, where $p(x) = \prod_{i=k+1}^{n}\tau_i$, $(i \geq k+1)$. τ_i can be uniquely taken from x_i^0 or Δx_i, i.e., from the following $2(n-k)$ variables: $x_{k+1}^0, \Delta x_{k+1}, x_{k+2}^0, \Delta x_{k+2}, \ldots, x_n^0, \Delta x_n$. Thus, the number of $p(x)$ is 2^{n-k}, and $P(X)$ is equal to their sum.

$$x_{i-effect} = \frac{v^0}{x_i^0}\Delta x_i + \sum_{k < j \neq i}\frac{v^0}{2x_i^0 x_j^0}\Delta x_i \Delta x_j + \sum_{k < j \neq r \neq i}\frac{v^0}{3x_i^0 x_j^0 x_r^0}\Delta x_i \Delta x_j \Delta x_r$$
$$+ \ldots + \frac{v^0}{(n-k)x_{k+1}^0 x_{k+1}^0 \ldots x_n^0}\Delta x_{k+1}^0 \Delta x_{k+1}^0 \ldots \Delta x_n^0, \quad \text{if } k < i \leq n \quad (20)$$

It should be pointed out that for the influence factors of the decomposition model, the change of dependent variable is caused by several factors, when the factors change at the same direction. The residual term is decomposed according to the principle of "average distribution". When the factors change in different directions, the changing direction that offsets each other must be considered by the residual items decomposition. The more the variables are, the more complex their changing directions are.

2.2. Transportation Energy Consumption Decomposition Model

Transportation energy consumption is connected with transportation turnover volume (the product of transportation volume and average distance), transportation structure (the transport structure usually refers to the transport volume structure. In a certain period, within the scope of a country or region, the proportion of various transport modes in the total passenger and freight transport volume or total turnover. It reflects the status and role of modes of transportation in the whole transportation system. Transportation volume share of mode i among all modes) and transportation energy intensity. The perfect complete decomposition model for explaining the change of transportation energy consumption can be written as follows.

$$E = \sum_i E_i^t = \sum_i \frac{E_i^t}{D_i^t} \times D_i^t = \sum_i \frac{E_i^t}{D_i^t} \times \frac{D_i^t}{D^t} \times D^t = \sum_i I_i^t \times S_i^t \times D^t \quad (21)$$

where $i = 1, 2, 3, 4, 5$ presents five transportation modes, namely, highway, railway, aviation, water transportation, pipeline, respectively; E is total energy consumption of the five transportation modes; E_i^t is energy consumption transportation mode i in year t; D_i^t is transportation turnover volume of mode i in year t; $I_i^t = E_i^t/D_i^t$ is transportation energy intensity of mode i in year t; $S_i^t = D_i^t/D^t$ is transportation structure share of mode i in year t.

It can be seen from Equation (21), transportation energy consumption can be decomposed into the common effect of three factors: transportation turnover volume, transportation structure and transportation energy consumption intensity. The impact of each factor on transportation energy consumption not only has a close relationship with the changes of the factors, but is also connected with the initial and final value of the other two factors.

The contribution of each influence factor to transportation energy consumption change can be seen as the product of five "three factors". Transportation energy consumption factor decomposition model can be constructed. $D_i^t = D_i^0 + \Delta D_i$, $S_i^t = S_i^0 + \Delta S_i$, $I_i^t = I_i^0 + \Delta I_i$, according to the three-factor decomposition model, the change of transportation energy consumption in the base year 0 and target year t, ΔE can be calculated as Equations (22) and (23).

$$\Delta E = E^t - E^0 = \sum_i I_i^t \times S_i^t \times D^t - \sum_i I_i^0 \times S_i^0 \times D^0 \quad (22)$$

$$\Delta E = \Delta E_t - \Delta E_0 = \Delta E_D + \Delta E_I + \Delta E_S \quad (23)$$

where ΔE_I, ΔE_S and ΔE_D are the contributions of transportation energy consumption intensity, transportation structure, and transport turnover volume, respectively.

Based on the perfect decomposition model, three influence factors of transportation energy consumption changing direction can be divided into two cases.

2.2.1. Change at the Same Direction

Three influence factors change at the same direction, i.e., "the three factor" increase ($\Delta D_i > 0$, $\Delta S_i > 0$, $\Delta I_i > 0$) or decrease ($\Delta D_i < 0$, $\Delta S_i < 0$, $\Delta I_i < 0$) simultaneously. According to the perfect decomposition model, the contribution of three factors to the transportation energy consumption can be determined as follows:

$$\Delta E_D = \sum \Delta D I_i^0 S_i^0 + \frac{\sum \Delta D}{2}(\Delta I S_i^0 + I_i^0 \Delta S) + \frac{\sum \Delta D \Delta I_i \Delta S_i}{3} \quad (24)$$

$$\Delta E_S = \sum \Delta S I_i^0 D_i^0 + \frac{\sum \Delta S}{2}(\Delta I D_i^0 + I_i^0 \Delta D) + \frac{\sum \Delta D \Delta I_i \Delta S_i}{3} \quad (25)$$

$$\Delta E_I = \sum \Delta I S_i^0 D_i^0 + \frac{\sum \Delta I}{2}(\Delta S D_i^0 + S_i^0 \Delta D) + \frac{\sum \Delta D \Delta I_i \Delta S_i}{3} \qquad (26)$$

2.2.2. Change at Different Directions

Three influence factors change at different directions, i.e., one factor decreases ($\Delta D_i \times \Delta I_i \times \Delta S_i < 0$) or two-factor decrease ($\Delta D_i \times \Delta I_i \times \Delta S_i > 0$), simultaneously. According to the perfect decomposition model, the contribution of three factors to the transportation energy consumption is as follows.

(i) When $\Delta D_i < 0$, $\Delta S_i > 0$ and $\Delta I_i > 0$ the formulas are as follows:

$$\Delta E_D = \sum \Delta D_i S_i^t I_i^t \qquad (27)$$

$$\Delta E_S = \frac{\sum \Delta S_i}{2} D_i^0 (I_i^t + I_i^0) \qquad (28)$$

$$\Delta E_I = \frac{\sum \Delta I_i}{2} D_i^0 (S_i^t + S_i^0) \qquad (29)$$

When $\Delta D_i > 0$, $\Delta S_i < 0$, $\Delta I_i > 0$ and $\Delta D_i > 0$, $\Delta S_i > 0$, $\Delta I_i < 0$, the formulas can also be obtained just by replacing the corresponding variables of the calculated formula.

(ii) When $\Delta D_i < 0$, $\Delta S_i < 0$ and $\Delta I_i > 0$, the formulas are as follows:

$$\Delta E_D = \frac{\sum \Delta D_i}{2} I_i^0 (S_i^t + S_i^0) \qquad (30)$$

$$\Delta E_S = \frac{\sum \Delta S_i}{2} I_i^0 (\sum D_i^T + \sum D_i^0) \qquad (31)$$

$$\Delta E_I = \sum D_i^0 \times S_i^0 \times \Delta I_i \qquad (32)$$

The cases $\Delta D_i > 0$, $\Delta S_i < 0$, $\Delta I_i < 0$ and $\Delta D_i < 0$, $\Delta S_i > 0$, $\Delta I_i < 0$ of the formulas can be similarly obtained.

3. Effective Verification and Case Study

3.1. The Effective Verification of the Perfect Decomposition Model

To examine the effectiveness of the proposed decomposition model, transportation energy consumption data based on the transportation sectors in China are decomposed from 1985 to 2012. To illustrate the remaining items and to omit for index, we use period wise decomposition. The change is only analyzed by the validation between the two base years, i.e., setting two different time intervals, 10 years (1985–1995) and 27 years (1985–2012). The results are presented in Table 1.

It can be seen from Table 1 that the total contribution of the perfect model and the complete decomposition model is identical ($\Delta E = \Delta E_D + \Delta E_S + \Delta E_I$), because the residual items are completely decomposed in both models. However, the Laspeyres model neglects the residual, therefore the results ΔE is different from the sum of influence factors change $\Delta E_D + \Delta E_S + \Delta E_I$. Compared with Laspeyres model and the complete decomposition model, the results of the proposed perfect decomposition model are more accurate and the method is more appropriate.

Table 1. The calculation results of the three decomposition model.

Time	1985–1995				1985–2012			
Model	ΔE	ΔE_D	ΔE_S	ΔE_I	ΔE	ΔE_D	ΔE_S	ΔE_I
Laspeyres model	2680.71	3069.77	538.38	−927.44	19,807.90	19,735.56	1442.69	−1370.36
Complete decomposition model	2306.64	2906.42	722.16	−1321.94	23,233.83	20,770.77	6205.35	−3742.29
Errors	—	5.32%	34.14%	42.54%	—	5.25%	330.12%	173.09%
proposed perfect decomposition model	2306.64	3397.81	716.47	−1807.63	23,233.83	26,504.38	5190.34	−8460.89
Errors	—	10.69%	33.08%	94.91%	—	34.30%	259.77%	517.42%

For the influence of each factor, compared with Laspeyres, the errors of three influence factors (transportation turnover volume, transportation structure and transportation energy consumption intensity) are 10.69%, 33.08% and 94.91%, respectively, from 1985 to 1995. When the analysis period is longer (from 1985 to 2012), the errors are larger, i.e., −34.29%, 259.77% and 517.42%, respectively. It can be seen that the errors caused by the negative factors are very obvious, and the perfect decomposition model takes into account the effects of the different factors, when dealing with the remaining items. The weight is more consistent with the actual situation, and the perfect decomposition model is necessary.

Compared to the complete decomposition model, the errors of three influence factors are 16.91%, 0.79% and 36.74% during the period from 1985 to 1995. When the analysis period is enlarged, the error will be greater, which are 27.60%, −17.66% and 126.09%, respectively, during the period from 1985 to 2012. The longer the analysis time is, the larger the error percentage is, because the residual processing, are even greater than the total change of a single factor. Therefore, the Lapsers index and its model of residual term ignore will produce a larger error. The changing directions of the factor are considered in the perfect decomposition model when dealing with the remaining items; therefore, decomposition results are more reasonable.

3.2. Perfect Decomposition Results Analysis

Index decomposition models can be divided into period wise and time series according to research objects. Time series decomposition can reflect energy changes trajectory within a certain period, and better explain the change mechanism of transportation energy consumption. Time series decomposition model is used to study transportation energy consumption, and conversion turnover is used in the decomposition process.

Based on the perfect decomposition model, we analyze the three factors (transportation turnover volume, transportation structure, transportation energy intensity) affecting transportation energy consumption change and calculate the change of related factors and the contribution rate with the analysis time period from 1985 to 2012. The decomposed results are shown in Table 2.

Table 2. The decomposed results.

Years	ΔE_D	ΔE_S	ΔE_I	ΔE	ΔTE	ΔTER
1985–1986	340.21	32.75	−141.23	231.72	108.49	2.56
1986–1987	431.09	181.26	−172.46	439.90	−8.80	−0.19
1987–1988	412.91	170.07	−280.40	302.58	110.33	2.22
1988–1989	196.77	−33.64	−51.38	111.75	85.02	1.68
1989–1990	−63.88	46.12	55.58	37.82	−101.70	−2.07
1990–1991	283.64	−37.40	−159.97	86.28	197.36	3.73
1991–1992	361.64	93.44	−231.47	223.61	138.03	2.53
1992–1993	388.21	50.38	−183.30	255.29	132.92	2.33
1993–1994	414.79	72.13	−63.15	423.77	−8.98	−0.15
1994–1995	167.67	116.10	−89.85	193.92	−26.25	−0.43
1995–1996	241.12	66.81	142.52	450.45	−209.33	−3.25
1996–1997	−376.25	448.84	−18.99	53.61	−429.86	−6.85
1997–1998	−42.96	187.87	195.32	340.23	−383.19	−5.75
1998–1999	209.89	123.32	378.85	712.06	−502.17	−6.93
1999–2000	1139.04	−299.36	−760.84	78.85	1060.19	11.92
2000–2001	−30.15	205.71	528.81	704.36	−734.51	−9.41
2001–2002	610.43	75.41	−107.62	578.22	32.20	0.35
2002–2003	552.78	−126.55	24.64	450.87	101.90	1.05
2003–2004	1719.26	−287.88	−184.78	1246.60	472.66	4.19
2004–2005	1225.18	−9.56	343.83	1559.46	−334.27	−2.78
2005–2006	1322.41	208.45	−479.51	1 OS 1.34	271.06	1.98
2006–2007	1794.73	308.25	−291.36	1811.62	−16.89	−0.11
2007–2008	1663.56	612.10	−102.06	2173.60	−510.04	−3.02
2008–2009	698.49	1017.68	−349.13	1367.05	−668.56	−3.69
2009–2010	2985.27	344.51	−526.76	2803.02	182.25	0.84
2010–2011	2759.44	318.17	−644.49	2433.12	326.32	1.34
2011–2012	1650.27	1438.64	23.83	3112.74	−1462.47	−5.70
1985–2012	26,504.38	5190.34	−8460.89	23,233.83	3270.55	30,395.08

From the decomposed results of the perfect model where transportation turnover volume is the main influence factor in determining the main trend of transportation energy consumption, the contribution is gradually strengthened. The transportation volume makes transportation energy consumption increase by 265.044 Mtce from 1985–2012, according to the perfect decomposition model. Except for a few years, the contribution rate of transportation turnover volume to transportation energy consumption growth is more than 80%.

The change in transportation structure, transportation energy, and consumption intensity saves transportation energy consumption by 32.706 (51.903 − 84.609 = −32.706) Mtce, with an energy saving rate of 10.76%. The average annual energy saving rate is 1.06% from 1985 to 2012.

The transportation energy consumption intensity is the main factor of energy saving. From 1985 to 2012, the change in energy intensity saves 84.609 Mtce. Except for 1989–1990, 1995–1996, 1997–1998, 1997–1999, and 2000–2001, 2001–2003, and 2004–2005, most of the other years have energy saving effects.

The change in transportation structures reduces the demand for energy in 1988–1989, 1990–1991, 1999–2000, 2002–2003, 2003–2004 and 2004–2005, and the energy demand has been increased in most of the other years. Energy demand increased by 51.903 Mtce due to the change of transportation structure from 1985 to 2012, and thus the transportation structure adjustment is the key to saving energy and has great potential to economize.

According to the above analysis, the structure adjustment has great energy saving potential. It is necessary to analyze the specific contribution of each transportation mode to energy consumption. It is transportation turnover volume, transportation structure and transportation energy consumption intensity of five transportation modes changes on the influence of transportation energy consumption, which are shown in Table 3.

Table 3. The contribution of five transportation modes for irallic mileage.

Years	Highway	Railway	Aviation	Water Transportation	Pipeline	Total
1985–1986	125.42	176.46	8.96	20.47	8.90	340.21
1986–1987	168.10	216.39	12.83	23.22	10.55	431.09
1987–1988	179.10	189.73	12.92	21.67	9.49	412.91
1988–1989	88.76	86.46	6.28	10.84	4.43	196.77
1989–1990	−29.16	−27.47	−2.10	−3.74	−1.41	−63.88
1990–1991	129.17	119.25	10.84	18.01	6.37	286.64
1991–1992	162.18	149.98	15.40	25.64	8.14	361.64
1992–1993	180.05	150.95	19.46	29.47	8.28	388.21
1993–1994	198.96	148.99	23.50	35.10	8.25	414.79
1994–1995	83.26	56.39	10.47	15.00	2.54	167.67
1995–1996	120.99	79.83	15.64	21.20	3.46	241.12
1996–1997	−195.97	−117.09	−25.54	−32.67	−4.98	−376.25
1997–1998	−23.73	−12.61	−3.22	−2.83	−0.57	−42.96
1998–1999	121.91	54.96	16.45	13.93	2.65	209.89
1999–2000	706.40	259.61	82.90	76.28	13.86	1139.04
2000–2001	−18.19	−6.47	−2.51	−2.62	−0.35	−30.15
2001–2002	394.87	123.22	58.37	26.93	7.05	610.43
2002–2003	364.39	102.70	53.19	26.16	6.33	552.78
2003–2004	1127.44	310.14	169.19	92.48	20.00	1719.26
2004–2005	795.29	200.77	129.56	84.43	15.13	1225.18
2005–2006	862.69	196.88	140.16	102.68	19.99	1322.41
2006–2007	1223.53	253.38	199.73	85.96	32.13	1794.73
2007–2008	1160.47	214.18	177.65	81.43	29.83	1663.56
2008–2009	503.70	80.31	70.08	33.05	11.36	698.49
2009–2010	2142.39	343.70	317.42	135.72	46.05	2985.27
2010–2011	1975.57	319.64	285.68	133.43	45.11	2759.44
2011–2012	1233.04	147.62	163.59	76.76	29.25	1650.27
1985–2012	12,016.48	10,303.26	1944.56	1702.93	537.14	26,504.38

From the decomposition results of this perfect method in the share of transport volume contribution, the energy demand is increased with the growth of each mode transportation volume. Railway transportation turnover volume plays a dominant role from the 1985–1987, and contribution rate is more than road transportation rate, at 51.87%, 50.20% and 45.95%, respectively. With the rapid development of highway, road transportation turnover volume increase during 1987–1988, the contribution share of highway is more than rail for the first time. Since 1995–1996, the contribution share is more than 50% and continues to increase. The highest contribution share reached 74.72% from 2010–2011. Certain volatility exists in other modes of transportation.

Highway is the dominant in the changes of the transportation structure, in the total contribution of the transportation structure, and the contribution of highway was more than 30% apart from 1988–1989. Therefore, the adjustment of the transportation structure is the key to reduce energy demand. Railway plays an obvious energy saving role in the transportation structure in two-thirds of the analysis period. The change of air transport structure over a few years has an energy saving effect. Pipeline plays a certain role in energy saving during more than half of the analysis period. The change of water transportation turnover volume saves energy over eight years.

Except for a few years, the five transportation modes play a certain important role in energy saving due to the lower energy intensity. Over the course of 27 years, road energy consumption intensity has an energy saving effect in 15 years, with the highest years saving 6.1414 million tons of standard coal (in 1999–2000). The energy consumption increase was promoted by the change of energy intensity in the rest of the years. The main reason is continuous increasing comfort requirements for highway transportation services, turning out the increase of highway energy intensity. The change of railway energy intensity on energy conservation effects in 23 years, the highest energy saving is in 2010–2011. The main reasons are the implementation of electrification railways instead of steam and diesel locomotives, which prompts the railway transportation energy consumption intensity to decrease. The energy consumption intensity of aviation changes have been saving energy in 18 years, with the highest energy saving is in 2009–2010. The energy consumption intensity of the pipeline has no change in some years without obviously energy saving. The year with the highest energy saving are 1993–1994.

In the long-term development, highway transportation will still account for a large proportion of the total transportation volume, but highway transportation is mainly based on regional short-distance transportation and passenger and freight distribution and plays a role in the connection. The railway will bear a large proportion of inter-regional and inter-city transport demand. Civil aviation mainly completes long-distance transportation and transportation of high-value-added products. The waterway undertakes the transportation of medium and long-distance bulk and cheap goods. However, from the analysis of the energy consumption intensity, the energy consumption intensity of railways is the lowest among various transportation modes. China's railway energy consumption accounts for only 8% of the total consumption of the national transportation industry, fully reflecting the comparative advantage of "low energy consumption and high efficiency", therefore, the railway is the best way to adapt to the development direction of China's energy structure in the transportation industry and plays a significant role in adjusting and optimizing the energy consumption structure in transportation.

4. Conclusions

A perfect model that decomposes the residual term is proposed on the basis of the Laspeyres Index Decomposition and complete decomposition method. This paper focuses on the residual terms in the exponential decomposition method. The existing complete decomposition model is improved, the improved decomposition model is summarized and deduced in detail, and the unified expression of the decomposition model is derived. The model is applied to build a complete decomposition model of the impact factors of transportation energy consumption in different directions. The decomposition model not only has the advantages of the existing decomposition methods but also can "perfect" decompose the remaining items, taking into account the direction of the change of the index influencing factors. This technique makes it possible to present symmetric decomposition without residuals. The perfect method decomposes the residual term completely according to the direction of index change. More accurate calculation results are obtained by comparing Laspeyres and the complete decomposition method. The validity of the perfect model is verified. Lastly, this decomposition model to transportation energy consumption is applied in China and the following conclusions have been drawn.

Transportation turnover volume is the main influencing factor that determines the main trend of transportation energy consumption. Except for a few years, the contribution rate of transportation turnover volume to transportation energy consumption growth is more than 80%.

Transportation energy consumption intensity is the main factor for energy savings. From 1985 to 2012, the change of energy intensity saves 84.609 Mtce. Except for a few years, five transportation modes play a key role in energy saving due to their lower energy intensity. Research on energy consumption intensity should focus on reducing energy consumption intensity of highway and aviation.

The transportation structure adjustment is the key to saving energy and has great potential to save energy. Highways account for absolute advantage. Reducing energy demand is mainly decided by the adjustment of highway transportation structures. Railways play an obvious energy saving role in structure share.

Author Contributions: Conceptualization, Y.Y., X.J. and C.S.L.; methodology, Y.Y. and X.J.; software, Y.Y.; validation, Y.Y., X.J. and C.S.L.; formal analysis, Y.Y., X.J. and C.S.L.; investigation, X.J.; resources, X.J.; data curation, X.J.; writing—original draft preparation, Y.Y., X.J. and C.S.L.; writing—review and editing, Y.Y.; visualization, Y.Y.; supervision, X.J. and C.S.L.; project administration, X.J. and C.S.L.; funding acquisition, X.J. and C.S.L. All authors have read and agreed to the published version of the manuscript.

Funding: This research was funded by the Fundamental Research Funds of the National Natural Science Foundation of China (U2034208) and the study of the spatio-temporal tunnel theory for railway transportation organization (04060075), Beijing Jiaotong University, China.

Institutional Review Board Statement: Not applicable.

Informed Consent Statement: Not applicable.

Data Availability Statement: Data is unavailable due to privacy or ethical restrictions.

Conflicts of Interest: The authors declare no conflict of interest.

References

1. Jenne, C.A.; Cattell, R.K. Structural change and energy efficiency in industry. *Energy Econ.* **1983**, *5*, 114–123. [CrossRef]
2. Doblin, C.P. Declining Energy Intensity in the U.S. Manufacturing Sector. *Energy J.* **1988**, *9*, 109–135. [CrossRef]
3. Howarth, R.B. Energy use in U.S. manufacturing: The impacts of the energy shocks on sectoral output, industry structure, and energy intensity. *J. Energy Dev.* **1991**, *14*, 175–191.
4. Marlay, R.C. Trends in industrial use of energy. *Science* **1984**, *226*, 1277–1283. [CrossRef]
5. Boyd, G.; McDonald, J.F.; Ross, M.; Hansont, D.A. Separating the Changing Composition of U.S. Manufacturing Production from Energy Efficiency Improvements: A Divisia Index Approach. *Energy J.* **1987**, *8*, 77–96. [CrossRef]
6. Boyd, G.A.; Hanson, D.A.; Sterner, T. Decomposition of changes in energy intensity: A comparison of the Divisia index and other methods. *Energy Econ.* **1988**, *10*, 309–312. [CrossRef]
7. Divisia, F. *L'indice Monétaire et la Théorie de la Monnaie*; Société anonyme du Recueil Sirey: Paris, France, 1926.
8. Howarth, R.B.; Schipper, L. Manufacturing Energy Use in Eight OECD Countries: Trends through 1988. *Energy J.* **1991**, *12*, 15–40. [CrossRef]
9. Howarth, R.B.; Schipper, L.; Duerr, P.A. Manufacturing energy use in eight OECD countries: Decomposing the impacts of changes in output, industry structure and energy intensity. *Energy Econ.* **1991**, *13*, 135–142. [CrossRef]
10. Li, J.W.; Shrestha, R.M.; Foell, W.K. Structural change and energy use: The case of the manufacturing sector in Taiwan. *Energy Econ.* **1990**, *12*, 109–115. [CrossRef]
11. Sun, J.W. Changes in energy consumption and energy intensity: A complete decomposition model. *Energy Econ.* **1998**, *20*, 85–100. [CrossRef]
12. Sun, J.W.; Ang, B.W. Some properties of an exact energy decomposition model. *Energy* **2000**, *25*, 1177–1188. [CrossRef]
13. Sun, J.W. *Quantitative Analysis of Energy Consumption, Efficiency and Savings in the World, 1973–1990*; Turku School of Economics Press: Turku, Finland, 1996; Volume A-4.
14. Liu, X.Q.; Ong, H.L. The application of the Divisia index to the decomposition of changes in industrial energy consumption. *Energy J.* **1992**, *13*, 161–177. [CrossRef]
15. Wood, R.; Lenzen, M. Aggregate measures of complex economic structure and evolution: A review and case study. *J. Ind. Ecol.* **2009**, *13*, 264–283. [CrossRef]
16. Ang, B.W. Decomposition of industrial energy consumption: The energy intensity approach. *Energy Econ.* **1994**, *16*, 163–174. [CrossRef]
17. Ang, B.W.; Lee, S.Y. Decomposition of industrial energy consumption: Some methodological and application issues. *Energy Econ.* **1994**, *16*, 83–92. [CrossRef]
18. Ang, B.W.; Choi, K.H. Decomposition of aggregate energy and gas emission intensities for industry: A refined Divisia index method. *Energy J.* **1997**, *18*, 59–73. [CrossRef]
19. Choi, K.-H.; Ang, B.W. Attribution of changes in Divisia real energy intensity index—An extension to index decomposition analysis. *Energy Econ.* **2012**, *34*, 171–176. [CrossRef]
20. Ang, B.W. Decomposition methodology in industrial energy demand analysis. *Energy* **1995**, *20*, 1081–1095. [CrossRef]
21. Ang, B.W.; Zhang, F.Q. A survey of index decomposition analysis in energy and environmental studies. *Energy* **2000**, *25*, 1149–1176. [CrossRef]
22. Ang, B.W. Decomposition analysis for policy making in energy: Which is the preferred method? *Energy Policy* **2004**, *32*, 1131–1139. [CrossRef]
23. Ang, B.W.; Liu, F.L. A new energy decomposition method: Perfect in decomposition and consistent in aggregation. *Energy* **2001**, *26*, 537–548. [CrossRef]

24. Ang, B.W.; Liu, F.L.; Chew, E.P. Perfect decomposition techniques in energy and environmental analysis. *Energy Policy* **2003**, *31*, 1561–1566. [CrossRef]
25. Ang, B.W.; Tian, G. Index decomposition analysis for comparing emission scenarios: Applications and challenges. *Energy Econ.* **2019**, *83*, 74–87. [CrossRef]
26. Wang, H.; Pan, C.; Ang, B.W.; Zhou, P. Does Global Value Chain Participation Decouple Chinese Development from CO_2 Emissions? A Structural Decomposition Analysis. *Energy J.* **2021**, *42*. [CrossRef]
27. Su, B.; Ang, B.W. Improved granularity in input-output analysis of embodied energy and emissions: The use of monthly data. *Energy Econ.* **2022**, *113*, 106245. [CrossRef]
28. Chung, H.S.; Rhee, H.C. A residual-free decomposition of the sources of carbon dioxide emissions: A case of the Korean industries. *Energy* **2001**, *26*, 15–30. [CrossRef]
29. Lenzen, M. Decomposition analysis and the mean-rate-of-change index. *Appl. Energy* **2006**, *83*, 185–198. [CrossRef]
30. Wood, R.; Lenzen, M. Zero-value problems of the logarithmic mean divisia index decomposition method. *Energy Policy* **2006**, *34*, 1326–1331. [CrossRef]
31. Lee, K.; Oh, W. Analysis of CO2 emissions in APEC countries: A time-series and a cross-sectional decomposition using the log mean Divisia method. *Energy Policy* **2006**, *34*, 2779–2787. [CrossRef]
32. Albrecht, J.; François, D.; Schoors, K. A Shapley decomposition of carbon emissions without residuals. *Energy Policy* **2002**, *30*, S0301–S4215. [CrossRef]
33. Wang, W.W.; Zhang, M.; Zhou, M. Using LMDI method to analyze transport sector CO_2 emissions in China. *Energy* **2011**, *36*, 5909–5915. [CrossRef]
34. Wang, L.; Li, H.M. Decomposition Analysis on Dematerialization for the Further Development of Circular Economy. *Bioinform. Biomed. Eng.* **2010**, *30*, 1–4. [CrossRef]
35. Zhang, M.; Li, G.; Mu, H.; Ning, Y. Energy and exergy efficiencies in the Chinese transportation sector, 1980–2009. *Energy* **2011**, *36*, 770–776. [CrossRef]
36. Zhang, M.; Mu, H.; Ning, Y. Accounting for energy-related CO_2 emission in China, 1991–2006. *Energy Policy* **2009**, *37*, 767–773. [CrossRef]

Disclaimer/Publisher's Note: The statements, opinions and data contained in all publications are solely those of the individual author(s) and contributor(s) and not of MDPI and/or the editor(s). MDPI and/or the editor(s) disclaim responsibility for any injury to people or property resulting from any ideas, methods, instructions or products referred to in the content.

Article

Setting Up and Operating Electric City Buses in Harsh Winter Conditions

Maarit Vehviläinen [1], Rita Lavikka [2,*], Seppo Rantala [3], Marko Paakkinen [3], Janne Laurila [1] and Terttu Vainio [3]

1. Climate and Environmental Policy Unit of the City of Tampere, 33101 Tampere, Finland; maarit.vehvilainen@tampere.fi (M.V.); janne.laurila@tampere.fi (J.L.)
2. VTT Technical Research Centre of Finland, 02150 Espoo, Finland
3. VTT Technical Research Centre of Finland, 33101 Tampere, Finland; seppo.rantala@vtt.fi (S.R.); marko.paakkinen@vtt.fi (M.P.); terttu.vainio@vtt.fi (T.V.)
* Correspondence: rita.lavikka@vtt.fi; Tel.: +358-50-384-1662

Abstract: The city of Tampere in Finland aims to be carbon-neutral in 2030 and wanted to find out how the electrification of public transport would help achieve the climate goal. Research has covered topics related to electric buses, ranging from battery technologies to lifecycle assessment and cost analysis. However, less is known about electric city buses' performance in cold climatic zones. This study collected and analysed weather and electric city bus data to understand the effects of temperature and weather conditions on the electric buses' efficiency. Data were collected from four battery-electric buses and one hybrid bus as a reference. The buses were fast-charged at the market and slow-charged at the depot. The test route ran downtown. The study finds that the average energy consumption of the buses during winter was 40–45% higher than in summer (kWh/km). The effect of cabin cooling is minor compared to the cabin heating energy needs. The study also finds that infrastructure needs to have enough safety margins in case of faults and additional energy consumption in harsh weather conditions. In addition, appropriate training for operators, maintenance and other personnel is needed to avoid disturbances caused by charging and excessive energy consumption by driving style.

Keywords: electric city bus; energy consumption; winter; weather; temperature; infrastructure; driving style; cooling; heating; emissions

Citation: Vehviläinen, M.; Lavikka, R.; Rantala, S.; Paakkinen, M.; Laurila, J.; Vainio, T. Setting Up and Operating Electric City Buses in Harsh Winter Conditions. *Appl. Sci.* **2022**, *12*, 2762. https://doi.org/10.3390/app12062762

Academic Editors: Chun Sing Lai, Adel Razek, Kim-Fung Tsang and Yinhai Wang

Received: 17 January 2022
Accepted: 4 March 2022
Published: 8 March 2022

Publisher's Note: MDPI stays neutral with regard to jurisdictional claims in published maps and institutional affiliations.

Copyright: © 2022 by the authors. Licensee MDPI, Basel, Switzerland. This article is an open access article distributed under the terms and conditions of the Creative Commons Attribution (CC BY) license (https://creativecommons.org/licenses/by/4.0/).

1. Introduction

Several cities across the world have sustainable mobility plans to reduce carbon dioxide (CO_2) emissions, pollution and traffic jams [1–3]. For example, the city of Tampere in Finland aims to be carbon-neutral in 2030 and wants to find out how the electrification of public transport would help achieve the climate goal. Public transportation, especially in the form of green solutions, such as electrification, walking, and cycling, can have an enormous effect on reducing CO_2 emissions [4]. The European Commission's 2016 strategy towards low emission mobility includes zero-emission vehicles, such as fully electric cars [5]. Research shows that electric buses produce up to 75% fewer emissions than conventional diesel buses [6]. However, fewer emissions are determined by the grid emissions of the used electricity [7]. Electric buses can also decrease the city transport noise [2,8]. Some studies have also mentioned that electric buses are more comfortable than buses with combustion engines [9].

Electric city buses are still a fairly new phenomenon in city transport. Less research has been conducted on testing electric city buses in various climatic zones. For example, in Finland, the temperature can vary from +35 °C to −35 °C [8]. It is of utmost importance for city traffic planners to understand how electric buses perform in different ambient temperatures [10]. This understanding forms the basis for making other crucial decisions related to electric city buses, such as investment costs, the number of buses, charging stations and routes [11,12]. It needs to be noted that the electric buses' procurement cost is still higher than buses with conventional combustion engines [13].

A research gap exists as less is known about the effects of ambient temperature change on the efficiency (vehicle fuel economy) and range of the city buses in cold climatic zones. In addition, less is known how to operate city buses in hard winter conditions. Therefore, this study aims to understand how to operate battery-electric buses in a city located at latitude 61.3 North, where temperature varies between +32 °C and −32 °C, and buses need to operate on snowy and icy street conditions. In addition, the study aims to understand how electric buses perform in such conditions.

This study collected and analysed weather and electric city bus data to understand the effects of ambient temperature, driving conditions and weather on the efficiency (vehicle energy economy) and range of the city buses in the city of Tampere, Finland.

2. Theoretical Background

2.1. Challenges and Opportunities for Wider Dissemination of Electric City Buses

Previous technology-driven research has covered various crucial aspects of electric buses, such as the performance of battery technologies [14,15], energy-efficient heating, ventilation, air-conditioning (HVAC) systems [16,17] and optimised charging infrastructure settings [18–23]. For example, Cho et al. [14] studied the time-dependent low-temperature power performance of a lithium-ion battery. Their study shows that the interfacial charge-transfer resistance of the anode (graphite) and the cathode (lithium cobalt dioxide) greatly impact the low-temperature power decline. Other non-technical studies have focused on incentives, such as contracting and financing mechanisms, to increase the adoption of electric buses in cities [2,13,24]. For example, Li, Castellanos, et al. [13] found three contracting and financing mechanisms to accelerate electric bus adoption: (1) public and private grants, (2) less costly sources of financing and (3) innovative ways of structuring contractual implementation.

Several studies have also examined the lifecycle assessment (LCA) of the energy and carbon dioxide emissions and calculated lifecycle costs (LCC) of city buses [6,7,12,25–27]. For example, Meishner and Sauer [12] conducted an economic comparison of four different battery charging methods based on the total cost of ownership (TCO), including all investment and operating costs in the bus service. They found that electric buses are economically competitive under favourable conditions. Topić et al. [28] developed a simulation tool to calculate the optimal type and number of buses and charges and predict the TCO of city bus fleets. On the other hand, Bi et al. [26] created an integrated LCA and LCC model to compare the lifecycle performance of plug-in charging versus wireless charging of an electric bus system. It turned out that the wireless charging bus system had the lowest LCC per bus kilometre and had the potential to reduce use-phase carbon emissions due to the light-weighting benefits of onboard battery downsizing compared to plug-in charging [26].

Based on the previous studies, the wider dissemination of battery electric vehicles (BEVs) in cities requires two decisions by authorities. The first is that the city decision makers identify the right contracting and financing mechanisms for replacing conventional buses with BEVs. The second decision is selecting the optimal infrastructure setting for electric city buses. For example, some buses can run almost the whole day with a big enough battery. In contrast, other buses are slow or fast charged in specific charging stations, overnight or in dedicated bus stops. Some researchers have created models for determining the optimal number and location of required charging stations for a bus network and the adequate battery capacity for each bus line [10,18,29,30]. Some studies also report efforts to quickly change the battery in a battery-changing station [20]. In addition, wireless charging technology for electric buses might be an option in the future [26].

2.2. The Effect of Ambient Temperature on Electric City Buses' Electric Consumption

It is already widely communicated that the range of electric vehicles varies with temperature [31]. Research results also confirm this observation. For example, previous studies have shown that the range of electric buses decreases a lot as the temperature drops below zero degrees [32]. This happens because the functioning of the lithium-ion battery varies with temperature [14]. In 2015, Graurs et al. [33] studied public electric bus energy consumption during the coldest months in Latvia (Jan-Mar) and found that the bus consumed 2.86 kWh per kilometre. In 2015, a study estimated that an electric bus with a light aluminium chassis (9000 kg curb weight) consumed on average 1 kWh/km [34]. Electric Commercial Vehicles project (2012–2016) reports that the voltage drop caused by the internal impedance and the applied current at cold temperatures is the reason for reduced battery capacity [8]. In addition, Henning et al. [32] report that temperature drops from around 10.0 °C to around 0 °C caused battery-electric buses to lose around 32.1% of their battery capacity. Another study found that ambient temperature impacted energy consumption a lot in the case of a DC/DC-converter, heat pump and drive motor [35]. However, power steering and an air compressor did not have an insignificant impact.

Ambient temperature and several other factors influence the range of electric buses, such as topography, the road pave (sand, concrete), and the road's surface conditions (wetness, snowfall, sleet, ice). Bartłomiejczyk and Kołacz [36] studied the relationship between ambient temperature and demand for heating power. They found that traffic congestion can result in a 60% overall increase in energy consumption. They also found that auxiliaries may consume 70% of the electric bus's energy during winter, whereas they generally consume almost 50% of total energy use [36]. In addition, the driving style affects the use of energy. For example, fast accelerations consume lots of energy [20]. Preheating the battery and indoor air before starting the bus, on the other hand, saves energy [16]. However, heating the bus cabin during the drive also reduces the range. Research shows that heating and cooling can consume 35% of all energy in electric cars [37]. While the energy consumption of an electric bus increases, the battery's state of charge and the travel range are reduced [38]. Therefore, researchers have proposed a fuzzy braking strategy of which electricity consumption was shown in a simulation platform to decrease by 9.8% compared to the normal braking energy management control strategy [39]. Then again, using a novel sorption air conditioner was shown to save cruising electric vehicles' mileage by 100 km [40].

The bus body and insulation materials also impact how efficiently a bus passenger cabin is kept warm or cool, which again affects the energy usage of the bus [32]. For example, Chiriac et al. [41] estimated that the average energy demand due to ambient heat loss of bus structure and opening the doors at the stops was 12–14 kWh. The bus was 12-m long, had 100 passengers and three doors, 150 kW electric traction motor and 33 kW installed power for the heating system [41].

3. Case Study

This paper adopts a case study methodology to describe the procurement and operational models used for setting up the needed technical infrastructure for electric city buses and operating those buses in a cold climatic zone to understand the effect of this climate on the energy use of buses. A single case study approach [42] was adopted because it well suits the study of a topic that is not yet well explored.

The city of Tampere in Finland aims to be carbon-neutral in 2030. Currently, the traffic is accountable for about a quarter of the city's carbon dioxide emissions. In 2019, an amendment to the Clean Vehicles Directive came into force. The amendment obliges the public sector to procure zero-emission vehicles. The city of Tampere is one of the lighthouse cities of the EU-funded STARDUST project where new solutions for reducing emissions are piloted. In Tampere, the focus is on reducing emissions from public transport and supporting light transport. Furthermore, the 2017 climate strategy of the Finnish Government necessitates a 50% reduction, compared to the year 2005, in transport greenhouse gas (GHG) emissions by 2030 [43]. Thus, the cities in Finland are encouraged to electrify public transportation.

Nysse is the Tampere regional public transport organisation in Finland. It had around 41,300,000 passengers in 2019. In 2020, the number of passengers was lower due to COVID-19, which made people remote work. Nysse serves an area that includes eight municipalities where approximately 390,000 inhabitants live. About 60% of bus services are outsourced to three major private operators, and in-house operator drives around 40% of bus services. Nysse has 280 buses in operation.

The climate in Tampere is cold and temperate can vary between +32 °C and −32 °C. The lowest average temperature is −8.2 °C in February, and the average temperature is 16.0 °C in July. The average annual temperature is 3.7 °C.

In Tampere, electric city buses were introduced at the end of 2016. A test system was created to collect data on electric buses. Initially, a meter was placed in one bus for sensing energy efficiency. Later on, measuring equipment was installed on three other electric buses. When problems in setting up the data collection were tackled, the data has been collected for more than two years, including a couple of winter seasons.

Since 2017, electric buses have been operating in the city of Tampere's bus route 2, which starts from Pyynikintori market and ends at Rauhaniemi. The round-trip length is around 8.8 km. The buses are fast-charged at the market and slow-charged at the depot. The buses spend the night inside a warm depot; thus, the buses' interiors are warm when they leave.

Data were collected from four battery-electric buses and one hybrid bus as a reference. The electric buses were model Solaris Urbino 12 low entry, with an autonomy of approximately 60 km. The electric buses used lithium-titanate (LTO) batteries of 3×25 kWh that last at least 10 years and demand 0.5 h for a total fast charge. The maximum speed of the buses is over 70 km/h, and the average consumption is 100–150 kWh each 100 km (without heating). Buses have 32 fixed seats, four-fold seats, and 46 standing spaces. The dimensions of the buses are the following:

- Wheelbase: 5900 mm
- Length: 12,000 mm
- Width: 2550 mm
- Height: 3300–3480 mm
- Curb weight: 14 t
- Max torque: 973 Nm
- Max power: 250 kW

Data was collected with WRM-247 of Wapice Ltd. for three years, between January 2017 and August 2021. The following data was collected by the meter in real-time: latitude, longitude, speed, energy consumption, charging time, temperature, battery's state of charge, battery power and distance travelled. The typical sampling frequency was 1 Hz.

WRM-247 devices allow remote management, measurement and control. They were purchased and installed by the STARDUST lighthouse project. One of the e-buses already used this device, so the other three e-buses and the hybrid bus were equipped with the same devices to achieve comparable data. The buses' mobile networks (3G/4G) were used as the connectivity layers towards the server that collects the data. The aim was to also have the same equipment on the charging platform to acquire more accurate information about the charging of the buses.

The data from the WRM-247 devices was sent to an IoT-Ticket instance, which is an IoT platform product by Wapice. City's Azure logic app checks the IoT-Ticket REST API for new data once every two hours and transfers all the new data to City's Azure blob storage. From there, the data is aggregated and stored in Azure SQL database. Data analysis from the stored data is carried out and presented in a dashboard using Microsoft Power BI. Datasets are shared via a REST API. Figure 1 illustrates the architecture of the e-bus data in the city's Azure.

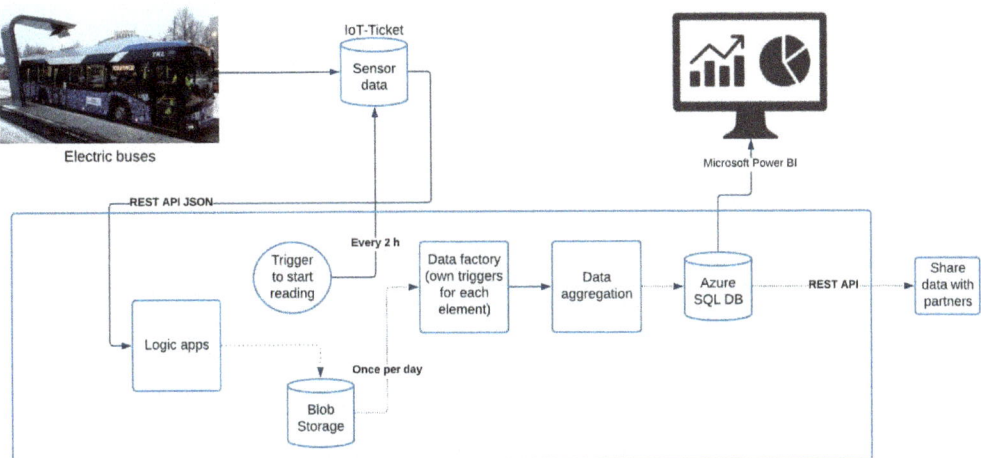

Figure 1. The architecture of the e-bus data in the city's Azure.

In addition, two interviews were conducted in December 2021 to understand the operation of the electric bus system. The purpose of the interviews was to form an overall picture of the operation and find out whether there were any challenges in the charging system or the operation of the buses related to winter conditions. Both interviewees work in the city of Tampere, and they are experienced experts in city transportation and urban environmental infrastructure systems such as maintenance. The interviews focused on specific questions that emerged during the setup phase of the charging system and measurement data-analysis process. Each interview lasted around one hour, and they were not recorded, but detailed notes were taken. The failure situations have also been collected and listed in an excel document throughout the operation period.

4. Results and Discussion

This section first reports the procurement and operational model of the electric city buses in Tampere, Finland. After that, the section presents the results regarding the energy consumption of the buses and explains the energy monitoring and estimation in detail. Finally, the section discusses the energy consumption of the buses and the charging strategy.

4.1. Procurement and Operational Model of Electric City Buses

Tampere has been travelling by public transport since 1948. The City of Tampere Transport Authority—now Tampere City Transport, better known as TKL—operated the public transport service for half a century. A new era of public transport started in April 2006, when it was organised with a subscriber-producer model and brand "Nysse" was born to Tampere public transport. In addition to customer service, planning and administration departments were located in the subscriber unit. TKL stayed to provide a transport service. At the same time, some bus lines began to be put out to tender for private transport producers. The name of the comprehensive service was Tampere Public Transport.

Several parties were involved in preparing and implementing the electric bus system procurement; Tampere city was the main implementer of the procurement project (Figure 2). The buses were acquired and operated by TKL. The public transport planning was responsible for the procurement of the charger. The City of Tampere's public transport unit acted as a subscriber to electronic transport. At the beginning of the project, the City of Tampere's ECO_2 project worked as a leader and coordinator. Inter-city co-operation was also used in the preparation of the acquisition. The collaboration during the acquisition preparation deepened with establishing a joint Forum eKEKO (Extended Management Team for Electric Bus Projects) headed by VTT. In the forum, Finnish cities working with electric buses and VTT as an organizer shared experiences about electric bus operation. Sharing experiences and deepening knowledge, especially with Helsinki and Turku, has been enlightening, especially from the operator's point of view.

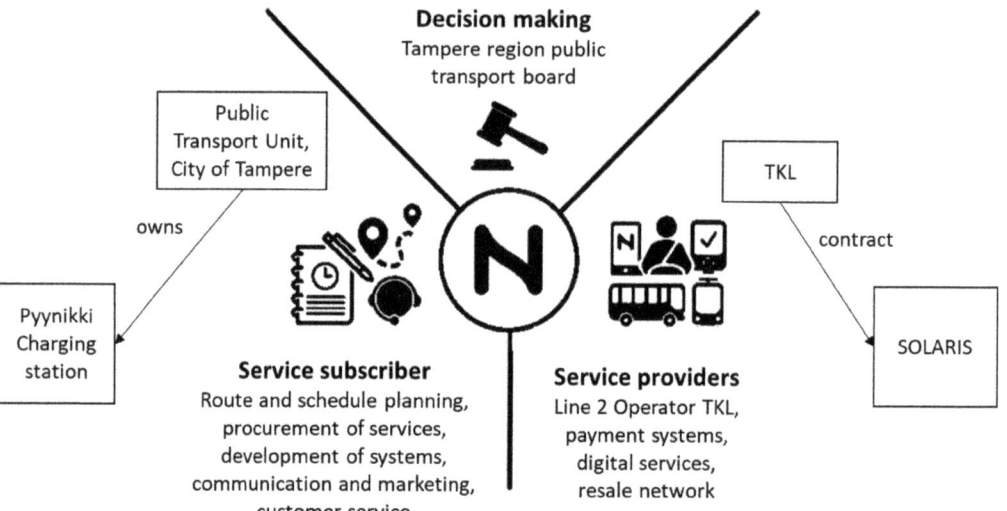

Figure 2. The parties involved in preparing and implementing the electric busses' acquisition.

Figure 2 illustrates the operational model of the electric bus system in the subscriber-producer framework. The city owns charging stations. The procurement process was felt to be easier to start when the city decided to make the charging station investment itself. The supplier is responsible for its maintenance for the five first years, and the charging station also has a warranty. TKL signed a contract with Solaris to lease the four e-buses and maintain the charging stations. The companies Ekoenergetyka and Schunk acted as charger suppliers. The buses used roof-mounted pantographs Schunk SL102.

The tender for the electric bus system of line 2 was opened on 1 July 2015, and the offer period ended on 30 September 2015. With the competition, Tampere became the first city in Finland to acquire an electric bus system through an open tender. Tenders were initially reviewed, and a supplier was selected in October 2015. Tampere decided to ask for a tender for a five-year leasing agreement. At the end of the deadline, Tampere could have decided to buy the buses in a case seen as feasible and reasonable.

The city bought a study that showed that choosing the line in economic terms necessitates making decisions on the following questions [44]:

- How many buses can use the same charger, and what the utilisation rate is? For example, charger costs can be shared between the buses that use it.
- How much is the annual mileage? For example, an electric bus saves more expenses compared to a diesel bus the more you drive it.
- What is the line terminus time? For example, optimal terminus times minimise the indirect costs for additional equipment and staff.
- How much does the terminal stop time shorten during peak hours?
- How long is the route? This information affects the battery dimensioning.

With these calculations, line 2 was identified to be well suited for electric buses. The line is relatively slow and contains a lot of traffic lights. Electric buses are very suitable for urban traffic on routes with lots of stops and traffic lights, allowing energy efficiency compared to diesel buses.

The selection of the charging type in the procurement affects the selected line. The study by Markkula and Vilppo states that charging at the terminus is the best option because then the battery size may be small [44]. If buses were to be charged only in the depot, the passenger space would be reduced because the battery needs to be bigger and thus requires space. The charging method selection was also in favour of line 2, which is suitable for an electric bus line due to its Pyynikintori terminal. Pyynikintori has several line terminuses, which allows the charging station to be used jointly on several bus lines. Using the charger on more than one line would increase the system's profitability. It was also considered whether the terminal has enough time to charge when choosing a line.

During the monitoring period, line 2 experienced minor route changes as the construction of a tram site progressed in the centre of Tampere. Figure 3 illustrates how the route has changed since 2018.

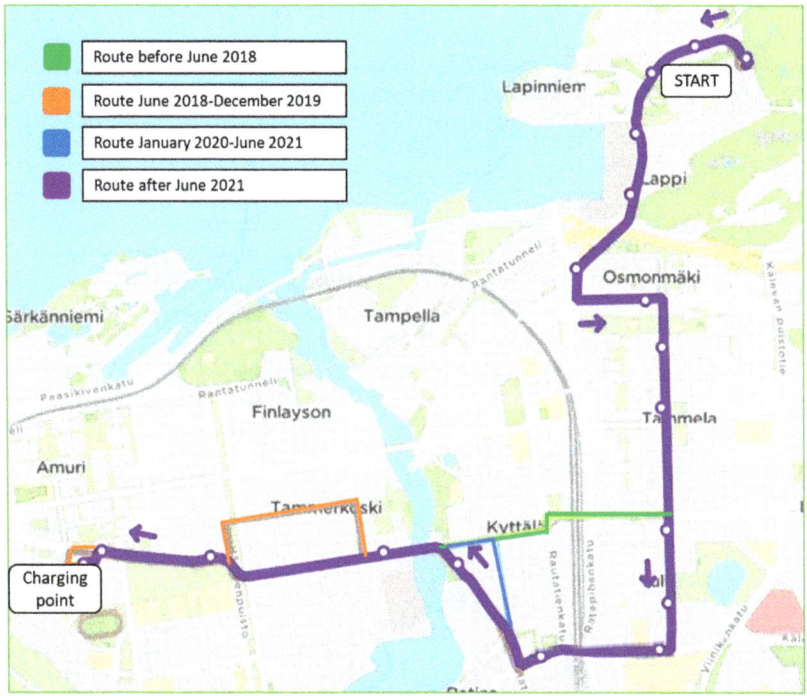

Figure 3. Line 2 changes due to the tram site.

During the procurement process, many issues related to the procurement method and the technology to be procured had to be resolved, confirming earlier findings that the operation of electric buses requires more planning than the operation of conventional buses [10]. For example, the choice of procurement method was already the subject of debate. The decision on the open procedure was questioned. The experience of Tampere concerning electric bus systems was still limited, which raised many additional questions during the acquisition. In this respect, the conciliation procedure would have been more forgiving. The market dialogue aimed to identify available solutions and meet the city's needs. The key issue for the procurement was whether the electric buses and chargers were to be purchased separately or together. The city ended up with a single-supplier model because it was perceived to simplify procurement, which already had sufficient uncertainties. This model could avoid possible disagreements between the bus and charger supplier.

Already in 2019, the buses had travelled 600,000 km. Theoretically, the mileage should have been 800,000 km. The lost driving time is due to, e.g., traffic accidents in cramped urban traffic that led to sheet metal crashes. According to Nysse's (Tampere regional public transport organization) own customer feedback survey on using the electric buses, the passengers have given positive feedback. The electric buses have a low noise level, and they are easily accessible because of low-floor vehicles, three wide doors and no steps on the aisle. The only negative feedback has been a large number of rear-facing seats in low-floor city buses.

The development of a sustainable public transport system requires a shift to emission-free bus transport, the development of smooth travel chains and new mobility services, and an overall improvement in service level to increase the modal share of public transport in line with the target set. Targets in the number of outsourced transport services using low emission fuel sources (bus and tramway line kilometres) are set 35% (2025) and 100% (2030). More than 700 tonnes of CO_2 emissions have been saved during the pilot. This is a conservative estimate because the saved CO_2 emissions had to be partially extrapolated to the monitoring time due to data interruptions. In spring 2021, Tampere started a tender for two different bus lines. The requirement was that buses must comply with the clean vehicles Directive (EU 2019/1161). The selected operator implements the requirements with an electric bus system. During spring 2022, there will be 26 new electric buses when the winning operator brings its buses into service.

4.2. Monitoring and Estimation of Energy Consumption

During the monitoring period between 2019 and the end of August 2021, the four electric buses travelled approximately 500,000 kilometres using an average of 1.43 kWh/km electricity. Figure 4 shows seasonal variation of energy efficiency between 2019–2021 in monthly intervals. However, it has to be taken into account that cooling systems use electricity during summer and heating during winter. These electricity consumptions could not be separated in our monitoring setup.

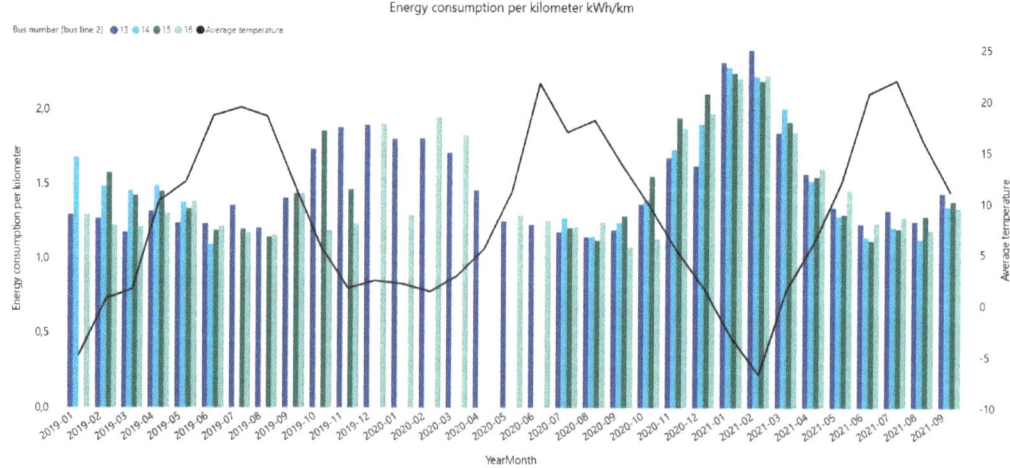

Figure 4. Example of created monitoring dashboards: energy consumption. Disclaimer: The comma is used as a decimal separator instead of a dot because the version of Power BI used in the visualization followed the grammar rules of the Finnish language.

The monitoring setup experienced fewer problems during the last two years. Thus, these years are more closely analysed. Figures 5 and 6 show summer, and Figures 7 and 8 wintertime, separately divided into weekly intervals. Summer is defined as the months between May and October. Respectively, winter is defined by the months between November and April.

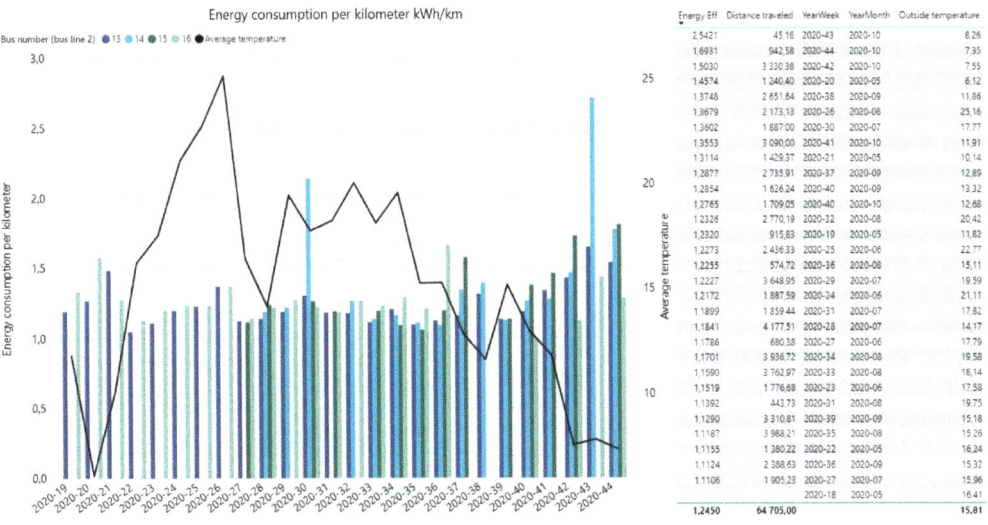

Figure 5. Energy efficiency and outside temperature per week during 1 May 2020–31 October 2020. Disclaimer: The comma is used as a decimal separator instead of a dot because the version of Power BI used in the visualization followed the grammar rules of the Finnish language.

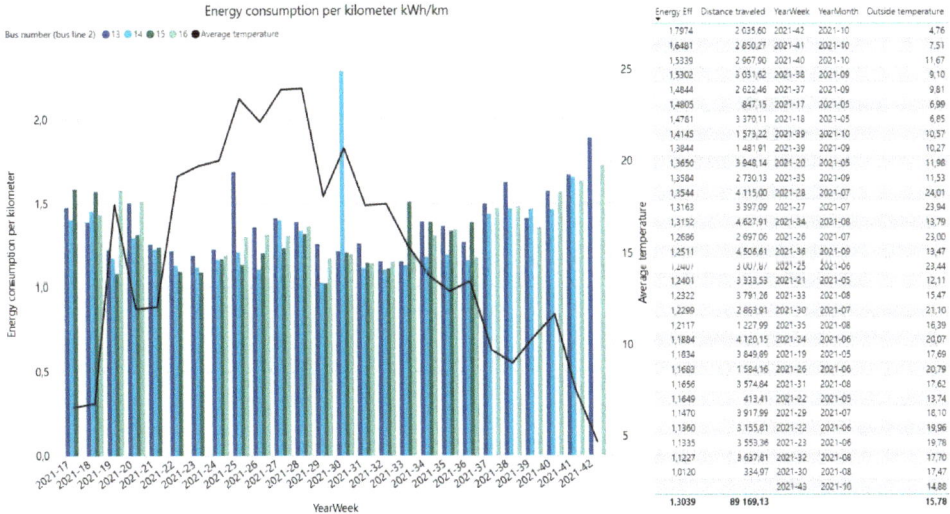

Figure 6. Energy efficiency and outside temperature per week during 1 May 2021–31 October 2021. Disclaimer: The comma is used as a decimal separator instead of a dot because the version of Power BI used in the visualization followed the grammar rules of the Finnish language.

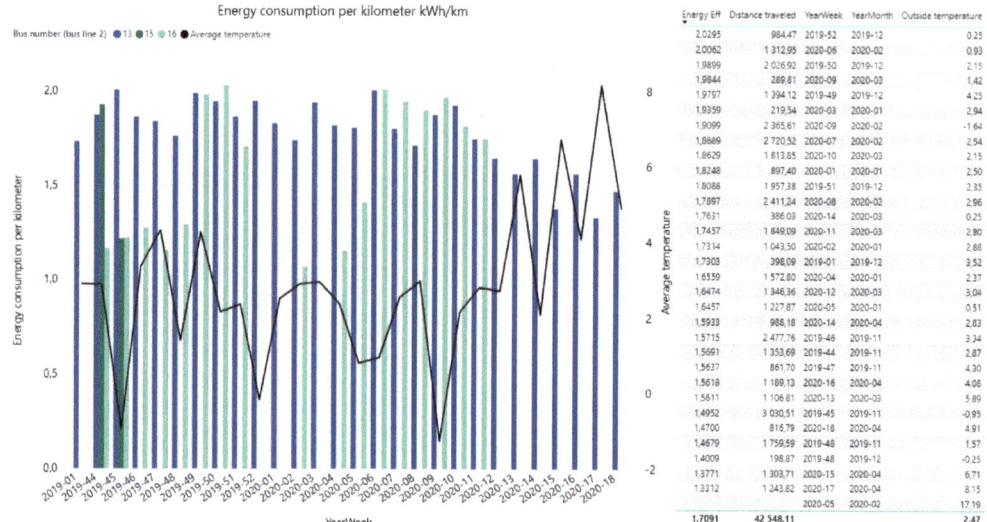

Figure 7. Energy efficiency and outside temperature per week during 1 November 2019–30 April 2020. Disclaimer: The comma is used as a decimal separator instead of a dot because the version of Power BI used in the visualization followed the grammar rules of the Finnish language.

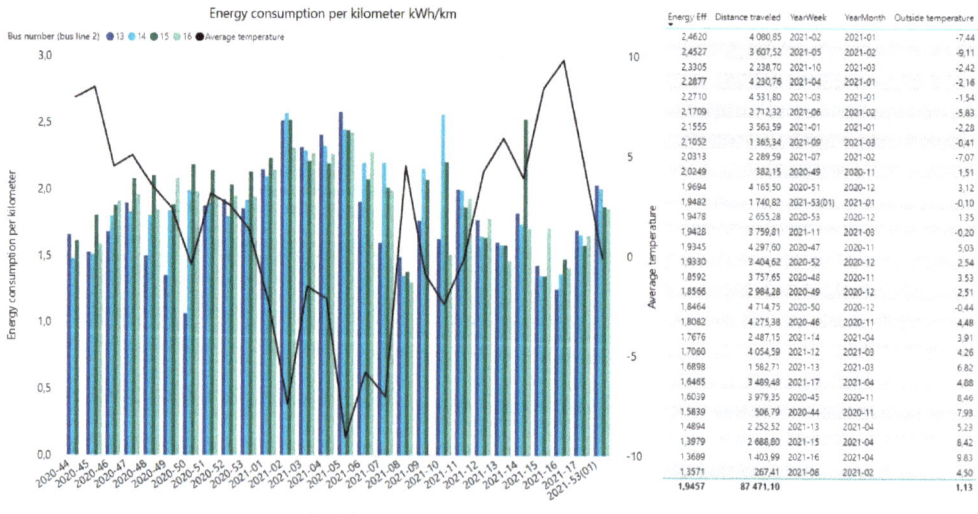

Figure 8. Energy efficiency and outside temperature per week during 1 November 2020–30 April 2021. Disclaimer: The comma is used as a decimal separator instead of a dot because the version of Power BI used in the visualization followed the grammar rules of the Finnish language.

The findings show that the energy consumption was 1.24 kWh/km during summer 2020 and 1.30 kWh/km during summer 2021. Then again, the energy consumption was 1.71 kWh/km during winter 2020 and 1.95 kWh/km during winter 2021.

The results of this study inform traffic planners on how electric buses perform in different environmental conditions. Several factors influence the energy consumption of electric city buses. The design considerations such as the total mass of the bus and the regeneration rate can significantly affect the energy efficiency. Several studies have been made where the driving range of different structure selection have been analysed by making simulation or analysing measured data [10,45,46]. This investigation focuses on the effect of environmental factors since the monitored buses are completely similar. In cold climatic zones, the temperature changes the most energy consumption. Still, the number of passengers, road topography, traffic congestion, driving style, and surface condition contribute to it, as previous studies have shown [16,20,36].

The previous results dealt with daily averages. It is necessary to analyse each driving from Pyynikintori to Rauhaniemi individually to obtain more detailed information on the effect of weather phenomena on consumption. Since the elevation variations along the route are about 28 m, the directions are analysed separately. Measurements from January 2019 to August 2021 have been selected for the study. There have been some changes to the route, but they are so minor that their effect is negligible. The lengths of the routes have ranged from 8.8 km to 10.3 km. The analyses have been performed only for working day hours from 6:30 to 22:30 and on a route where the doors have been open for more than half a minute to obtain comparable results.

Figure 9 shows electrical energy consumption as a function of temperature so that each blue dots represent one drive between the start and end station. Since 2019, it has been possible for the operators to choose between diesel fuel and electricity for heating the battery and the interior. The measurements show that most drivers had opted for fuel heating. This option was removed from 2020 onwards, and the fuel heater was controlled automatically; it was activated only when the ambient temperature was below −15 °C. This can be clearly seen from the figure. The figure also has a polynomial curve fit to data, which have been carried out in two separate cases due to the diesel heating. Below 0 °C, the fitting is carried out to temperature data between −15 and 10 °C. At temperatures above zero degrees, a fitting was made to those values. The energy consumption (EC) is shown in Equation (1), where T is the temperature in degrees Celsius.

$$EC = \begin{cases} 5.4 \times 10^{-5} T^3 - 2.7 \times 10^{-4} T^2 - 0.05 T + 1.6, & T \geq 0\,°C \\ -0.04 T + 1.6, & T < 0\,°C \end{cases} \quad (1)$$

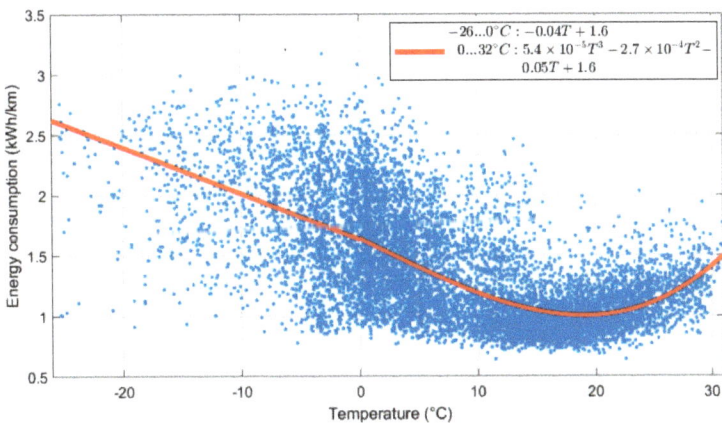

Figure 9. Energy consumption as a function of temperature (blue dots—samples; red line—polynomial curve fit).

The graph shows that the average energy consumption increased by about 0.4 kWh when the temperature decreased to 10 °C. Air conditioning increases consumption when the temperature is above 15 degrees. Its effect is approximately equivalent to an increase in consumption at temperatures below zero.

Two outlier groups are interesting: (1) Although there is considerable frost, there is small consumption, and (2) high consumption near zero degrees. A common feature of the first case is the short duration when doors are open, meaning few passengers and little heat escaping from the doors. There might be several reasons for the second group: snow, slush or slippery road. In any case, they have the doors open for a long time, indicating a lot of passengers.

Table 1 shows energy consumption (kWh/km) for different cases for 2019 and 2020–2021 separately due to the change in the heating mentioned above. The data is the same as in Figure 9 except that the cumulative passenger entry and exit time shall be at least one minute. There are some high values for 2019 since the winter was rather cold and snowy. The highest six values occurred when the temperature was between −1 and 5 °C. This reflects the fact that snowy weather, particularly snowmelt, increases consumption significantly. Winter 2020 was warm and had hardly any snow, but 2021 had cold and snowy winter. The median energy consumption was 0.8 kWh/km greater when the temperature was below zero than over zero. Again, there were some high consumptions near zero degrees, which can be observed from the figure. The median difference between driving the route when the snow was melting and without melting was 0.2 kWh/km. The distribution of melting cases was twofold: either it had little effect, or the consumption increased greatly. Understandably, a small melt has little effect, but if the vehicle is driven in slush ice, consumption increase considerably.

Table 1. Energy consumption (kWh/km) for specific conditions.

	2019				2020–2021			
	Samples	Median	Mean	Max.	Samples	Median	Mean	Max.
Doors open 1–3 min	3354	1.1	1.2	3.1	7592	1.2	1.3	3.0
Doors open > 3 min	2839	1.2	1.3	3.5	1536	1.4	1.5	3.2
Temperature ≥ 0 °C	4638	1.1	1.3	3.5	7282	1.1	1.2	3.0
Temperature < 0 °C	1555	1.2	1.4	3.5	1846	1.9	1.9	3.2
Snowing > 0 cm/h	551	1.2	1.4	3.0	539	1.8	1.8	3.0
Snowing > 1 cm/h	239	1.2	1.4	2.8	229	1.9	1.9	2.9
Snow melt or sublimation	1220	1.2	1.3	3.5	1260	1.7	1.7	3.1
Temperature −1–10 °C without snow melt	1602	1.2	1.4	3.4	2158	1.4	1.5	3.0
Temperature −1–10 °C and snow melt	877	1.2	1.3	3.5	977	1.6	1.6	3.1

The results show that the weather and climate affect the operation of buses and the entire electric bus system. Four electric buses travelled approximately 500,000 kilometres (2019–2021) using an average of 1.43 kWh/km electricity. However, this number also includes the energy use by cooling systems during summer and heating during winter, as they could not be separated in the monitoring setup. The average energy consumption during thermal winter was 2.1 kWh/km and 1.2 kWh/km during thermal summer. Thermal winter starts when the average temperature is below 0 degrees Celsius at least five days in a row. Thermal summer starts when the temperature is over +10 degrees Celsius at least five days in a row. In Tampere, the thermal winter was 7 December 2020–22 March 2021, and thermal summer was 10 May 2021–14 September 2021.

4.3. Experiences in Operating and Charging Electric Buses

The electric motor does not generate waste heat the same way as a diesel motor, and separate heating must be provided. When electric buses started to operate in Tampere, the heating system worked with diesel fuel. The heating system was changed in winter

2019 to an electricity-based system where the water circulation system is electrically heated. In the beginning, a bus driver could select if the electric heating is used instead of a fuel heater. It was noticed that drivers tended to select the fuel heater, as they wanted to avoid charging as it was difficult. Therefore, the operation was changed so that the electric heating is the default, and the fuel heater activates if the outside temperature decreases under −15 degrees Celsius. The effect can be seen when comparing the winter energy consumption between 2020 and 2021.

The study finds that the effect of cooling is minor compared to the effect of heating energy needs. The data includes two different periods in which the buses were using different control strategies for indoor heating. The second winter period shows the increase in electrical energy consumption when using a full electric indoor heating instead of using an auxiliary fuel heater for heating when a heat pump cannot produce enough heating power. When considering local emissions and total greenhouse gas emissions of an electric bus system, one option to minimise the emissions is to go for full electric heating. Still, the penalty in cold climates is the increased battery energy consumption at very low temperatures, which needs to be taken into account in the electric bus charging design—either the battery capacities need to be increased to be able to handle the additional consumption, or opportunity charging needs to be arranged to be able to charge the buses more often.

A more detailed analysis of energy consumption would have required data that was not accessible because there was no mention of ownership or access to the data in the model leasing agreement. This must be considered in future agreements in other areas than e-mobility.

Another area that could use detailed data is the passenger number and its effect on energy consumption. Currently, only passengers boarding the bus can be entered into the information system. Therefore, the exact number of passengers is not known. The city of Tampere has carried out pilots to monitor the number of passengers to monitor, improve and optimise the occupancy rate, but this is still a clear area for development and research; how many passengers there are and how their number affects energy consumption. The graphs (Figures 10 and 11) show the number of passengers per hour made by one bus. Winter weather is not attractive for cycling or walking. Passenger number is smaller during summertime compared to winter. The bus runs a round trip from the departure stop to the charging station (one end) and back during the hour. It can be estimated that the passenger number in the bus simultaneously during peak hours is ~50.

The energy consumption of electric buses as a whole has been lower than expected, but the differences per driver have been surprisingly large. For the project, it was impossible to monitor driver-specific energy consumption more accurately since it would have required an act on co-operation [47].

The study also reveals that when setting up and operating charging systems with automated charging devices (pantographs), the effects of the weather must be considered when selecting and preparing the location of the charging point. The buses were charged with a bus-mounted pantograph, where an automated charging connector rises from the bus roof to connect with a receptacle mounted on a charging mast or pole (Figure 12). This connection has some tolerances for misalignment, but snow build-up on the driving tracks during winter has shown in practice that these tolerances are not enough to maintain a reliable connection in all weather conditions without additional measures.

Positioning the bus under the charging system pantograph was a difficult task in the beginning since the bus needed to be in an exact correct spot to initiate the charging. Therefore, paint markers on the curb were used. The bus's front door was aligned with the markers when the bus was in a correct position (Figure 13). Another challenge was the alignment of the bus lateral distance from the kerb.

A defrost system was built at the Pyynikki charging point to prevent a hard snow ridge from building on the charging point driving tracks. The defrosting system caused decreasing soil bearing capacity, and buses driving to the same spot for charging caused a depression to the charging area, causing problems in the charging connection. A heated concrete foundation was built, and the area was paved again with new asphalt. There have been no problems with durability since the latest repair.

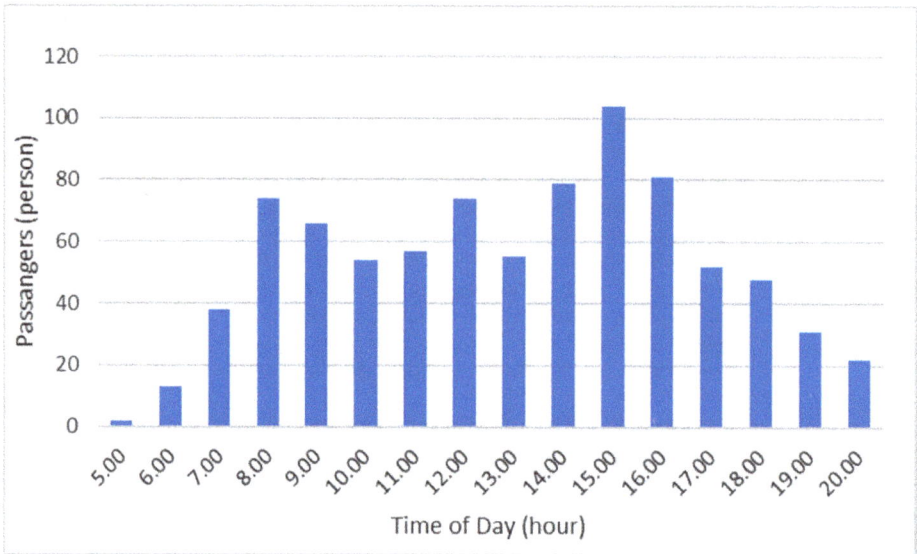

Figure 10. Passengers on one bus shift (13 February 2020).

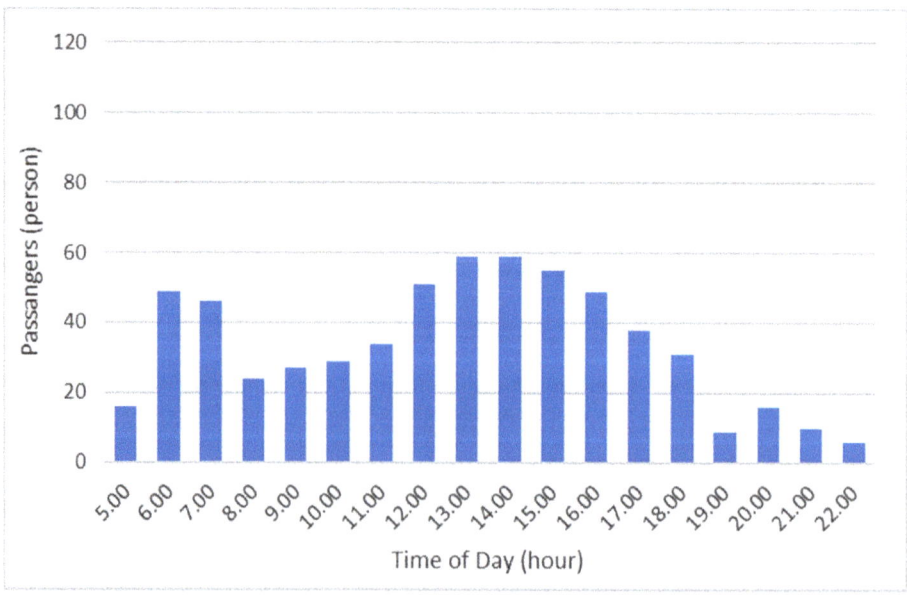

Figure 11. Passengers on one bus shift (11 July 2019).

Figure 12. Heated foundation for the charging point to remove snow build-up on driving tracks.

Figure 13. Paint markers on the kerb to position the bus for charging.

In addition to the equipment that affects the operation of the electric bus system, people and the operation of electric buses also notice the great importance of the way drivers drive, for example, in energy consumption, which raises the importance of training the drivers.

Operating electric buses is generally different from operating diesel or hybrid buses. There was some training included in the contract with Solaris, but common thought amongst drivers has been that there should have been more operator training. The charging was perceived to be difficult for drivers, and the driving style affected the electricity consumption greatly. Driver assistance tools could be one solution to help drivers better operate electric buses [48].

5. Conclusions

The data shows the increased energy consumption of electric buses in cold climatic conditions. During thermal winter, the average energy consumption was 2.1 kWh/km and 1.2 kWh/km during thermal summer. Thermal winter starts when the average temperature is below 0 degrees Celsius at least five days in a row. Thermal summer starts when the temperature is over +10 degrees Celsius at least five days in a row. When comparing the best-case energy consumption in summer, with energy consumption of roughly 1.1 kWh/km with the hot summer weeks of around 1.35 kWh/km and cold winter weeks with the highest energy consumption of almost 2.5 kWh/km, one can see that the effect of cooling is minor compared to the effect of heating energy needs.

When using a fuel heater, the energy consumption from the battery can be reduced, with the best results being roughly 1.35 kWh/km. This was also indicated by the operators' behaviour when they could prioritise the diesel heater to avoid charging. However, using a fuel heater comes with the cost of local emissions, even though the fuel would be from sustainable sources. When using electric heating, local emissions can be minimised. Even with diesel heating during colder months, the greatly increased energy consumption from the traction battery needs to be taken into account in the charging design—either the battery capacities need to be increased to be able to handle the additional consumption, or opportunity charging needs to be arranged to be able to charge the buses more often. Driving in the heavy slush ice, in particular, increases consumption considerably.

The comments from the interviews highlight the systemic nature of the electrification of transport. The design of an electric bus system, especially in cold climate conditions, needs to address appropriate energy transfer to the buses, without affecting the operation, in all conditions. The system needs to have enough safety margins in case of faults and for additional energy consumption in harsh weather conditions. Charging equipment and locations need to withstand the continuous loading of soil on same positions and maintain the potentially needed opportunity charging systems within their operating tolerances. Finally, appropriate training for operators, maintenance, and all relevant personnel can help avoid disturbances caused by charging and excessive energy consumption by driving style.

The motivation for this test was to determine whether the electrification of public transport helps achieve the carbon neutrality goal of Tampere. This goal was met, but the test raised several technical issues, such as the energy consumption of different devices and the impact of driving styles. The depth of analysis was limited because the test project was not granted access to the leasing bus internal data collection system. In addition, some issues during the analysis phase were caused by the synchronisation of time series data from different sources, which should also be taken care of when setting up a data collection system.

Promoting alternative propulsion for transport and procuring an electric bus system are ways to achieve the city's climate goals. In addition to meeting the emission targets, there is a desire to try new promising technology in public transport, which will lead to cost savings in the longer term. The acquisition was also based on the goal of making the electric bus line an innovation platform for intelligent transport, which can be used to test and put into practice products and services related to intelligent transport. On the operational side, the experience was gained, and an overview of operations was obtained in winter conditions. From the point of view of innovation and information, leasing buses is not ideal unless the contract ensures that the collected data can be utilised with sufficient precision for the subscriber. Unfortunately, the agreement did not take a sufficient position on data ownership. There was no agreement on data collection, which prevented a more detailed analysis of energy consumption. With a view to future agreements, data ownership must be considered, and transparency of operations is important for development and scientific studies.

Further research could collect more data to conduct a more thorough operational analysis using the charger and charging process data, including the vehicle alignment to the charging point. For example, the number of unsuccessful charging attempts could have pointed out areas of improvement in the charging process or training. The effect of drivers' driving style on energy consumption is an interesting area for study, but such research must take into account GDPR and the required anonymisation of data. Tampere collects a lot of data related to traffic and distributes it as open data [49]. Combining this data with more accurate data collected from buses creates the basis for new studies and the ability to find correlations between conditions and different parameters.

Author Contributions: Conceptualization, M.V., R.L., S.R., M.P., J.L. and T.V.; methodology, R.L. and J.L.; formal analysis, S.R. and J.L.; data curation, S.R.; writing—original draft preparation, R.L.; writing—review and editing, M.V., R.L., S.R., M.P., J.L. and T.V.; visualization, J.L., M.V. and S.R. All authors have read and agreed to the published version of the manuscript.

Funding: This research was funded by STARDUST, a city lighthouse project, which has received funding from the European Union's Horizon 2020 research and Innovation programme under grant agreement N°774094.

Institutional Review Board Statement: Not applicable.

Informed Consent Statement: The interviewees have been explained the purpose of the interview and the study, and they have given their consent to the interview. Participation in this research project have been voluntary and interviewees did not receive any payments for participating in this research interview. Sensitive information was not discussed in the interviews. Interviews were conducted by teams discussions, and no audio or video was recorded, only written notes were collected.

Data Availability Statement: The city of Tampere provides the following data publicly available for reuse: the hourly energy consumption and distance traveled of the three electric buses and one hybrid bus. The data, provided in an Excel format, and the data's metadata can be downloaded from the IDA research data storage service, which is organised by the Finnish Ministry of Education and Culture and maintained by CSC (IT Center for Science Ltd.). The dataset's metadata can be downloaded as a JSON file. Two datasets are available in the IDA service: (1) *Ebus distance data*, which shows how many kilometres the four electric city buses have travelled in Tampere, is available: https://etsin.fairdata.fi/dataset/199dff1c-389d-4069-a55d-c2d9e2379a2d (accessed on 3 March 2022). The columns of the 4.68 MB Excel file are: busId, nodeId, variable, unit, date, time, number start, number (end), actual, updated_timestamp. (2) *Ebus electricity consumption data*, which shows the electrical energy consumption of the four electric buses in Tampere, is available: https://etsin.fairdata.fi/dataset/cbea812a-9093-11c6-8ce9-041770b41c7d (accessed on 3 March 2022). The columns of the 1.93 MB Excel file are: date, time, busId, nodeId, SUM, Updated_date. The datasets are available under the creative commons license (CC BY 4.0), which allows the user to (1) Share—copy and redistribute the material in any medium or format and (2) Adapt—remix, transform, and build upon the material for any purpose, even commercially, provided that the dataset is given appropriate credit and any changes are indicated.

Acknowledgments: The research was part of STARDUST, a city lighthouse project, which has received funding from the European Union's Horizon 2020 research and Innovation programme under grant agreement N°774094. The funder had no role in the design of the research; in the collection, analyses, or interpretation of data; in the writing of the manuscript, or in the decision to publish the research. We would like to express our great appreciation to the following parties: Wapice Oy Ltd. for their support with monitoring sensors and data collection. Tampere University of Applied Sciences (TAMK) for the support for data collection and maintenance work, especially the assistance given by Jukka Pellinen and the team has been essential. TKL and especially Kalle Keinonen for the help, expertise and enthusiasm. Mika Heikkilä from the city of Tampere for long-term cooperation. Persons in the city of Tampere public transport organization, especially Juha-Pekka Häyrynen and his expertise.

Conflicts of Interest: The authors declare no conflict of interest. The funders had no role in the design of the study; in the collection, analyses, or interpretation of data; in the writing of the manuscript, or in the decision to publish the results.

References

1. Hipogrosso, S.; Nesmachnow, S. Analysis of sustainable public transportation and mobility recommendations for Montevideo and Parque Rodó neighborhood. *Smart Cities* **2020**, *3*, 479–510. [CrossRef]
2. Pagliaro, M.; Meneguzzo, F. Electric bus: A critical overview on the dawn of its widespread uptake. *Adv. Sustain. Syst.* **2019**, *3*, 1800151. [CrossRef]
3. Li, Z.; Li, C.; Liao, K.; Yin, Z. Rethinking impact evaluation and carbon reduction analysis on electric bus vehicles in China. In Proceedings of the IOP Conference Series: Earth and Environmental Science, Banda Aceh, Indonesia, 26–27 September 2018; Volume 121.
4. Toledo, A.L.L.; Rovere, E.L. La Urban mobility and greenhouse gas emissions: Status, public policies, and scenarios in a developing economy city, Natal, Brazil. *Sustainability* **2018**, *10*, 3995. [CrossRef]
5. European Commission A European Strategy for Low-Emission Mobility. Available online: https://ec.europa.eu/transport/themes/strategies/news/2016-07-20-decarbonisation_en (accessed on 15 December 2021).
6. Lajunen, A.; Lipman, T. Lifecycle cost assessment and carbon dioxide emissions of diesel, natural gas, hybrid electric, fuel cell hybrid and electric transit buses. *Energy* **2016**, *106*, 329–342. [CrossRef]
7. Mulley, C.; Hensher, D.A.; Cosgrove, D. Is rail cleaner and greener than bus? *Transp. Res. Part D Transp. Environ.* **2017**, *51*, 14–28. [CrossRef]
8. Pihlatie, M.; Pippuri-Mäkeläinen, J. *Electric Commercial Vehicles (ECV)—Final Report*; VTT Technology: Tampere, Finland, 2019; p. 348.
9. UITP (The International Association of Public Transport). *ZeEUS eBus Report # 2 ZeEUS eBus Report # 2—An Updated Overview of Electric Buses in Europe*; ZeEUS EU-Funded R&D Project; UITP: Brussels, Belgia, 2017.
10. Ranta, M.; Pihlatie, M.; Paakkinen, M. Analysis of key performance indicators of electric bus systems in Helsinki and comparison to simulated results. In Proceedings of the 30th International Electric Vehicle Symposium (EVS30), Stuttgart, Germany, 9–11 October 2017; European Association for Electromobility (AVERE): Stuttgart, Germany, 2017; pp. 3300–3311.
11. Rogge, M.; van der Hurk, E.; Larsen, A.; Sauer, D.U. Electric bus fleet size and mix problem with optimization of charging infrastructure. *Appl. Energy* **2018**, *211*, 282–295. [CrossRef]
12. Meishner, F.; Sauer, D.U. Technical and economic comparison of different electric bus concepts based on actual demonstrations in European cities. *IET Electr. Syst. Transp.* **2020**, *10*, 144–153. [CrossRef]
13. Li, X.; Castellanos, S.; Maassen, A. Emerging trends and innovations for electric bus adoption—A comparative case study of contracting and financing of 22 cities in the Americas, Asia-Pacific, and Europe. *Res. Transp. Econ.* **2018**, *69*, 470–481. [CrossRef]
14. Cho, H.; Choi, W.; Go, J.; Bae, S.; Shin, H. A study on time-dependent low temperature power performance of a lithium-ion battery. *J. Power Sources* **2012**, *198*, 273–280. [CrossRef]
15. Horita, T.; Shimazaki, J. Performance evaluation of lithium-ion batteries on small electric bus. *Electr. Eng. Japan* **2017**, *198*, 68–76. [CrossRef]
16. Suh, I.S.; Lee, M.; Kim, J.; Oh, S.T.; Won, J.P. Design and experimental analysis of an efficient HVAC (heating, ventilation, air-conditioning) system on an electric bus with dynamic on-road wireless charging. *Energy* **2015**, *81*, 262–273. [CrossRef]
17. Zhang, Z.; Wang, J.; Feng, X.; Chang, L.; Chen, Y.; Wang, X. The solutions to electric vehicle air conditioning systems: A review. *Renew. Sustain. Energy Rev.* **2018**, *91*, 443–463. [CrossRef]
18. Kunith, A.; Mendelevitch, R.; Goehlich, D. Electrification of a city bus network—An optimization model for cost-effective placing of charging infrastructure and battery sizing of fast charging electric bus systems. *Int. J. Sustain. Transp.* **2017**, *11*, 707–720. [CrossRef]
19. An, K. Battery electric bus infrastructure planning under demand uncertainty. *Transp. Res. Part C* **2020**, *111*, 572–587. [CrossRef]
20. Kim, J.; Song, I.; Choi, W. An electric bus with a battery exchange system. *Energies* **2015**, *8*, 6806–6819. [CrossRef]
21. Bagherinezhad, A.; Palomino, A.D.; Li, B.; Parvania, M. Spatio-temporal electric bus charging optimization with transit network constraints. *IEEE Trans. Ind. Appl.* **2020**, *56*, 5741–5749. [CrossRef]
22. Chen, H.; Hu, Z.; Zhang, H.; Luo, H. Coordinated charging and discharging strategies for plug-in electric bus fast charging station with energy storage system. *IET Gener. Transm. Distrib.* **2018**, *12*, 2019–2028. [CrossRef]
23. Jahic, A.; Eskander, M.; Schulz, D. Charging schedule for load peak minimization on large-scale electric bus depots. *Appl. Sci.* **2019**, *9*, 1748. [CrossRef]
24. Pucci, P. Spatial dimensions of electric mobility—Scenarios for efficient and fair diffusion of electric vehicles in the Milan Urban Region. *Cities* **2021**, *110*, 103069. [CrossRef]
25. Ayetor, G.K.; Mbonigaba, I.; Sunnu, A.K.; Nyantekyi-Kwakye, B. Impact of replacing ICE bus fleet with electric bus fleet in Africa: A lifetime assessment. *Energy* **2021**, *221*, 119852. [CrossRef]
26. Bi, Z.; De Kleine, R.; Keoleian, G.A. Integrated life cycle assessment and life cycle cost model for comparing plug-in versus wireless charging for an electric bus system. *J. Ind. Ecol.* **2016**, *21*, 344–355. [CrossRef]

27. Thein, S.; Chang, Y.S. Decision making model for lifecycle assessment of lithium-ion battery for electric vehicle—A case study for smart electric bus project in Korea. *J. Power Sources* **2014**, *249*, 142–147. [CrossRef]
28. Topić, J.; Soldo, J.; Maletić, F.; Škugor, B.; Deur, J. Virtual simulation of electric bus fleets for city bus transport electrification planning. *Energies* **2020**, *13*, 3410. [CrossRef]
29. Ranta, M.; Pihlatie, M.; Pellikka, A.-P.; Laurikko, J.; Rahkola, P.; Anttila, J. Analysis and comparison of energy efficiency of commercially available battery electric buses. In Proceedings of the 2017 IEEE Vehicle Power and Propulsion Conference (VPPC), Belfort, France, 11–14 December 2017; IEEE: Belfort, France, 2017; pp. 1–5.
30. Anttila, J.; Todorov, Y.; Ranta, M.; Pihlatie, M. System-level validation of an electric bus fleet simulator. In Proceedings of the 2019 IEEE Vehicle Power and Propulsion Conference (VPPC), Hanoi, Vietnam, 14–17 October 2019; Institute of Electrical and Electronics Engineers: Hanoi, Vietnam, 2019.
31. Bullis, K. Electric Vehicles out in the Cold. Available online: https://www.technologyreview.com/2013/12/13/175150/electric-vehicles-out-in-the-cold/ (accessed on 19 February 2021).
32. Henning, M.; Thomas, A.; Smyth, A. *An Analysis of the Association between Changes in Ambient Temperature, Fuel Economy, and Vehicle Range for Battery Electric and Fuel Cell Electric Buses*; Urban Affairs at EngagedScholarship@CSU: Cleveland, OH, USA, 2019; Volume 12316.
33. Graurs, I.; Laizans, A.; Rajeckis, P.; Rubenis, A. Public bus energy consumption investigation for transition to electric power and semi-dynamic charging. *Eng. Rural Dev.* **2015**, *14*, 366–371.
34. Vilppo, O.; Markkula, J. Feasibility of electric buses in public transport. *World Electr. Veh. J. Vol.* **2015**, *7*, 357–365. [CrossRef]
35. Taavetinkangas, J. The Effect of Ambient Temperature on the Consumption of Electric Buses. Bachelor's Thesis, Turku University of Applied Sciences, Turku, Finland, 2018.
36. Bartłomiejczyk, M.; Kołacz, R. The reduction of auxiliaries power demand: The challenge for electromobility in public transportation. *J. Clean. Prod.* **2020**, *252*, 119776. [CrossRef]
37. Evtimov, I.; Ivanov, R.; Sapundjiev, M. Energy consumption of auxiliary systems of electric cars. *MATEC Web Conf.* **2017**, *133*, 06002. [CrossRef]
38. Cigarini, F.; Fay, T.A.; Artemenko, N.; Göhlich, D. Modeling and experimental investigation of thermal comfort and energy consumption in a battery electric bus. *World Electr. Veh. J.* **2021**, *12*, 7. [CrossRef]
39. Ye, L.; Liang, C.; Li, X.; Li, D. Energy efficiency improvement of eddy-current braking and heating system for electric bus based on fuzzy control. *IET Electr. Syst. Transp.* **2020**, *10*, 385–390. [CrossRef]
40. Jiang, L.; Wang, R.Z.; Li, J.B.; Wang, L.W.; Roskilly, A.P. Performance analysis on a novel sorption air conditioner for electric vehicles. *Energy Convers. Manag.* **2018**, *156*, 515–524. [CrossRef]
41. Chiriac, G.; Lucache, D.D.; Nițucă, C.; Dragomir, A.; Ramakrishna, S. Electric bus indoor heat balance in cold weather. *Appl. Sci.* **2021**, *11*, 11761. [CrossRef]
42. Yin, R.K. *Case Study Research: Design and Methods, Essential Guide to Qualitative Methods in Organizational Research*; Sage Publications: Thousand Oaks, CA, USA, 2009.
43. Huttunen, R. *Government Report on the National Energy and Climate Strategy for 2030*; Publications of the Ministry of Economic Affairs and Employment: Helsinki, Finland, 2017.
44. Markkula, J.; Vilppo, O. *Tampereen Bussiliikenteen Sähköistäminen (Electrification of Tampere Bus Traffic, in Finnish)*; City of Tampere: Tampere, Finland, 2014.
45. Miri, I.; Fotouhi, A.; Ewin, N. Electric vehicle energy consumption modelling and estimation—A case study. *Int. J. Energy Res.* **2021**, *45*, 501–520. [CrossRef]
46. Kivekas, K.; Lajunen, A.; Baldi, F.; Vepsalainen, J.; Tammi, K. Reducing the energy consumption of electric buses with design choices and predictive driving. *IEEE Trans. Veh. Technol.* **2019**, *68*, 11409–11419. [CrossRef]
47. Ministry of Economic Affairs and Employment of Finland, Act on Co-Operation within Undertakings Is Not just a Law on Terminating Employment 2021. Available online: https://tem.fi/en/negotiation-obligation (accessed on 1 March 2022).
48. Halmeaho, T.; Antila, M.; Kataja, J.; Silvonen, P.; Pihlatie, M. Advanced driver aid system for energy efficient electric bus operation. In Proceedings of the 1st International Conference on Vehicle Technology and Intelligent Transport Systems—VEHITS, Lisbon, Portugal, 20–22 May 2015; pp. 59–64.
49. The City of Tampere Open Data Tampere—Open Data from Tampere Region. Available online: https://data.tampere.fi/en_gb/ (accessed on 5 February 2022).

Article

Parameter Optimization and Tuning Methodology for a Scalable E-Bus Fleet Simulation Framework: Verification Using Real-World Data from Case Studies

Mohammed Mahedi Hasan [1,2], Nikos Avramis [3], Mikaela Ranta [4], Mohamed El Baghdadi [1,2] and Omar Hegazy [1,2,*]

1. ETEC Department & MOBI Research Group, Vrije Universiteit Brussel, Pleinlaan 2, 1050 Brussels, Belgium
2. Flanders Make, 3001 Heverlee, Belgium
3. TNO Automotive, Automotive Campus 30, 5708 JZ Helmond, The Netherlands
4. VTT Technical Research Center, Vuorimiehentie 3, P.O. Box 1000, 02044 Espoo, Finland
* Correspondence: omar.hegazy@vub.be; Tel.: +32-488-819-954

Abstract: This study presents the optimization and tuning of a simulation framework to improve its simulation accuracy while evaluating the energy utilization of electric buses under various mission scenarios. The simulation framework was developed using the low fidelity (Lo-Fi) model of the forward-facing electric bus (e-bus) powertrain to achieve the fast simulation speeds necessary for real-time fleet simulations. The measurement data required to verify the proper tuning of the simulation framework is provided by the bus original equipment manufacturers (OEMs) and taken from the various demonstrations of 12 m and 18 m buses in the cities of Barcelona, Gothenburg, and Osnabruck. We investigate the different methodologies applied for the tuning process, including empirical and optimization. In the empirical methodology, the standard driving cycles that have been used in previous studies to simulate various use case (UC) scenarios are replaced with actual driving cycles derived from measurement data from buses traversing their respective routes. The key outputs, including the energy requirements, total cost of ownership (TCO), and impact on the grid are statistically compared. In the optimization scenario, the assumptions for the various vehicle and mission parameters are tuned to increase the correlation between the simulation and measurement outputs (the battery SoC profile), for the given scenario input (the velocity profile). Improved simple optimization (iSOPT) was used to provide a superfast optimization process to tune the passenger load in the bus, cabin setpoint temperature, battery's age as relative capacity degradation (RCD), SoC cutoff point between constant current (CC) and constant voltage charging (CV), charge decay factor used in CV charging, charging power, and cutoff in initial velocity during braking for which regenerative braking is activated.

Keywords: e-bus powertrain; tuning and optimization; iSOPT; digital twins; internet-of-things

1. Introduction

Automotive system engineering has come a long way since Henry Ford spearheaded the assembly line process a century prior, resulting in sharp increases in productivity and manufacturing efficiency and corresponding decreases in the price of the manufactured vehicle [1]. The evolution in automotive system engineering in the 21st century saw the advent of Industry 4.0, empowered by the very high-speed internet (Internet 2.0), resulting in paradigm shifts in manufacturing production operations by merging the boundaries of the physical and virtual worlds [2]; the current state of the art (SotA) includes the Internet-of-things (IoT), cloud-connected processes, and digital twins (DT) technology. A DT model can have various levels of fidelity [3] in the virtual domain, but they are all tuned to accurately reflect a physical object or system. A DT model relies on the real-time measurement of data from numerous sensors installed in the physical system to continuously train itself to behave as its physical counterpart to corresponding input stimuli [4]. A fully trained and tuned DT offers several advantages, including quicker iterative testing of the virtual model, using multiple copies for evaluation of different aspects of the vehicle at a fraction of the cost and time. For a city bus operator (CBO) and electricity distribution system operator (DSO), the virtual models can substitute for their real counterparts during fleet use case (UC) simulations to determine the real-world feasibility of electrification of the bus routes.

In [5], a simulation framework developed for the European Commission's Horizon 2020 project ASSURED was used to investigate the UCs of single buses and fleets of buses in various cities to determine their energy expenditure and TCO. The simulation framework was also used to study the reduction in energy utilization possible by applying different energy saving (ECO) strategies, and various optimization scenarios were investigated to determine the charging infrastructure that will minimize the fleet TCO (for the CBO) and load on the grid due to fleet charging (for the DSO). However, due to the lack of measurement data during the research conducted using various assumptions, including the use of a standard (hybrid SORT) driving cycle as the input scenario, constant average vehicle speed profile and a randomized passenger profile throughout the simulation period of one day, these assumptions naturally were not consistent with real-world conditions, including traffic situations on the road, and did not differentiate between peak and non-peak hours for passenger commutes. The hybrid SORT driving cycle can only be applied repetitively, synchronized to a constant average vehicle speed, throughout the simulation implying a constant traffic situation throughout the day. Furthermore, although the passenger profile was randomized, the output of the randomizer tended towards a full bus with time??, resulting in energy requirements that were aggressive. Similarly, the charging scenario assumed constant duration spacing in between two charging events, which resulted in a more simplified charging strategy. Finally, the results of the simulation framework were not validated using actual measurements; thus, the output of the simulation framework could only be taken as estimates.

In this research, the measurement data from the electric buses in the cities of Barcelona, Osnabruck, and Gothenburg are used to tune and validate the simulation framework. Furthermore, the study investigates the differences in energy consumption between the standard and actual driving scenarios, and finally an optimization was performed to determine the optimal charging strategy, given variable durations between two charging events, based on the input scenario. The objectives of this research are twofold: one is to validate the simulation framework, so that it can be used to investigate different scenarios with a high degree of confidence in its results; and two is to lay the framework for the creation of a DT of the electric bus for future research. Section 2 introduces the simulation framework and the necessary modification that enables it to work with actual measurements. Section 3 reports on the energy requirements from the vehicle demonstrations in cities. The tuning methodologies used to ensure that the output of the simulation framework matches the measured output, given similar inputs, are described in Section 4. Section 5 describes the optimization procedure for the charging strategy for bus fleets, whose driving scenarios were constructed using the actual driving scenarios. Finally, Section 6 concludes with how this research can be used to construct a DT from the simulation framework.

2. The Simulation Framework

A low fidelity (Lo-Fi) simulation framework illustrated in [5] was used to evaluate the energy expenditure for fleets of vehicles and impact on the electricity grid for a given mission profile in this research, with modifications in the framework to accept measurement data as the scenario input. An offline scenario input process was developed for the framework in this research, meaning that the simulation is not occurring parallelly in real-time using the measurement data taken during the bus demonstrations. Rather, the measurement data from the sensors are stored and later input to the simulation.

The simulation framework is based on basic electrical, mechanical, kinematic, and thermal equations needed to represent the charging infrastructure and forward-facing electric bus (e-bus) powertrain model, as shown in Figure 1. Unlike a high fidelity (Hi-Fi) simulation model, where the simulation model is based on detailed physical equations of the actual system and uses small timesteps to ensure very high accuracy of the simulation output, the Lo-Fi framework uses look-up tables (LuTs) to define the efficiency maps of the various electronic, mechanical, electromechanical, and electrochemical devices integrated within the powertrain; and basic equations that model the overall energy transfer behavior of each component. The timestep in a Lo-Fi model is large to ensure high simulation speed at the cost of accuracy. Thus, even though a Lo-Fi model cannot simulate transient behaviors, they can be used to get a rapid estimate of the steady-state behavior. Therefore, Lo-Fi models can be used to simulate large time ranges covering the lifetime of the e-bus or large fleets of e-buses within a reasonable timeframe. Furthermore, a Lo-Fi model can be used to perform a fleet-level energy management and charging strategy (EM&CS) optimization, which require very fast simulation speeds. The Simulink framework was designed to use the measurement data from the demonstrations as inputs: the design of the energy storage system (ESS) block allowed comparison to be made between the simulated and measured battery SoC values for validation purposes, while the energy management system (EMS) block was designed to allow the model to be tuned to minimize the difference between the simulated and measured values. More details of the tuning process are provided in Section 4.

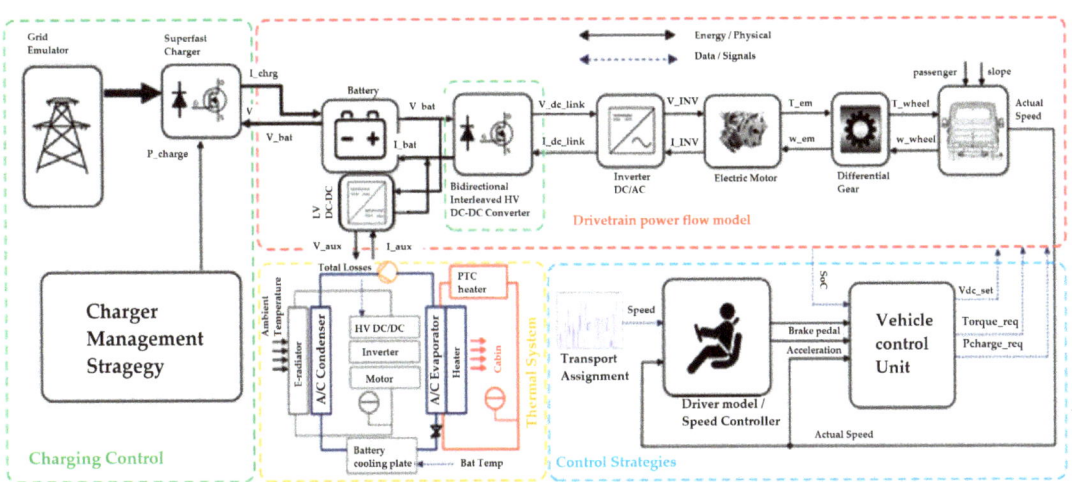

Figure 1. Overview of the simulation framework illustrating the forward-facing e-bus powertrain and grid infrastructure.

Inputs to the Simulation Framework

The framework was designed to accept measurement data from the bus as inputs in an offline process. In ASSURED, the various OEMs and CBOs involved in the demonstrations were responsible for the data collection process and then forwarding those data to the simulation team. However, different OEMs and CBOs used different data logging and data processing techniques. Therefore, it was not possible to apply a standardized methodology for data collection, making the offline validation the most suitable option. Table 1 gives a concise overview of the measurement data collected in each city. As can be seen, the collected data seems rather arbitrary; it is due to different stakeholders being involved in the data collection process. However, each stakeholder was required, at minimum, to provide the vehicle's speed profile (to be used as the simulation input) and battery SoC profile (to be compared with the simulation output), at a data logging frequency of 1 Hz, to ensure reasonable tuning and validation of the simulation framework. Beyond these constraints, each stakeholder communicated, according to their data sharing policies, a subset of the following parameters: energy usage rate, mileage, charging state, charging time, ambient temperature, road inclination, GPS coordinates, and altitude.

Table 1. Overview of the measured data collected.

City and Bus Type	Measured Parameter (Unit)	Logging Frequency
BCN, 12 m *	Speed profile (km/h) Measured energy (kWh) State of charge (%) GPS coordinates (°)	0.5 Hz
BCN, 18 m *	Speed profile (km/h) State of charge (%)	20 Hz
OSN, 12 m *	Speed profile (km/h) Measured energy (kWh) State of charge (%) GPS coordinates (°)	0.5 Hz
OSN, 18 m *	Speed profile (km/h) Mileage (km) State of charge (%) Charging state (-)	20 Hz
GOT, 12 m *	Speed profile (km/h) State of charge (%) Mileage (km) Charging time (s) Road inclination (°) Ambient temperature (°C)	10 Hz

* BCN: Barcelona, OSN: Osnabruck, GOT: Gothenburg; 12 m and 18 m refers to the bus length.

Sensor data in vehicles are mainly communicated via the CAN bus network and logged via CAN-based dataloggers attached to the vehicle's CAN network and wirelessly communicated to a central server via the GSM (3G/4G) or Wi-Fi. The data is then decoded from the CAN message format (.blf), which is binary, into a more user readable format, including comma separated values (.csv), excel (.xlsx), or a simple text (.txt) file, using a CAN database (.dbc) file structure. The next step is to convert them into a common format, the MATLAB data (.mat) file, after which the parameter values are brought to a common sampling rate of 10 Hz, using up-and-down sampling techniques; wherein 10 Hz was chosen as a simulation time step of the Lo-Fi model. The data is then pre-processed to remove noise from the data, especially those which were measured via the GPS module, since GPS user accuracies, even with augmentation and when operated in wide open areas, are in "meters" for horizontal (i.e., longitude and latitude) measurements, and much worse for vertical (i.e., altitude) measurements [6]. In an urban setting featuring many obstacles (i.e., buildings, bridges etc.) and a multipath signal environment due to reflected signals, these accuracies are further degraded. Finally, the data is thoroughly checked to ensure that the speed and acceleration do not exceed the vehicle maximum for those parameters, and that the road inclination and difference in altitude between the lowest and highest point of the route were within known ranges.

3. Use Case Demonstration Overview

Numerous demonstration runs were conducted in the cities of Barcelona and Osnabruck using 12 m and 18 m e-buses, and in the city of Gothenburg using 12 m e-bus. For the 12 m bus, the demonstrations took place at two different months of the year to account for variation in weather. Table 2 provides the details for all the demonstrations considered for simulation and analysis. The simulations were run for approximately the same duration as their standard driving cycle counterparts in [5]; thus, some of the scenarios described in Table 2 were repeated until the desired timeframe was achieved. The complete specifications of the scenarios of the three routes, the 12 m and 18 m bus, as well as the climate profile for each city, used as inputs for the simulations are presented in [5], while the exact maps of the demonstration routes are shown in the Appendix A.

Table 2. Overview of the demonstration scenarios.

City and Bus Type	Demonstration Month	Operational Scenario	Route
BCN, 12 m *	December	25.7 km in 160 min	H16
	February	16.7 km in 68 min 13.3 km in 31 min 26.3 km in 126 min	
BCN, 18 m *	June	109.8 km in 558 min	
OSN, 12 m *	March	49.1 km in 243 min 64.0 km in 310 min 63.2 km in 357 min	N5
	May	88.7 km in 473 min	
OSN, 18 m *	April	88.1 km in 252 min	
GOT, 12 m *	May	168.3 km in 784 min 99.5 km in 434 min	R55
	October	148.4 km in 575 min	

* BCN: Barcelona, OSN: Osnabruck, GOT: Gothenburg; 12 m and 18 m refers to the bus length.

The measurement data gathered from the demonstrations were used to improve the UCs that were simulated using the standard driving cycles. Comparing the kinematic characteristics between the actual and standard driving cycles, very striking differences can be seen in their respective profiles. All measurements from the demonstrations exhibited accelerations whose ranges were higher than what was assumed when simulating the UCs using the standard driving cycle. Similarly, the maximum measured velocity from the demonstrations were higher than the maximum velocities assumed in the standard driving cycle, except in the case of the Osnabruck 12 m bus. Finally, in Barcelona, the average velocity measured during the demonstrations were higher than what was assumed in the standard driving cycle, while those of Osnabruck and Gothenburg were lower. From these facts, it can be assumed that the energy requirements for the buses subject to the measured driving cycles will be higher. Table 3 details the characteristics of the measured driving cycle from the demonstrations as well as the standard driving cycle, while Figure 2 illustrates this difference visually. As can be seen from the figure, the standard driving cycle is composed of clean and repeating patterns, while the actual measurements look random and somewhat noisy.

Table 3. Comparison between the demonstration and the standard driving profile characteristics.

City and Bus Type	Demonstration Profile Characteristics	Standard Profile Characteristics
BCN, 12 m * (4 demos)	Avg. vel. 9.65~26.2 km/h Max. vel. 59.8~78.4 km/h Max acc. 1.30~2.06 m/s^2	BCN, Route H16, All buses: Avg. vel. 9.52 km/h Max. vel. 29.8 km/h Max. acc. 0.51 m/s^2
BCN, 18 m * (1 demo)	Avg. vel. 11.8 km/h Max. vel. 72.0 km/h Max acc. 2.36 m/s^2	
OSN, 12 m * (4 demos)	Avg. vel. 10.6~12.8 km/h Max. vel. 43.2~59.0 km/h Max acc. 1.30~3.51 m/s^2	OSN, Route N5, All buses: Avg. vel. 19.8 km/h Max. vel. 61.9 km/h Max. acc. 1.06 m/s^2
OSN, 18 m * (1 demo)	Avg. vel. 21.0 km/h Max. vel. 67.3 km/h Max acc. 2.39 m/s^2	

Table 3. *Cont.*

City and Bus Type	Demonstration Profile Characteristics	Standard Profile Characteristics
GOT, 12 m * (3 demos)	Avg. vel. 7.99~12.6 km/h Max. vel. 70.9~82.4 km/h Max acc. 4.99~5.33 m/s²	GOT, Route R55, 12 m bus: Avg. vel. 18.3 km/h Max. vel. 57.2 km/h Max. acc. 0.98 m/s²

* BCN: Barcelona, OSN: Osnabruck, GOT: Gothenburg; 12 m and 18 m refers to the bus length.

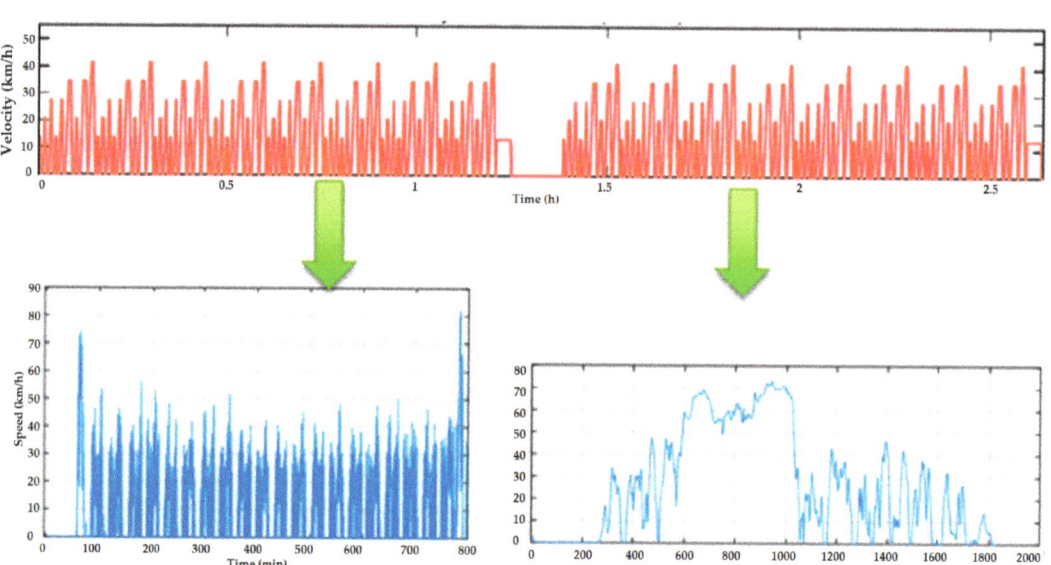

Figure 2. Comparison between the measured driving cycle from the demonstrations and standard driving cycles.

3.1. Simulation Output

Figure 3 compares the energy requirements determined from the simulation of the measurement data of the demonstrations to those from the UC simulations in [5], while Figure 4 illustrates the effects of the various ECO-features in reducing the energy consumption of the bus. For the remainder of the article, the baseline energy requirement is defined as the average energy requirement found from the UC simulations in [5] using the standard driving cycle. Figure 3 shows, as expected, that in Barcelona the energy requirements are significantly higher for the demonstrations compared with the baseline. However, the opposite is true for Osnabruck, where the energy requirements are significantly lesser than the baseline. This can be explained by the simple fact that in Barcelona, the average and maximum speeds of the buses in the demonstration are much higher than the baseline. Thus, the buses in the demonstration experience higher aerodynamic drag, leading to greater energy requirements compared with the baseline. In the case of Osnabruck, the opposite was true; for the 12 m bus, the average and maximum speeds of the demonstrations were less than those of the baseline, thus lesser energy was required than for the baseline. In the case of the 18 m bus, the average and maximum speeds are comparable between the demonstrations and the baseline; thus, the energy requirement between the baseline and demonstration is similar. For Gothenburg, the average velocity is less than the average velocity of the baseline, even if the maximum velocity is higher. Thus, the bus expends less energy on average compared to the baseline. From the results, it can also be deduced that normal acceleration and deceleration have a low impact on the rate of energy expenditure of the vehicle; this can be explained by the fact that the vehicle is an electric bus with an efficient energy recovery system (via regenerative braking), thus 70% to 80% of the traction energy expended during acceleration is recovered during braking [7]. The amount of energy recovered depends on several factors including the momentum of the vehicle during braking, SoC of the battery, and capability of the battery to accept the power influx. For small EVs such as the Renault Zoe, the cutoff velocity beyond which energy recovery can efficiently occur during braking is 5 m/s [7], but for heavy-duty vehicles such as buses, the regeneration can occur from a lower velocity due to their larger masses resulting in greater braking momentum. Thus, regenerative braking in an urban scenario with low speeds and heavy traffic is more suitable for electric buses and trucks.

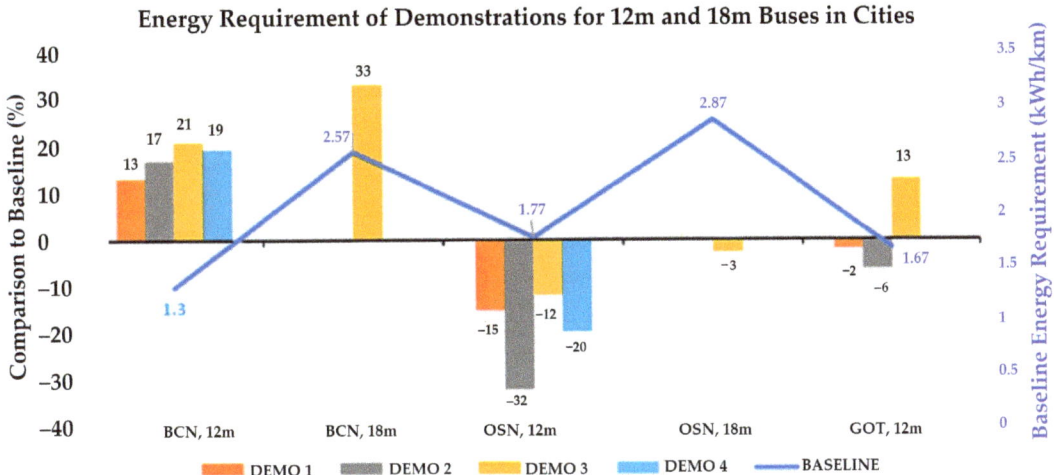

Figure 3. Comparison of the energy requirement for 12 m and 18 m buses in Barcelona, Osnabruck, and Gothenburg between the standard and actual driving cycles.

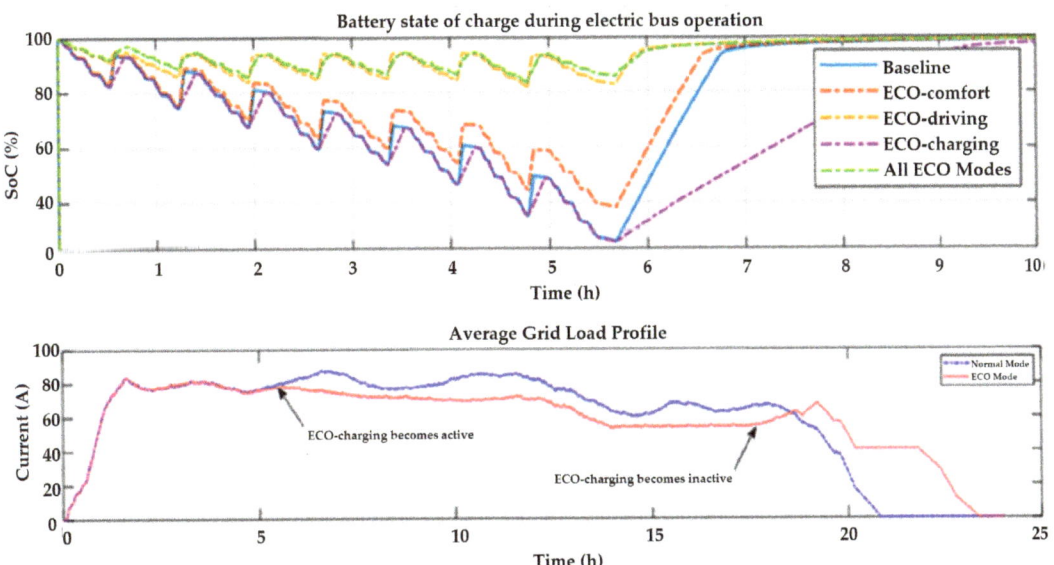

Figure 4. Effects of the ECO-features of the energy requirements (**top row**) and impact on the electricity grid (**bottom row**).

3.2. Energy Reduction Using ECO-Features

Three energy management techniques are considered to reduce the energy requirements of the buses, namely, ECO-comfort [8], ECO-driving [9], and ECO-charging [10,11]. ECO-comfort optimizes the thermal management system of the bus responsible for the cabin and battery cooling systems, ECO-driving optimizes the EMS of the bus responsible for vehicle traction and regeneration, and ECO-charging optimizes the charging management system of the vehicle responsible for battery charging. Figure 4 highlights the effects of the ECO-features on the SoC; and details about the functionality of the three ECO-algorithms are presented in the Appendix B. Based on the SoC profile shown in the top row of Figure 4, ECO-driving has a significant effect on energy reduction, as seen from the smaller drop in the battery SoC with ECO-driving compared with the baseline. This is because the baseline driving profile featured aggressive driving, i.e., high speed (max. velocity of 18.7 m/s) and acceleration (max. acceleration of 2.39 m/s^2), and ??these see the highest reduction in the energy requirement due to the application of ECO-driving. There is modest energy savings due to ECO-comfort, as it was simulated for moderate springtime weather conditions. ECO-charging does not change the energy requirement of the vehicle compared to the baseline, but as can be seen from the bottom row of Figure 4, it does spread out the charging duration, resulting in a lower average load on the electricity grid; this is important during fleet charging so as not to put undue stress on the electricity grid. Overall, the 12 m bus saw an average reduction of 0.4 kWh/km from the baseline energy requirements, while the 18 m bus had a reduction of almost 1.8 kWh/km from the baseline. On average, at least three quarters of the reduction was achieved due to ECO-driving, while barely 2% is due to ECO-charging, as shown in Figure 5.

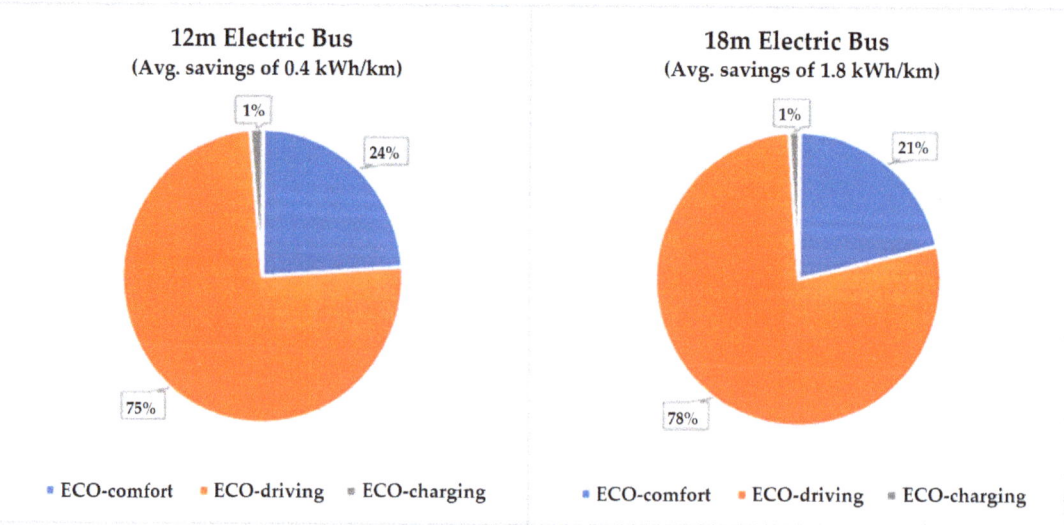

Figure 5. Breakdown of energy savings due to ECO-features for 12 m and 18 m electric bus.

Table 4 shows that there is a high correlation between the amount of energy savings due to ECO-driving and average speed of the vehicle in the baseline scenario. There is also a link between the size of the bus and possible energy savings. However, the data also show that there is no link between the top speed of the vehicle in the baseline scenario and possible energy reduction; this may be because the vehicle does not spend sufficient time at its top speed for it to matter. These results prove that there is a lot of room for improvement when it comes to driving behavior and low-speed driving is recommended for optimum traction energy utilization.

Table 4. Energy reduction possible due to ECO-driving.

City, Bus Type and Demo Number	Speed (Mean and maximum)	Baseline Energy Requirement	Energy Savings
BCN 12 m, Demo 1 *	2.68 m/s, 21.9 m/s	1.47 kWh/km	21.8%
BCN 12 m, Demo 2 *	4.08 m/s, 16.6 m/s	1.52 kWh/km	22.8%
BCN 12 m, Demo 3 *	7.28 m/s, 20.4 m/s	1.57 kWh/km	27.5%
BCN 12 m, Demo 4 *	3.48 m/s, 21.4 m/s	1.50 kWh/km	20.9%
BCN 18 m, Demo 1 *	4.28 m/s, 20.0 m/s	3.42 kWh/km	39.5%
OSN 12 m, Demo 1 *	3.37 m/s, 16.0 m/s	1.50 kWh/km	24.2%
OSN 12 m, Demo 2 *	3.44 m/s, 12.0 m/s	1.20 kWh/km	17.6%
OSN 12 m, Demo 3 *	2.95 m/s, 16.4 m/s	1.56 kWh/km	21.9%
OSN 12 m, Demo 4 *	3.13 m/s, 14.7 m/s	1.42 kWh/km	18.7%
OSN 18 m, Demo 1 *	5.82 m/s, 18.7 m/s	2.79 kWh/km	47.4%
GOT 12 m, Demo 1 *	3.50 m/s, 22.9 m/s	1.83 kWh/km	7.4%
GOT 12 m, Demo 2 *	2.22 m/s, 21.6 m/s	1.75 kWh/km	8.1%
GOT 12 m, Demo 3 *	3.30 m/s, 19.7 m/s	2.11 kWh/km	5.3%

* BCN: Barcelona, OSN: Osnabruck, GOT: Gothenburg; 12 m and 18 m refers to the bus length.

4. Validation of the Simulation Framework

This section focuses on the methodology followed to validate the simulation framework through real measurement data from the demonstrations. The measurements were also used to improve the inputs to the simulation model to have a better representation of the UCs; these improved inputs are then used for the simulation. Measurement data from Osnabruck and Gothenburg were used in the validation process. The quality of the data from the two sources were different. The Gothenburg dataset consists of continuous measurement values sampled at 20 Hz directly from the vehicle's CAN-bus. The data from Osnabruck, extracted from the CBO's cloud server, were only available at intermittent intervals. Thus, the two cases were handled differently.

The validation and tuning process addressed the following features:

- The EMS: The energy recovery system was tuned to align the traction energy profile with the measurement data. The regenerative braking system (RBS) is a proprietary system for many OEMs; thus, assumptions were made during model development.
- The charging management system (CMS): The cutoff between the constant current (CC) mode and the constant voltage (CV) mode, and the current decay parameter during the CV mode were tuned based on the measurement data. These parameter values are also not forthcoming by the OEMs.
- The passenger load estimation: Passenger load inside the bus is the one aspect that could not be automatically measured and requires manual counting; thus, it is usually ignored. Instead, some simulations involved an intricate passenger model based on the passenger appearance rate at each bus stop as a function of time [12], which is modeled on actual bus traffic data by the CBO. Others use agent-based modeling whereby each passenger is a unique object that has "preferences", such as drop in point, drop off point, and waiting time [13]. In [14], a cellular automata model is utilized to study behavioral characteristics of bus passengers boarding and alighting behavior. There are also certain cases where a fixed load was assumed within the bus, based on passenger load factor [15], when the passenger load is ancillary to other considerations. The UC simulations carried out in [5] assumed a random passenger profile as a function of time within the bus cabin; however, for this validation, the passenger inside the bus was estimated based on the measured SoC profile.

4.1. Tuning and Validation Methodology

The tuning was performed by using optimization to directly determine the parameters' values of the powertrain module (e.g., EMS, CMS, BMS) that needs to be tuned, to minimize the normalized root mean squared error (NRMSE) between the simulated and measured outputs.

$$C_{total} = \frac{\sqrt{\frac{\sum_{i=1}^{n}(SOC_{SIM,i} - SOC_{MEAS,i})^2}{n}}}{\max(SOC_{MEAS}) - \min(SOC_{MEAS})} + \text{constraint penalty} \quad (1)$$

The cost function, C_{total}, shown in (1), gives an estimate of the deviation between the simulated and measured SoC of the battery. The closer the value of the cost function is to zero, the closer the match between the two SoC signals. To achieve a minimum value of C_{total} in the optimization process, not only must the two SoC signals match as closely as possible, but the simulation must also not violate any of the constraints elaborated on in Section 4.2. The output score calculated by the NRMSE ranges from 0 (perfect match between simulated and measured signals) to 1 (implying the maximum mismatch between the two signals), thus any penalty applied has values greater than 1. The magnitude of the penalty depends on the extent of the violation of a given constraint.

By the standard definition [16], the tuning methodology described in this study is an example of the offline tuning process because the tuning occurs in the simulation model using saved data, i.e., the measured input during the demonstration was not processed in real-time but cached for later processing and simulation. Instead, for this study, a different definition is used to differentiate between an offline and an online tuning process. The online tuning process is defined as the tuning that occurred while the simulation was still ongoing, whereas in offline tuning, the tuning occurred in an iterative process between separate simulations, after each simulation had finished running in its entirety. For the online tuning process, the total time duration of the simulation was split into several "windows"; the tuning occurred in between each time window, and its result was applied to the next window until no further improvement could be seen, i.e., it converged. If the convergence occurred before the end of the simulation, the tuning was considered completed; otherwise, the simulation was repeated with the latest tuned configuration as the starting condition. As expected, the online process is faster due to the small dataset involved in the tuning process, so the tuning completes quicker.

The online tuning process was applied during the optimization; the simulation time duration was split into variable-sized windows based on the driving cycle. As there were no discernable patterns, the split was made according to different categories of driving, such as constant speed driving or driving with frequent accelerations, as shown in Figure 6. The tuning algorithm assessed various parametric configurations within a window sample to minimize the NRMSE within that window, before applying the best possible configuration to the next window and repeating the process, as in [16].

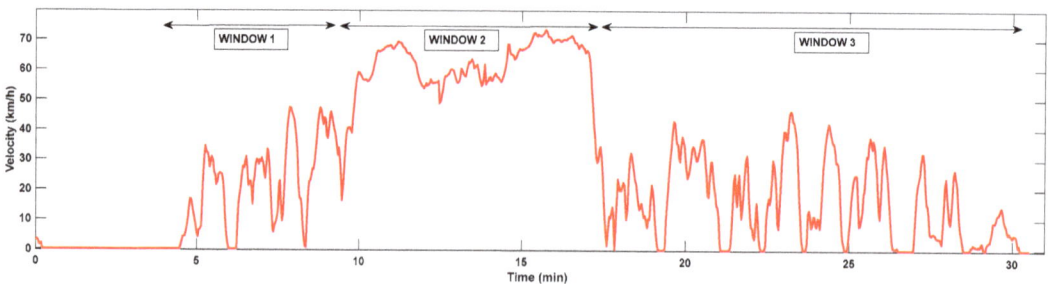

Figure 6. Tuning the simulation using variably sized windows applied based on the driving cycle.

4.2. Optimization Based Tuning Process

In [17], constrained minimization was used in the tuning process of the controller to allow the controller to become flexible, so it can respond in a robust fashion to changes in the inputs, and be used for different purposes by optimally retuning the control parameters subject to different constraints. In [18], a constrained nonlinear optimization was carried out using a sequential quadratic programming (SQP) algorithm to tune PID gains to allow the controller to adapt to changes in the plant; this not only offered superior performances when compared to traditional PID tuning, the tuning process was much quicker. Similarly, linear programming was utilized in [19] to tune the weights of a symmetric finite impulse response (FIR) filter of low-bandwidth controllers for a linear time and spatial invariant (LTSI) systems; a hybrid genetic algorithm (GA) followed by constrained nonlinear minimization was used in [20] to optimize in real time the autopilot gain of an unmanned aerial vehicle (UAV); the GA ensured a global minimum, but without running the GA process to its conclusion, and the f_{mincon} function utilized to finetune the results of the GA at a higher speed. In this study, the meta-heuristic algorithm, improved simple optimization (iSOPT) [21], was used to tune the EMS and CMS of the electric bus powertrain model, to ensure a global minimum within the fastest possible time, so that the tuning can be carried out in real-time.

The set of parameters that were tuned for the EMS are:
- Cutoff velocity for regenerative braking activation
- Passenger load in the bus (broad categories: full load, half load, driver only)

The set of parameters that were tuned for the thermal management system (TMS) are:
- Cabin setpoint temperature

The set of parameters that were tuned for the CMS are:
- Cutoff SoC between CC and CV charging mode
- The current decay factor for CV charging mode
- The charging duration and power
- Initial Battery ageing

The final two parameters that were tuned are the passenger load estimate in the bus and the cabin setpoint temperature. Thus, a total of seven parameters makes up the solution space. An initial population size of 11 with random combinations of the seven parametric values was generated, and the algorithm described in [5] is followed till its conclusion. The maximum number of iterations was set to 50. The optimization is handled via MATLAB scripts, which populates the variables of the Simulink model with updated values every iteration while simulating the demonstration scenario.

The following constraints were applied to the optimization, and a penalty was added to the optimization score if one or more of these constraints were exceeded in any way:
- The current decay factor, cutoff velocity, and cutoff SoC were positive
- The cutoff SoC was below 100%
- The RCD was below 25%
- The charging duration exceeds 1 min and charging power was positive
- The battery SoC should not drop below 10% during the simulation

The advantage of using optimization techniques to tune the model is that it preserves the integrity of the model, with the only factor being changed is the set of parameter values of the respective modules that are being tuned. The improvements of the optimization methodology followed in this research compared with [5] are twofold. The first is an improvement in speed of optimization. In all cases, it is noticed that $T_{opt} < n * T_{sim}$, where n was the number of iterative simulations required during the optimization before convergence and T_{sim} is the duration of one complete simulation. This is because using the methodology in [5], we would have needed to run the complete simulation 'n' times before convergence, but with the window technique presented in this article, we only needed to run the 1st and the 2nd windows 'n' times, and the subsequent windows needed to be run less than 'n' times, as the parameters values have already become optimal by that point. The second improvement was the fact that the optimization process could be made online in the traditional sense [16] by focusing on optimizing the model using the measurement data dump from a previous time window, while the measurement is in progress for the current time window. This is a necessary first step to overcome in the process to develop a real-time DT of the system, which is the end goal of this research track.

5. Validation Results

5.1. Osnabruck

The demonstration for Osnabruck city took place in the months of March, April, and May using an 18 m bus type. Measurement data are available for a total of 9 days, with 2 days each in March and May, and the rest in April. The demonstrations focus on different charging characteristics, with the March and April demonstration clearly focusing on low-power depot charging, and the May demonstration focusing on the high-power opportunity charging. The measurement data provided included the time, speed, and distance travelled data taken at 5-minute intervals. The sampling rate of the provided data is not sufficient to perform simulation and, therefore, each five-minute interval was replaced by the standard SORT driving cycle whose mean velocity was adjusted to match the measured speed value if the distance covered by the adjusted driving cycle was less than or equal to the actual distance traversed during that five-minute interval. If, on the other hand, the adjusted driving cycle covered a larger distance than the actual measured value, the simulation was conducted assuming a constant velocity for that five-minute interval. The measurement also consisted of the battery SoC level at different points during the demonstration. These SoC values are used to verify the simulation results by comparing the simulated SoC values with the actual demonstration SoC values at the same point in time. The simulation assumptions were tuned to give the scenario configuration that provides a simulation with the closest match between the simulated SoC values and measured SoC values.

Figure 7 illustrates the driving and charging scenario constructed from the demonstration data provided for March 29th and 31st, April 7th, 12th & 13th, and 20th & 21st, and May 6th and 12th. The driving and charging scenario will be shown within the same plot. There is no charging taking place between the 20th and 21st; the vehicle is switched off and restarted the next day. The estimated (average) power of the charger used during the March and April demonstrations is 18 kW, thus making it an AC charger in the depot; the estimated power of the charger used for the May demonstrations is 290 kW, thus making it a DC fast charger used for opportunity charging. The charging duration is determined by the type of charger, with opportunity charging active for 10 minutes, while the depot charging is active for hours. The total increase in battery SoC during charging was used to estimate the rated power of the charger.

Table 5 lists the estimates of the driving scenario that gave the closest SoC match between the simulated values and demonstration measurements. Based on these estimates, Figure 8 shows the validation output of the simulation framework.

Table 5. Estimation of the driving scenario configuration for the Osnabruck demonstration.

Parameter to Be Estimated	March	April	May 6th	May 12th
Passenger load	Driver only	Full load	Full load followed by driver only	
Cabin setpoint temperature	20 °C	15 °C		20 °C
Charging type	Depot		Opportunity	
Charging power	18 kW		290 kW	
Battery capacity	120 kWh			
Initial battery age	Relative capacity degradation of 20%			
RBS cutoff velocity	RBS active when vehicle speed above 1.5 m/s			

5.2. Gothenburg

The demonstration for Gothenburg city took place in the month of May and October using 12 m bus. Measurement data are available for a total of 3 days, with 2 days in May and 1 day in October. The demonstrations focus on different charging characteristics, with the May demonstration clearly focusing on shorter duration opportunity charging in the constant voltage (CV) mode, and the October demonstration focusing on the longer duration opportunity charging in the constant current (CC) mode. The duration of the May demonstration was between 12 h to 14 h per day, while the October demonstration was limited to below 3 h. The Gothenburg demonstration had access to continuous driving cycle data; thus, the actual speed measurements were used as inputs after suitable preprocessing. Furthermore, the Gothenburg demonstration also had access to the road inclination profile and ambient temperature profile, in addition to the velocity profile, as inputs. Thus, more relevant simulations could be produced for the validation process. The speed tracking and battery SoC level were validated by comparing the measured values against the simulated values. Table 6 lists the estimates made for the simulations, which achieved a high correlation between the simulated and measured SoC.

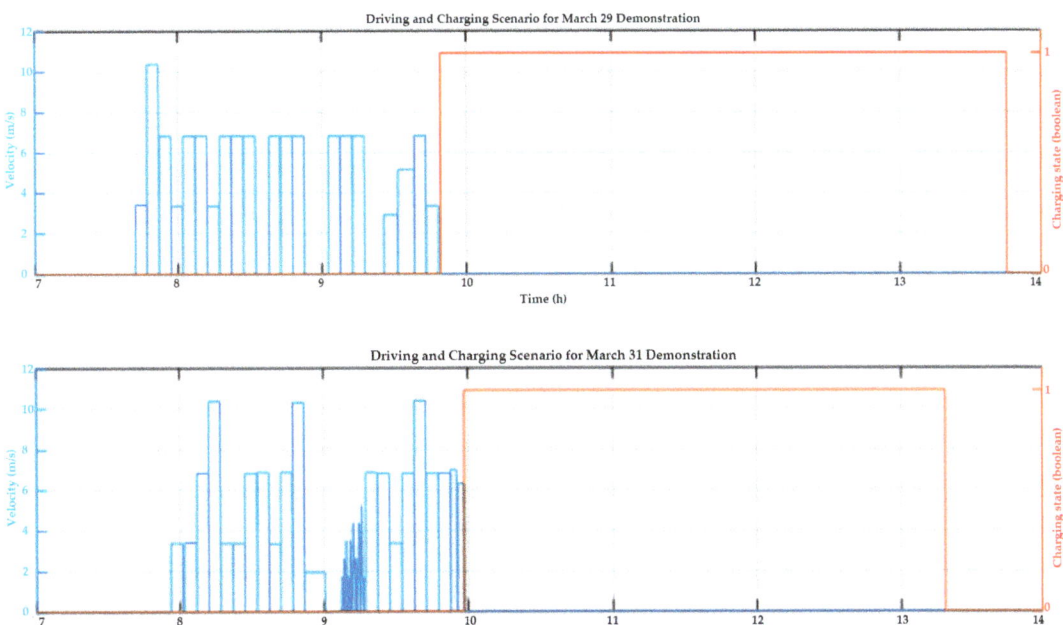

Figure 7. *Cont.*

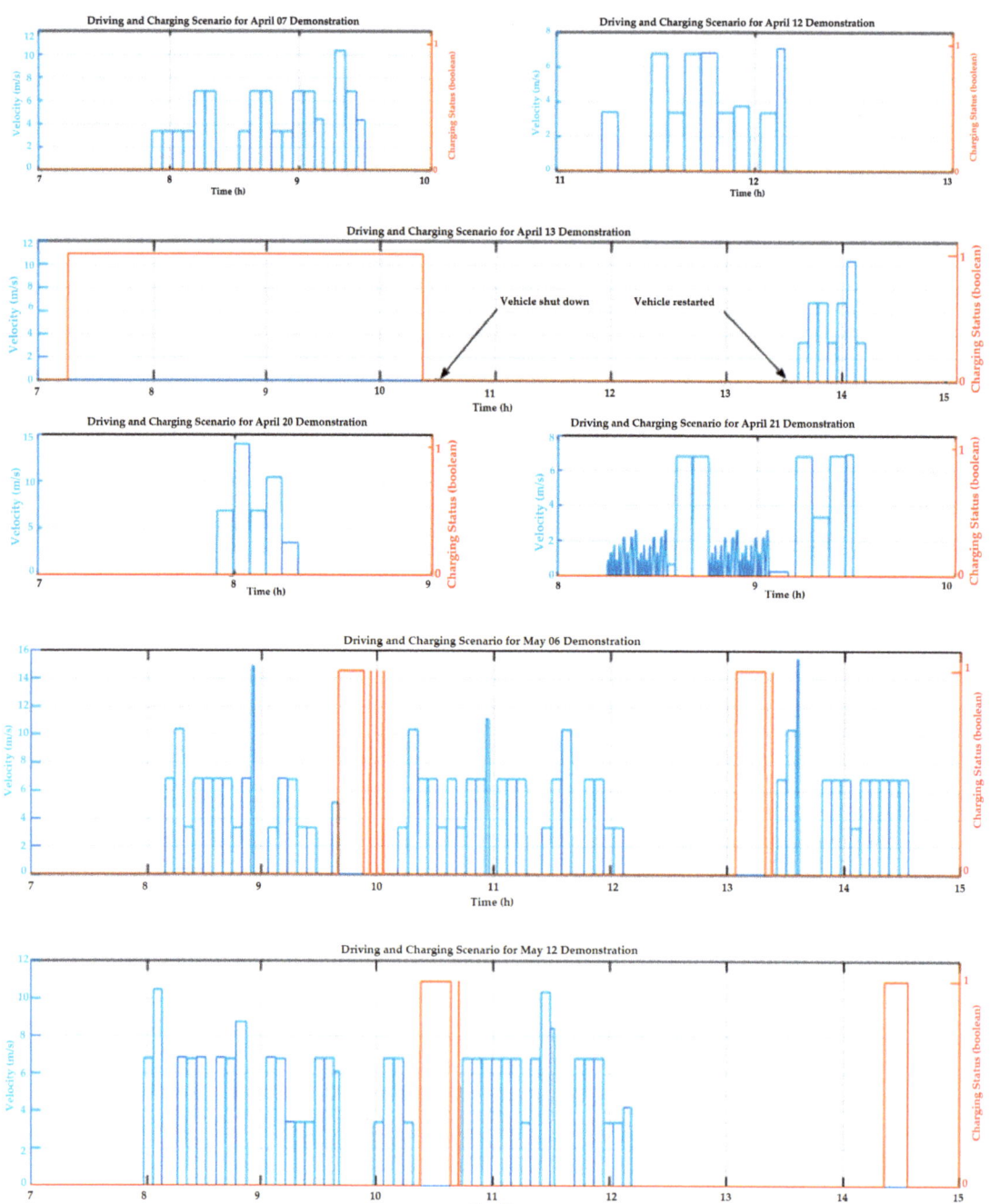

Figure 7. Osnabruck driving scenarios during the 18 m electric bus demonstrations.

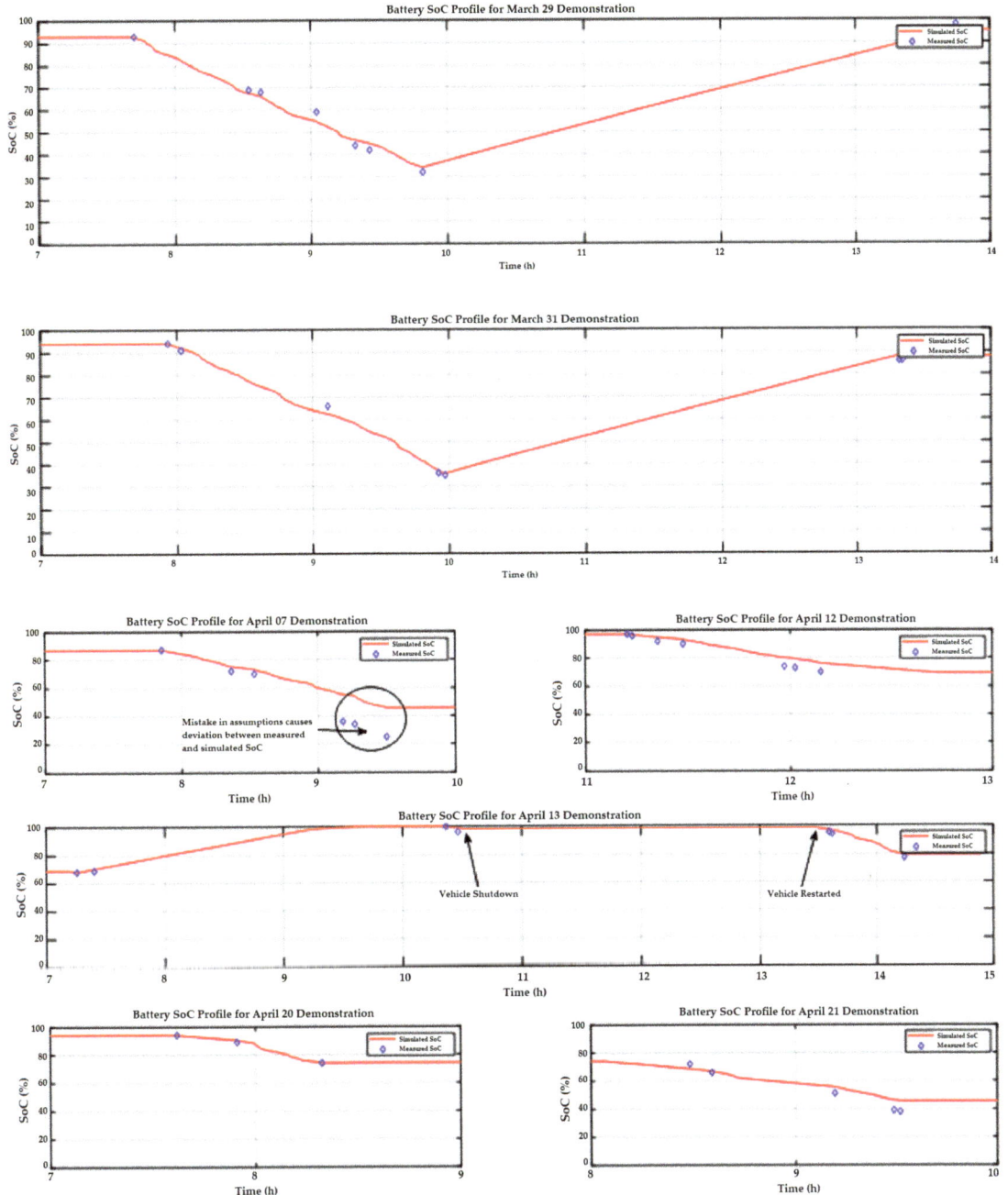

Figure 8. *Cont.*

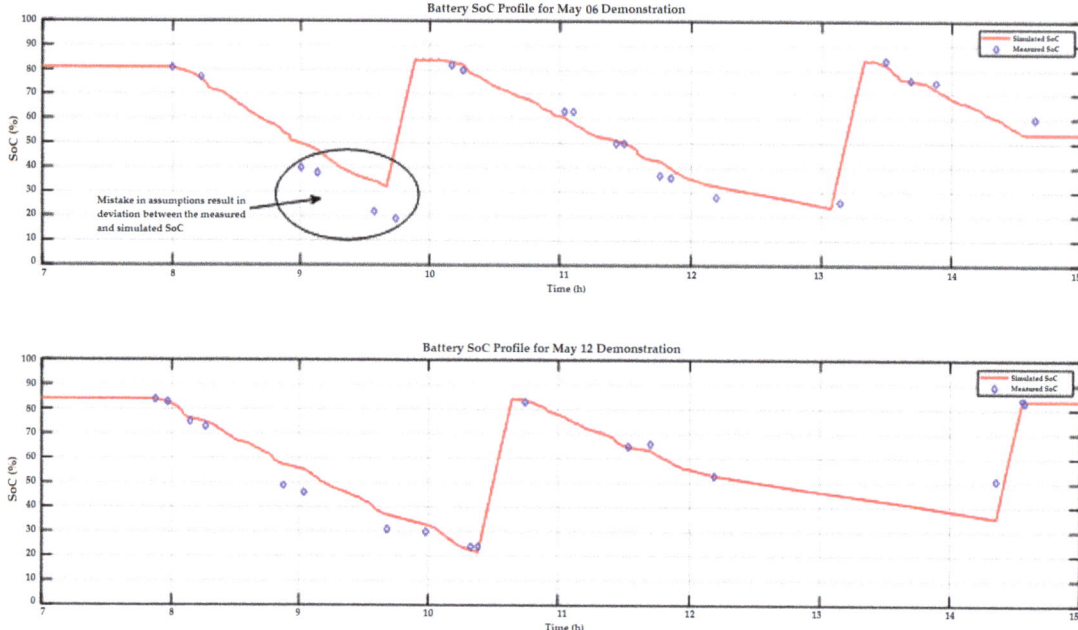

Figure 8. The SoC output profile of the 18m bus subject to the driving and charging scenario in Figure 7 using the estimates given in Table 5, and showing the correlation between the simulated outputs and measured values.

Table 6. Estimation of the driving scenario configuration for Gothenburg demonstration.

Parameter to Be Estimated	March	April	May 12th
Passenger load	Only driver initially, full load between 1 h and 4 h, then half load until 13 h, then driver only until end	Driver only	Only driver when idle (i.e., at end of the route or during charging), full load when bus is moving
Cabin setpoint temperature	20 °C		
Charging type	Opportunity		
Charging power	450 kW (current decay has a β = 0.23 in CV mode, which is activated when SoC > 87.5%)		
Battery capacity	200 kWh		
Initial battery age	New batteries with no degradation		
RBS cutoff velocity	RBS active when vehicle speed above 1.5 m/s		

Figure 9 shows the results of the October 13th demonstration based on the assumptions listed in Table 6. The total demonstration was conducted over 2.5 h with the bus standing idle for the first 40 min. The bus charged using an opportunity charger, with a rated power of 450 kW, at the 1.5 h mark. The simulation tracks the speed accurately with minimal deviation between the simulated and measured outputs. The battery SoC is also tracked accurately; however, there is a deviation between the measured and simulated outputs when the bus is standing idle. The energy usage during that time is very high according to the measured SoC values, which cannot be reasonably explained, unless the speed signal is missing/corrupted.

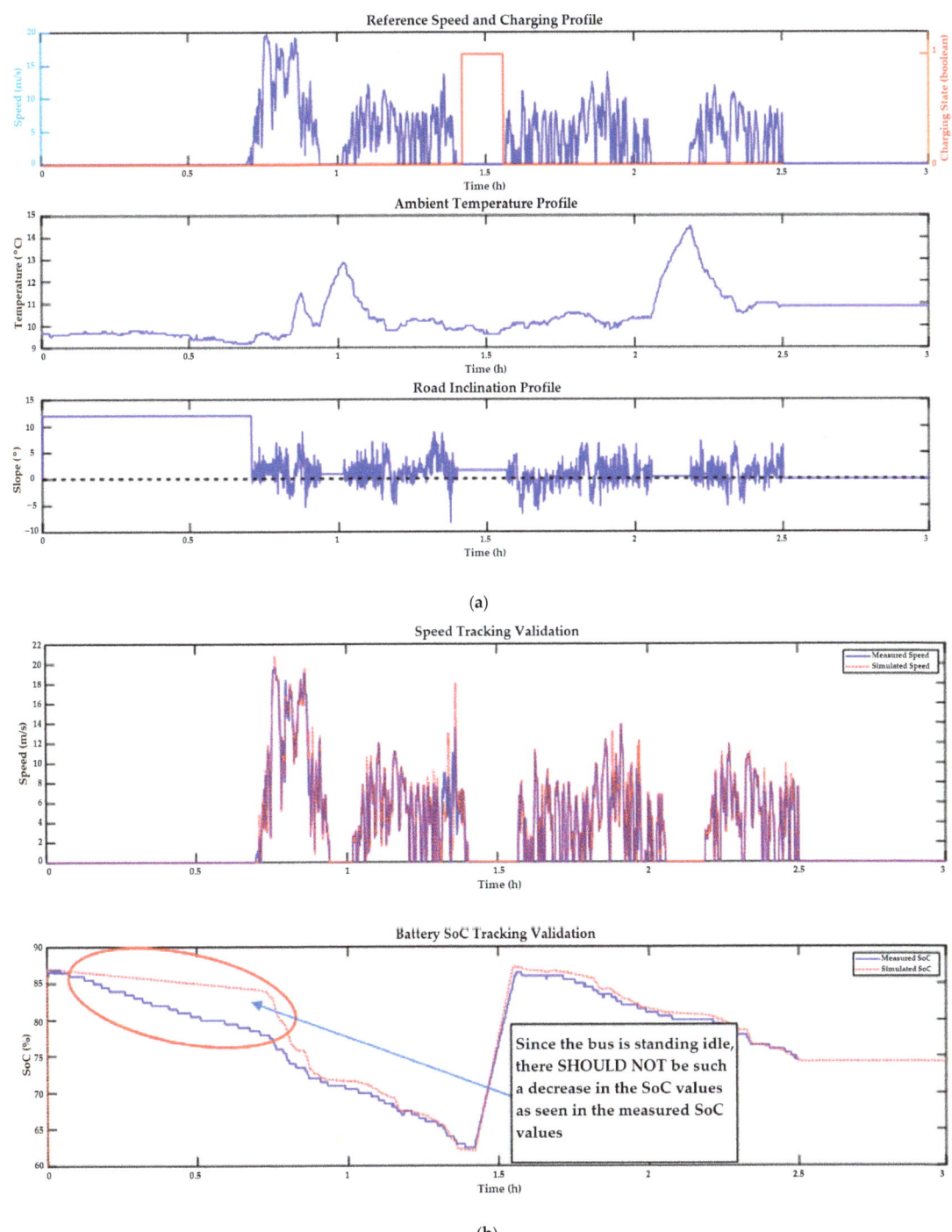

Figure 9. Validation of the October 13th demonstration of 12 m bus in Gothenburg city using the estimates in Table 6. (**a**) Scenario inputs, (**b**) Scenario outputs and validation.

Figure 10 shows the results of the May 27th demonstration based on the assumptions listed in Table 6. The total demonstration was conducted over 13.5 h with the bus standing idle for the first 1 h. The bus was charged using an opportunity charger, with a rated power of 450 kW at 22 different instances. The first charging instance occurs entirely in the CC mode and the second charging instance happens partially in both CC and CV modes, while the remaining charging occurs entirely in the CV mode. The charging current decay (β) of 0.23, when battery SoC exceeds 90%, accurately models the measured charging current. There is a deviation between the measured SoC and simulated SoC at two points; one when the bus was standing idle and the measured SoC showed greater than expected energy usage for an idle vehicle, and the other when the reference speed of the bus was 83 km/h, which exceeded the modeled maximum speed of the bus of 80 km/h.

Figure 11 shows the results of the May 29th demonstration based on the assumptions listed in Table 6. The total demonstration was conducted over 12.5 h; however, the measurements are only available after the 5 h mark. The bus was charged using an opportunity charger, with a rated power of 450 kW at 14 different locations; all the charging events were short in duration. For this demonstration, all charging events occurred entirely in the CV mode. Unlike the other demonstrations, which were modeled with high passenger loads, this one is modeled with only the driver to account for the minimal energy utilization observed. There is a deviation between the measured SoC and simulated SoC at a few locations; the deviations are most likely due to inaccurate battery models for LFP battery chemistry above 90% SoC. The deviations in the beginning can be explained by the fact that the measurements prior to the 5 h mark are not presented; thus, it was not possible to determine the state of the bus prior to the start of the simulation. The deviation at the end was most likely due to a more efficient energy recovery process during regenerative braking than was accounted for in the vehicle model.

There is a deviation between the measured SoC and simulated SoC at a few locations in Figure 11; the deviations are most likely due to inaccurate battery models for LFP battery chemistry above 90% SoC. The deviations in the beginning can be explained by the fact that the measurements prior to the 5 h mark are not presented; thus, it was not possible to determine the state of the bus prior to the start of the simulation. The deviation at the end was most likely due to a more efficient energy recovery process during regenerative braking than was accounted for in the vehicle model.

One of the clear outcomes of the validation process was an accurate determination of the current decay factor (β) during the CV mode of charging and the CC/CV cutoff SoC value. It is understood that after the bulk charging phase of a battery in CC mode, the charging switches to the CV mode, where the current reduces to a trickle. This reduction of the current was modeled as an exponential decay once the battery SoC exceeds 87.5%. The decay amount is given as:

$$\text{Iout} = \begin{cases} |I_{max\_cc} \times e^{\beta \times (SoC - 90)}, & SoC > 87.5 \\ |I_{max\_cc}, & \text{otherwise} \end{cases} \tag{2}$$

where I_{max_cc} is the maximum charging c-rate during the CC charging mode (-3C for an LFP battery chemistry), and β is the decay factor; it was found that a β of 0.23 models the charging current that gives the closest correlation between the measured and simulated battery SoC profile during charging.

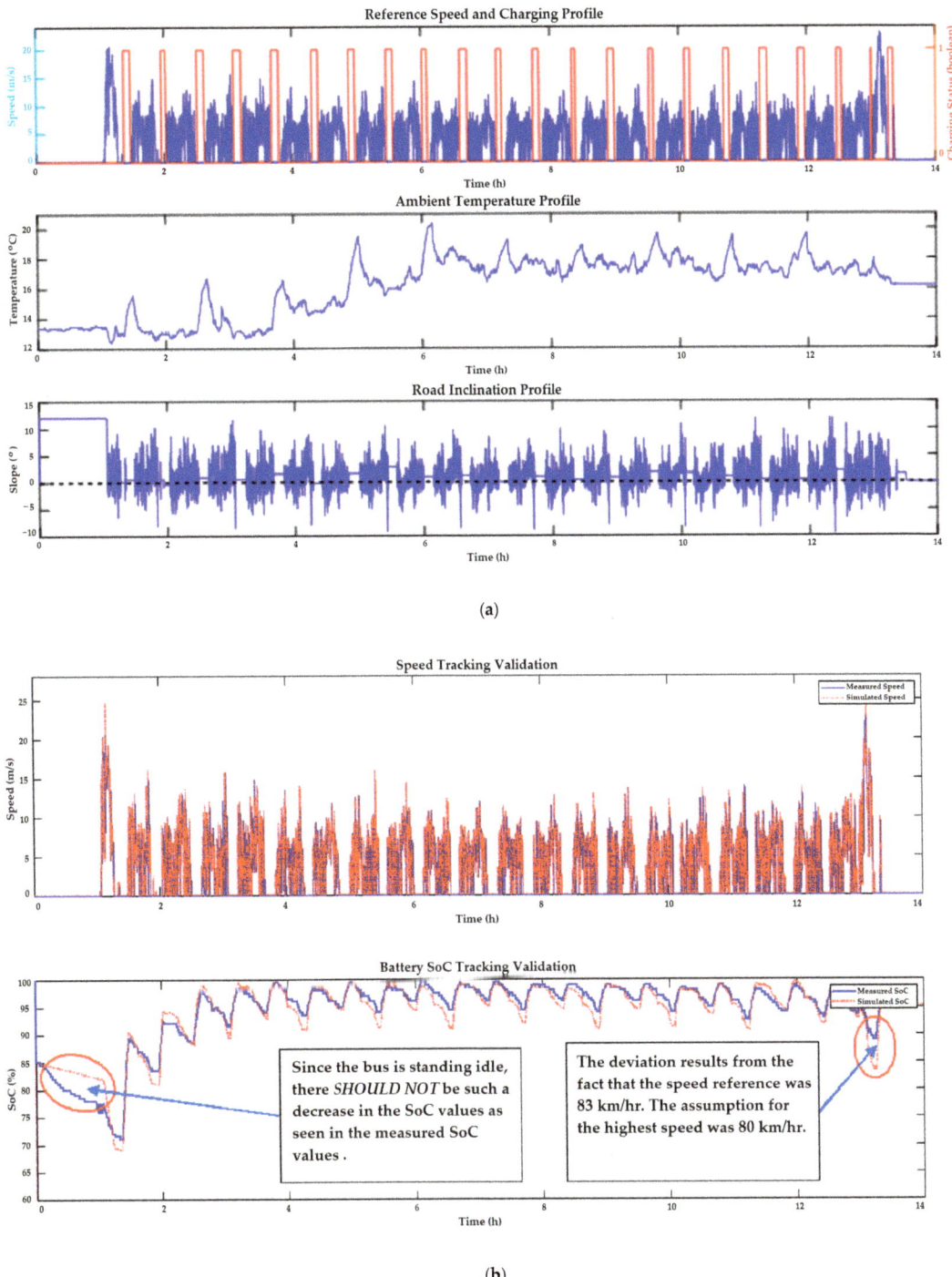

Figure 10. Validation of the May 27th demonstration of 12 m bus in Gothenburg city using the estimates in Table 6. (**a**) Scenario inputs, (**b**) Scenario outputs and validation.

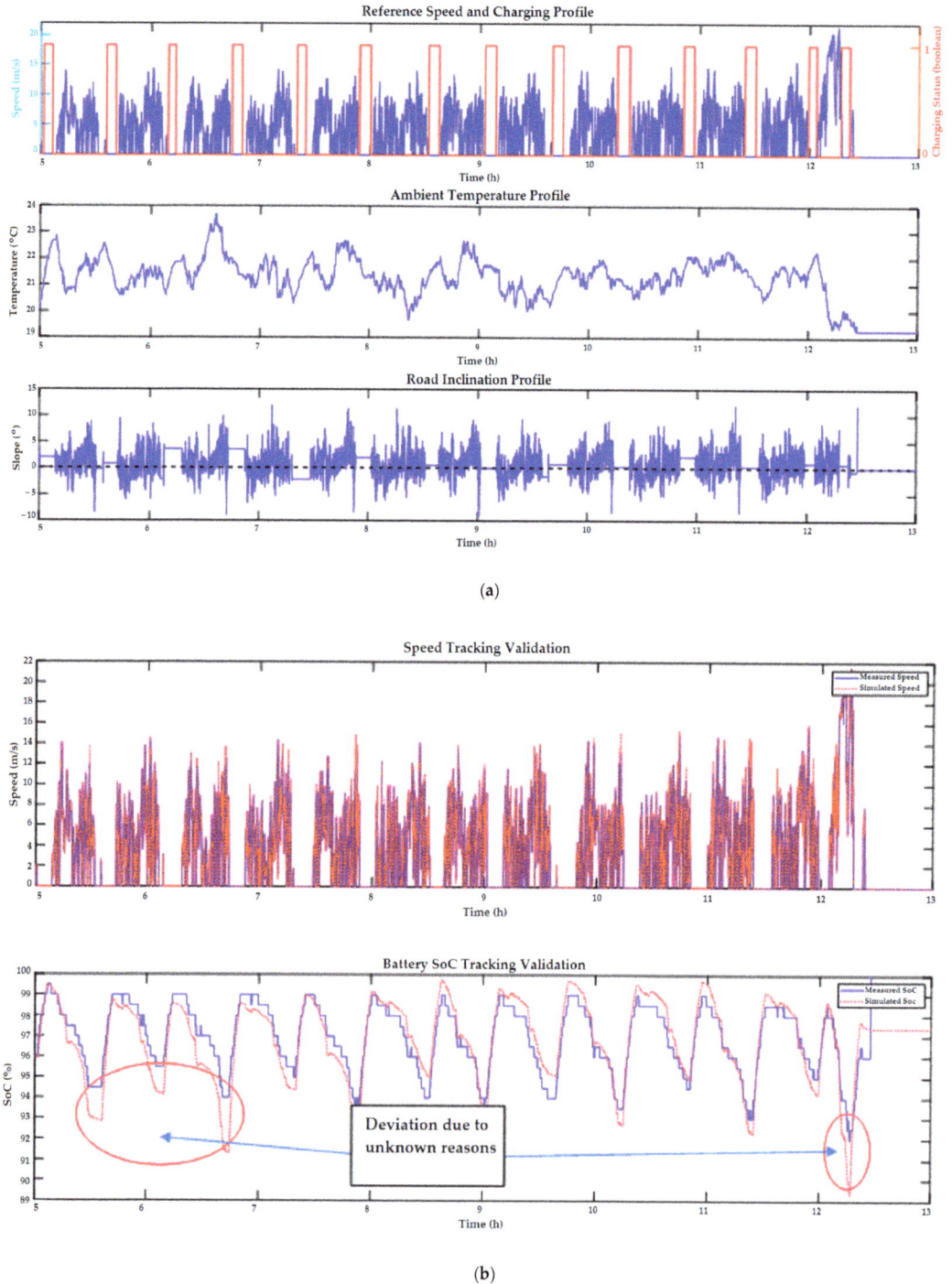

Figure 11. Validation of the May 29th demonstration of 12 m bus in Gothenburg city using the estimates in Table 6. (**a**) Scenario inputs, (**b**) Scenario outputs and validation.

6. Conclusions

This study presents a methodology for improving the accuracy of a Lo-Fi model of the electric bus powertrain using measurement data from 12 m and 18 m electric bus demonstrations in cities. First, a qualitative comparison is made of the bus's energy requirements between the baseline UC simulations, which used a standard driving profile, and the actual driving profile from the demonstrations. The results show that in Barcelona, the energy requirements of the 12 m buses were 17.5% higher, while those of the 18 m buses were 33% higher, when using the driving profile of the demonstration. For Osnabruck, the energy requirements were 20% lower for the 12 m buses when using the driving profile of the demonstrations, while the 18 m buses had similar energy requirements to the baseline. This is because the Barcelona demonstrations had a higher average velocity compared with the baseline, while the Osnabruck 12 m bus demonstrations had a lower average velocity. The magnitude of the acceleration and deceleration had less effect on the energy requirements of an electric powertrain, since energy expended during accelerations are recovered during decelerations. Only in cases where the driving profile showed many hard decelerations did the energy requirement become higher; this was because during hard decelerations, the bus requires friction brakes to decelerate in addition to the electric motor, leading to less energy recovered via regeneration.

Next, the measurement data of the vehicle's speed profile from the demonstration were used as inputs to the simulation framework, and the simulation results of the battery SoC profile were compared to measured battery SoC profiles from the demonstrations. A tuning methodology, based on iSOPT optimization, combined with splitting the simulation into smaller time windows during optimization, was used to minimize the NRMSE between the simulated and measured battery SoC signals and ensure that there is a high degree of correlation between them. The results show that the tuning process based on the window technique applied to the optimization process successfully synchronized the simulation and measurement outputs quicker than the technique presented in [5]. In rare cases, deviations are encountered between the simulated and measured output. Of these, the deviations that describe a situation that is physically impossible, based on the data provided, are ignored. Other deviations result from limitations in the assumptions made during the design of the simulation framework, and those were fixed by correcting the assumptions. However, in two cases, deviations occurred for which no suitable explanation could be determined, and those would require further research to fix. Overall, the optimization achieved more than 90% correlation between the simulated and measured SoC profile.

The techniques utilized in this research will be refined further in future research to perform real-time tuning of the platform with the aim of deploying a cloud-based DT of the electric bus that will be able to make predictions in real-time based on the measurement data from the real vehicle. To achieve that goal requires two systems working in synergy: first, it would be necessary to invest in CAN dataloggers with WiFi or 3G/4G capability that will capture the sensor data from the vehicle's CAN network and periodically transmit these measurements to a cloud server. Then, a highspeed simulation model needs to be deployed in the cloud server that will periodically take in these measurements data as inputs and quickly simulate the outputs and tune itself using appropriate tuning techniques to minimize the error between the simulated and measured outputs. The key will be to reduce the simulation time needed during the tuning process (whether via machine learning or optimization), so the model can tune itself in real time. This requires further improvements to the optimization technique and utilizing machine learning using artificial neural networks. Machine learning algorithms are also able to adapt to changes in behavior over time. Once the error has been reduced below an acceptable threshold, then many virtual copies of the DT can be deployed in the cloud to act as virtual testbeds for a myriad of different tests, or to simulate fleets of such vehicles to investigate the charging infrastructure requirements in city bus routes and depots.

Author Contributions: Conceptualization, M.M.H. and N.A.; methodology, M.M.H., N.A. and M.R.; software, M.M.H. and N.A.; validation, M.M.H.; formal analysis, M.M.H.; investigation, M.M.H. and N.A.; resources, N.A.; data curation, N.A.; writing—original draft preparation, M.M.H.; writing—review and editing, N.A., M.R., M.E.B. and O.H.; visualization, M.E.B. and O.H.; supervision, O.H.; project administration, M.E.B. and O.H.; funding acquisition, O.H. All authors have read and agreed to the published version of the manuscript.

Funding: This research was funded by the European Commission—Innovation and Networks Executive Agency, Grant number 769850, under the title of ASSURED—H2020-GV-2016-2017/H2020-GV-2017.

Data Availability Statement: Data used in this article are private and can be found in the project deliverables for those having access.

Acknowledgments: The authors acknowledge Flanders Make for their support to this research group. The authors acknowledge the OEMs and CBOs for providing measurement data from demonstrations for analysis in this article.

Conflicts of Interest: The authors declare no conflict of interest. The funders had no role in the design of the study; collection, analyses, or interpretation of data; writing of the manuscript; or decision to publish the results.

Appendix A. Input Scenario for Bus Demonstrations

Appendix A.1. Barcelona City

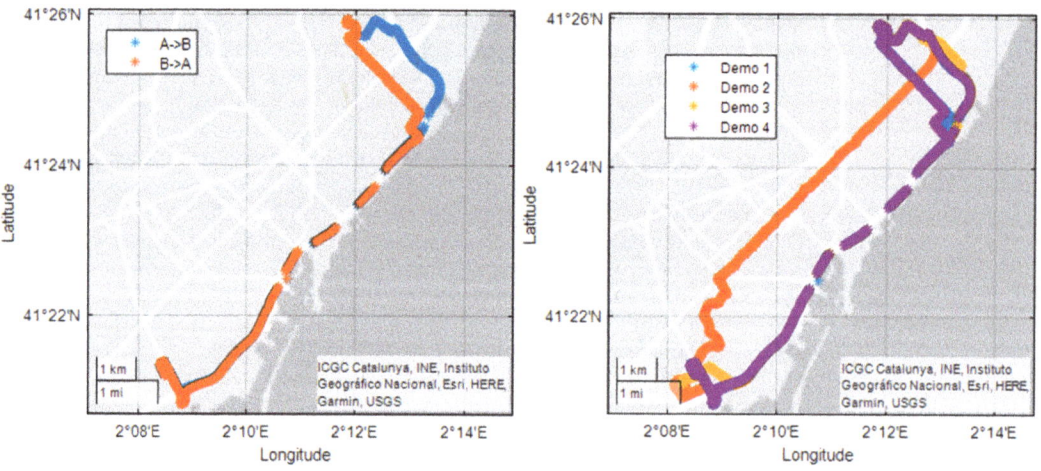

Figure A1. Route map of H16 for the demonstration.

Appendix A.2. Osnabruck City

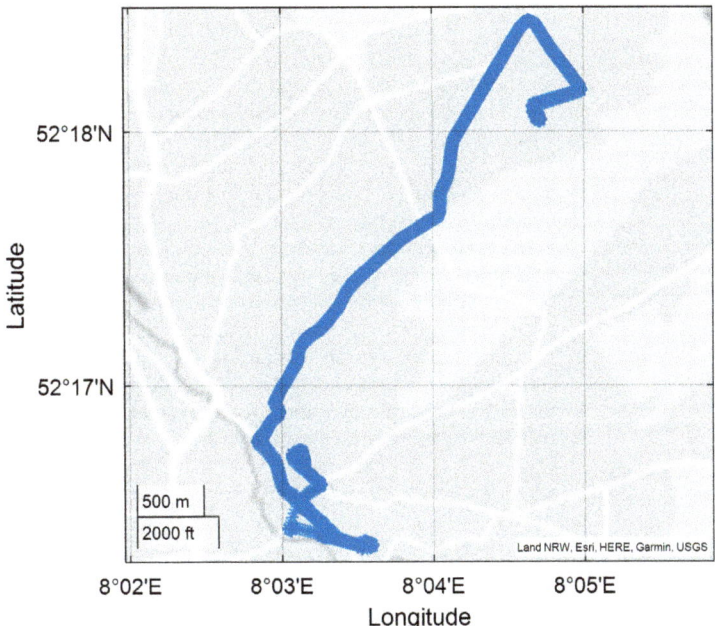

Figure A2. Route map of N5 for the demonstration.

Appendix A.3. Gothenburg City

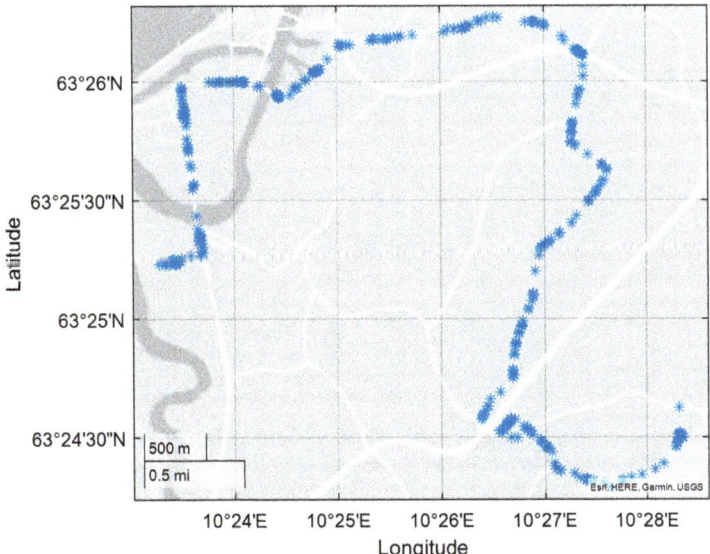

Figure A3. Route map of R55 for the demonstration.

Appendix B. Review of Energy Management (ECO) Features

Appendix B.1. ECO-Driving Functionality

ECO-driving transforms the driving cycle into an eco-friendlier profile that limits the maximum acceleration and speed of the vehicle resulting in less tractive energy requirements; furthermore, it also optimizes the energy recovery during regeneration by keeping the EM in the optimum power band to recover the maximum power. As can be seen from Figure A4, the velocity profile is smoothened by application of a ramp to the acceleration. The velocity modification ensures smoother changes in velocity and removes discontinuity in the acceleration. The top velocity and acceleration are also limited to save energy. The overall driving behavior is gentler, with minimal hard accelerations and braking. This is important because, unlike normal braking action, hard braking is not as efficient at energy recovery as a large portion of the braking power needs to be diverted to the friction brakes, rather than the electric motor, to cope with the braking load. This is why applying ECO-driving to an aggressive driving style results in significant energy savings. Therefore, good driving behavior is a requirement for proper regenerative braking action and is a core component of ECO-driving. The ECO-driving method also ensures that regardless of the velocity modification, the distances traveled between the ECO and non-ECO version remains synchronized. This distance synchronization is important to convince many CBOs to adopt ECO-driving principles for their routes, as they can still maintain their default bus schedules even while limiting top speed and acceleration.

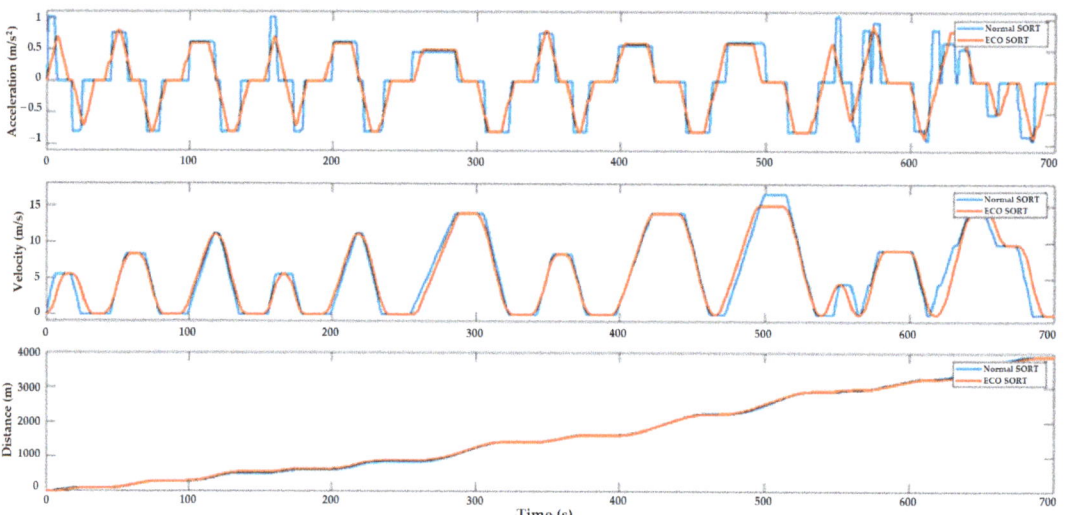

Figure A4. Velocity modification for Eco-friendly profile generated for a standard driving cycle.

Appendix B.2. ECO-Comfort Functionality

Figure A5 shows how the ECO-comfort functionality dynamically alters the cabin setpoint temperature throughout the day. The dynamic temperature setpoint of the ECO-comfort depends on the passenger count inside the bus as well as the ambient temperature. The temperature setpoint is devised to save the energy required for climate control at the expense of slightly reduced passenger comfort. This means a little less cooling inside the bus during summers and a little less heating inside the bus during winters. As well as dynamic temperature setpoints, ECO-comfort also uses pre-conditioning to reduce the energy requirement needed for heating or cooling when the bus is in motion. Pre-conditioning means to utilize the thermal management system to track the setpoint temperature of the bus while it is connected to the grid for charging; thus appropriating the energy from the grid instead of the battery. The energy reduction by ECO-comfort is highly dependent on the climate, e.g., for a hot climate, the maximum energy reduction due to ECO-comfort is achieved during mid-summer, while for colder climates, the maximum energy reduction is attained in mid-winter.

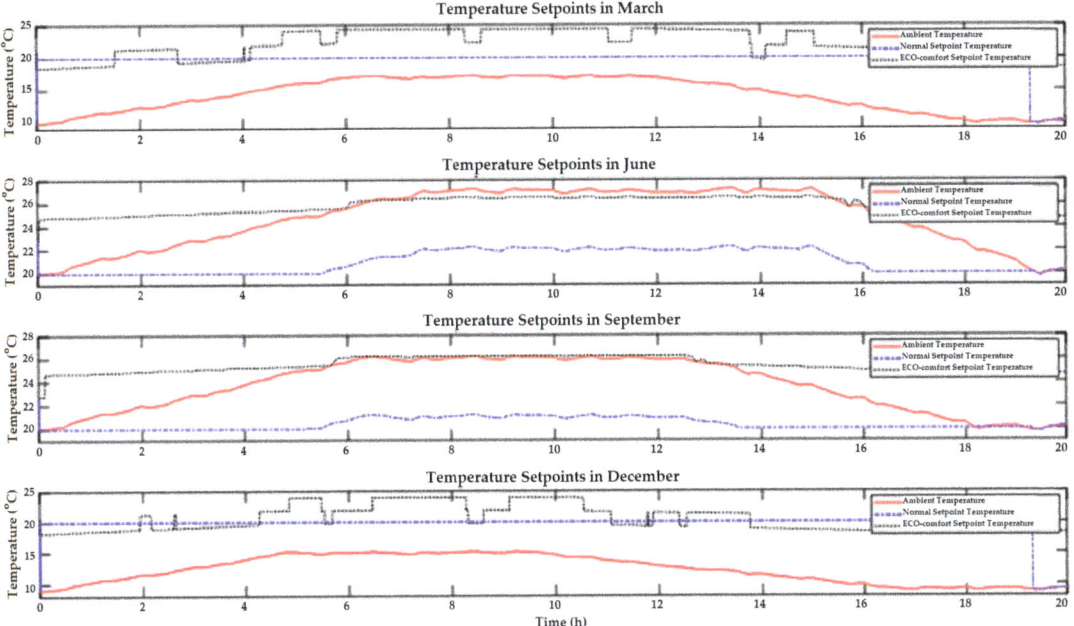

Figure A5. The daily dynamic cabin setpoint temperature for a 12 m bus.

Appendix B.3. ECO-Comfort Functionality

Figure A6 shows the ECO-charging functionality, which makes use of pulsed charging, instead of continuous charging. Since the charging is pulsed, the battery has a chance to cool down in between the charging pulses; this reduces the temperature increase during charging, and necessitates less cooling by the HVAC system. At the same time, this also results in low c-rate charging on average, thus improving battery longevity. The disadvantage of this charging method is that the battery will take longer to charge; to mitigate this, either the charging duration needs to be increased, which is not always possible due to bus scheduling constraints, or the battery size needs to be increased so that the battery can deliver the range required during its scheduled operational period. Thus, ECO-charging prevents excessive battery heating during charging, has minimal effect on the vehicle's energy requirements, and lowers the load on the electricity grid.

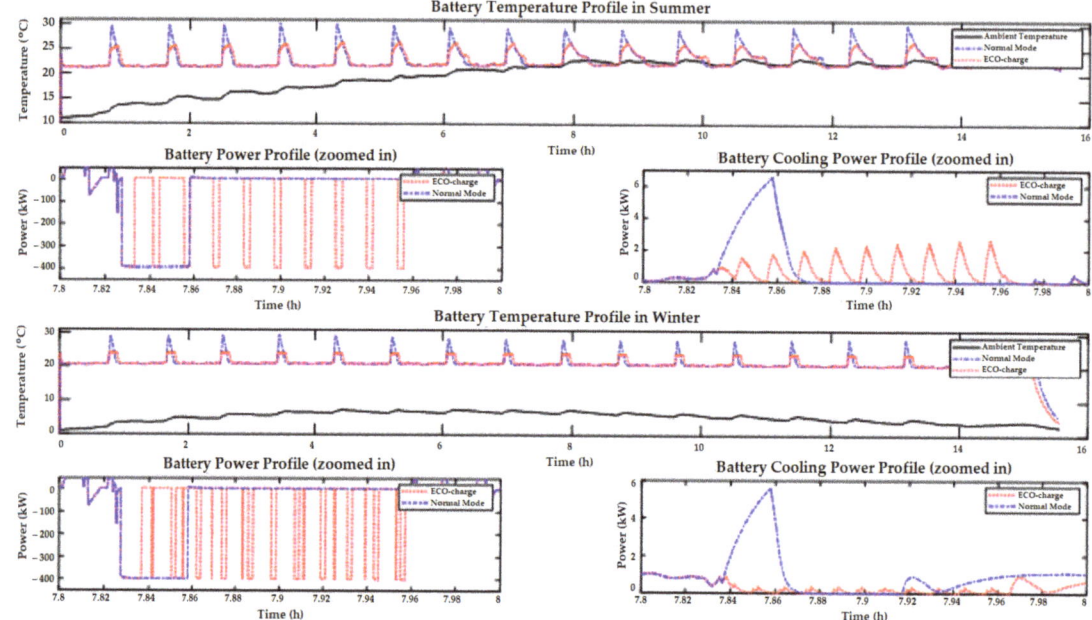

Figure A6. ECO-charging profile highlighting the effects of pulsed charging functionality.

References

1. Swamidass, P.M. (Ed.) Moving Assembly Line. In *Encyclopedia of Production and Manufacturing Management*; Springer: Boston, MA, USA, 2000; pp. 465–466. ISBN 978-1-4020-0612-8.
2. Automotive Product Development Cycles and the Need for Balance with the Regulatory Environment. Center for Automotive Research, updated: 20 September 2017. Available online: https://www.cargroup.org/automotive-product-development-cycles-and-the-need-for-balance-with-the-regulatory-environment/ (accessed on 16 June 2022).
3. Chakraborty, S.; Mazuela, M.; Tran, D.; Araujo, J.A.C.; Lan, Y.; Alacano, A.; Garmier, P.; Aizpuru, I.; Hegazy, O. Scalable Modeling Approach and Robust Hardware-In-The-Loop Testing of an Optimized Interleaved Bidirectional HV DC/DC Converter for Electric Vehicle Drivetrains. *IEEE Access* **2020**, *8*, 115515–115536. [CrossRef]
4. Grieves, M.; Vickers, J. Digital Twin: Mitigating Unpredictable, Undesirable Emergent Behaviors in Complex Systems. In *Transdisciplinary Perspectives on Complex Systems: New Findings and Approaches*; Kahlen, F.J., Flumerfelt, S., Alves, A., Eds.; Springer: Cham, Switzerland, 2017; pp. 85–113. ISBN 978-3-319-38754-3.
5. Hasan, M.M.; Avramis, N.; Ranta, M.; Saez-De-Ibarra, A.; El-Baghdadi, M.; Hegazy, O. Multi-Objective Energy Management and Charging Strategy for Electric Bus Fleets in Cities Using Various ECO-Strategies. *Sustainability* **2021**, *13*, 7865. [CrossRef]
6. GPS Accuracy, GPS.gov, updated: 03 March 2022. Available online: https://www.gps.gov/systems/gps/performance/accuracy/ (accessed on 20 June 2022).
7. Doyle, A.; Kolhe, M.L.; Muneer, T. *Electric Vehicles: Prospects and Challenges*; Elsevier: Amsterdam, Netherlands, 2017; Chapter 2; pp. 93–124. ISBN 978-0-12-803021. [CrossRef]
8. Hasan, M.M.; Maas, J.; el Baghdadi, M.; de Groot, R.; Hegazy, O. Thermal Management Strategy of Electric Buses Towards ECO-Comfort. In Proceedings of the 8th Transport Research Arena Conference (TRA 2020), Helsinki, Finland, 27–30 April 2020.
9. Hasan, M.M.; el Baghdadi, M.; Hegazy, O. Energy Management Strategy in Electric Buses for Public Transport Using ECO-Driving. In Proceedings of the 15th International Conference on Ecological Vehicles and Renewable Energies (EVER 2020), Monte-Carlo, Monaco, 28–30 May 2020.
10. Hasan, M.M.; Ranta, M.; El Baghdadi, M.; Hegazy, O. Charging Management Strategy Using ECO-Charging for Electric Bus Fleets in Cities. In Proceedings of the 2020 IEEE Vehicle Power and Propulsion Conference (VPPC 2020), Gijon, Spain, 26–29 October 2020.
11. Hasan, M.M.; Saez-de-Ibarra, A.; El Baghdadi, M.; Hegazy, O. Analysis of the Peak Load Reduction using ECO-charging Strategy for e-Bus Fleets in Gothenburg. In Proceedings of the 2021 IEEE Vehicle Power and Propulsion Conference (VPPC 2021), Gijon, Spain, 26–29 October 2021.

12. Hiroi, K.; Arai, T.; Kawaguchi, N. Simulation for Passengers Convenience using Actual Bus Traffic Data. In *Intelligent Transport Systems for Everyone's Mobility*; Mine, T., Fukuda, A., Ishida, S., Eds.; Springer: Singapore, 2019; pp. 175–194. ISBN 978-981-13-7433-3. [CrossRef]
13. Schelenz, T.; Suescun, A.; Karlsson, M.; Wikstrom, L. Decision Making Algorithm for Bus Passenger Simulation during the Vehicle Design Process. *J. Transp. Policy* **2013**, *25*, 178–185. [CrossRef]
14. Xue, Y.; Zhong, M.; Xue, L.; Zhang, B.; Tu, H.; Tan, C.; Kong, Q.; Guan, H. Simulation Analysis of Bus Passenger Boarding and Alighting Behavior Based on Cellular Automata. *Sustainability* **2022**, *14*, 2429. [CrossRef]
15. Shen, X.; Feng, S.; Li, Z.; Hu, B. Analysis of Bus Passenger Comfort Perception Based on Passenger Load Factor and In-Vehicle Time. *Springer Plus* **2016**, *5*, 62. [CrossRef] [PubMed]
16. Grefenstette, J.J. Optimization of Control Parameters for Genetic Algorithms. *IEEE Trans. Syst. Man Cybern.* **1986**, *16*, 122–128. [CrossRef]
17. Kamwa, I.; Trudel, G.; Lefebvre, D. Optimization-Based Tuning and Coordination of Flexible Damping Controllers for Bulk Power Systems. In Proceedings of the 1999 IEEE International Conference on Control Applications, Kohala Coast, HI, USA, 22–27 August 1999.
18. Neto, C.A.; Embirucu, M. Tuning of PID Controllers: An Optimization-Based Approach. *IFAC Proc. Vol.* **2000**, *33*, 367–372. [CrossRef]
19. Gorinevsky, D.; Boyd, S.; Stein, G. Optimization-Based Tuning of Low-Bandwidth Control in Spatially Distributed Systems. In Proceedings of the American Control Conference, Denver, CO, USA, 4–6 June 2003.
20. Ahsan, M.; Rafique, K.; Mazah, F. Optimization Based Tuning of Autopilot Gains for a Fixed Wing UAV. *Int. J. Comput. Syst. Eng.* **2013**, *7*, 781–786.
21. Thomas, J.; Mahapatra, S.S. Improved Simple Optimization (iSOPT) Algorithm for Unconstrained Non-Linear Optimization Problems. *Perspect. Sci.* **2016**, *8*, 159–161. [CrossRef]

Disclaimer/Publisher's Note: The statements, opinions and data contained in all publications are solely those of the individual author(s) and contributor(s) and not of MDPI and/or the editor(s). MDPI and/or the editor(s) disclaim responsibility for any injury to people or property resulting from any ideas, methods, instructions or products referred to in the content.

Article

Research on a Day-Ahead Grouping Coordinated Preheating Method for Large-Scale Electrified Heat Systems Based on a Demand Response Model

Guodong Guo * and Yanfeng Gong

State Key Laboratory of Alternate Electrical Power System with Renewable Energy Sources, North China Electric Power University, Changping District, Beijing 102206, China
* Correspondence: gbjdsf@163.com; Tel.: +86-183-22595221

Abstract: In recent years, the increasing winter load peak has brought great pressure on the operation of power grids. The demand response on the load side helps to alleviate the expansion of the power grid and promote the consumption of renewable energy. However, the response of large-scale electric heat loads to the same electricity price curve will lead to new load peaks and regulation failure. This paper proposes a grouping coordinated preheating framework based on a demand response model, which realizes the interaction of information between the central controller and each regulation group. The room thermal parameter model and the performance map of the inverter air conditioner/heat pump are integrated into the demand response model. In this framework, the coordination mechanism is adopted to avoid regulation failure, an edge computing structure is applied to consider the users' preferences and plans, the grouping and parallel computing structure is proposed to improve the computing efficiency. Users optimize their heat load curves based on a demand response model, which can consider travel planning and ensure user comfort. The central controller updates the marginal cost curve based on the predicted scenario set to coordinate the regulation groups and suppress the new peaks. The simulation results show that the proposed method can promote the consumption of renewable energy through coordinated preheating and reduce the system energy consumption cost and user bills. The parallel computing structure within the regulation group also ensures the computing efficiency under large-scale loads.

Keywords: demand response; coordinated preheating; inverter air conditioner; equivalent thermal parameter model; smart grid

1. Introduction

In recent years, the peak power consumption in winter has become more and more obvious. The maximum load in winter exceeds that in summer in many southern provinces in many southern provinces of China [1], and Texas in the United States has also set a record for the peak power consumption in winter in 2021 [2]. Different from the central heating in the north, the household independent heating mode is more common in hot summer and cold winter regions, with the characteristics of intermittent heating on demand [3]. It is important to promote heating electrification to achieve carbon neutrality in winter heating systems. The heat pump, hot air blower/cooling and heating air conditioner follows the reverse Carnot cycle and transfers more heat with less energy consumption, which is highly energy efficient. Jiang [4] pointed out that the air source heat pump is the most important way of heating electrification. The European Commission has also set a heat pump development target, that is, 40% of residential buildings and 65% of commercial buildings are expected to achieve electric heating by 2030 [5]. With the increase in power demand in the winter peak period [6], efficient demand management technology can help to reduce system costs, promote the consumption of renewable energy, and achieve carbon neutrality.

Residential air conditioning plays an important role in demand response resources. A demand response model for single residential buildings was established in [7], in which heat pumps preheat at low electricity prices to reduce power demand at high electricity prices. The home energy management system with integrated intelligent heat load has been studied in [8,9] and the corresponding optimization problems were found, with the purpose of minimizing energy consumption cost and ensuring user comfort. Comfort is actually guaranteed by the precooling/preheating. Large scale day-ahead heat load regulation is an attractive solution for power system scheduling due to its economy and security, but there are many difficulties to be solved. The assessment of the maximum adjustable load of the air conditioning cluster can be used for real-time regulation [10,11], but it cannot be used for day ahead scheduling because of the neglect of load-time coupling. For example, the time point and adjustment amount of preheating/precooling cannot be determined by the above methods. Equivalent energy storage models and cluster models for air-conditioning clusters are studied for day-ahead scheduling [12,13], in which temperature setpoints are regarded as consistent and fixed. In fact, the temperature setpoints of users are different and time-variant, and are related to their respective travel schedules. Most importantly, the scheduling results are not practical due to the lack of instructions specific to each user. In view of the above problems, some studies have been carried out from the characteristics of intermittent heating/cooling [14–16] and the respective thermal parameters and thermal demands of households [17]. However, the above methods are only applicable to a single residential or commercial building and will encounter difficulties in solving large-scale problems. It is worth noting that if all participants adjust based on the same electricity price curve, it will fundamentally change the marginal cost curve of the system, which will lead to failure of the adjustment. A new coordinated preheating scheme based on game theory was proposed in [18] to ensure the effectiveness of large-scale family collective preheating, which effectively takes into account changes in marginal costs in the coordination process. However, households need to iterate one by one, and the computation time is linearly related to the household scale, which is unacceptable in large-scale problems. Intermittent heating, personalized thermal demand, and computational efficiency are the three major difficulties in large-scale day-ahead thermal load regulation.

Accurate house thermal models [19] and air conditioning models have a significant impact on the conditioning effect. In terms of air conditioning, constant frequency air conditioning has been gradually replaced by inverter air conditioning (IAC). The performance of IAC is related to the indoor/outdoor temperature and the compressor speed, and its steady-state model can be used to study the coupled dynamic characteristics of the room and IAC. Research and experiments on steady-state models are abundant [20–22], but their computation time is unacceptable for scheduling problems. In [16,23], the performance map based on the steady-state model was obtained for the direct control of IAC.

This paper proposes a day-ahead group coordinated preheating method based on demand response model for large-scale electric heating load, which is carried out in a framework composed of a central controller and several regulation groups. Under this framework, users can reduce electricity bills through the proposed demand response model, in which personalized settings such as travel schedules and user temperature demand curves can be fully considered. The central controller updates the marginal cost curve after each round of regulation and transmits it to the next group, and the interaction avoids new peaks. In each round of adjustment, households in the regulation group solve their respective optimization problems in parallel, ensuring computational efficiency. In addition, a room equivalent thermal parameter (ETP) model and an IAC model are established to form a single household demand response model. In order to quickly obtain the performance parameters of IAC under given conditions, this paper develops a performance map based on the steady-state model of IAC, which can be applied to the direct control of the compressor frequency to determine the power consumption. And the mapping can be easily transformed into piecewise linear constraints and added to the optimization problem.

This article is organized as follows. In Section 3, a detailed ETP model of residence and the performance map based on the IAC model are established. In Section 4, a single household demand response model is given, and the grouping coordinated preheating framework is proposed. In Section 5, the performance of the proposed methods are presented through numerical simulation. Finally, the main conclusions are discussed in Section 6.

2. Methodology

2.1. Research Objectives

The preheating model is an economical and efficient load side regulation method to realize economic savings of a single household. However, regulation failure will occur such as new peaks and increasing system costs when preheating without coordination, which is due to the lack of interaction between global interests and demand scheduling. The response of large-scale electric heat loads to the same electricity price curve will lead to new load peaks and regulation failure. And most of the existing studies take the thermal comfort into consideration by reducing the temperature deviation with the desired value. However, the user's temperature demand is time-variant, which is related to the user's travel planning. In other words, many existing studies cannot take into account users' personal preferences and travel plans. Finally, considering the practicability of the model, the solution time of large-scale preheating planning must meet the scheduling requirements, which is extremely challenging. The research objectives of this paper are summarized as follows:

(1) To solve the problem of regulation failure under large-scale preheating, such as new peaks and increasing system costs.
(2) To consider the temperature preferences and travel planning of each user, and formulate customized heat consumption plan for each user.
(3) To ensure that the running time of the whole preheating framework can meet the scheduling time requirements.

2.2. Research Method

In order to solve the problem of regulation failure under large-scale preheating, a coordinated preheating mechanism is proposed which links the global interests with the demand side response. To formulate a customized heat consumption plan for each user according to their temperature preferences and travel planning, a kind of edge computing and central regulation framework is applied to the coordination mechanism. In addition, to meet the scheduling time requirements, a grouping and parallel computing structure is proposed. According to the above logic, the research architecture is shown in Figure 1.

To formulate the single household regulation model with IAC, a detailed ETP model of residence and the performance map of the IAC model need to be established, which can link the indoor temperature change with the user energy consumption scheduling. Then the grouping coordinated preheating mechanism based on edge computing and central regulation framework is proposed, which links the single household regulation model and the changes of marginal cost curve.

2.3. Simulation Parameters

The simulation is implemented on the python platform. CoolProp is called in the IAC system simulation to obtain physical properties, and Gurobi is called as the solver for the single household demand response optimization. The building parameters and thermal parameters are shown in Table 1. User status division and distribution of parameters can be seen in Section 5.

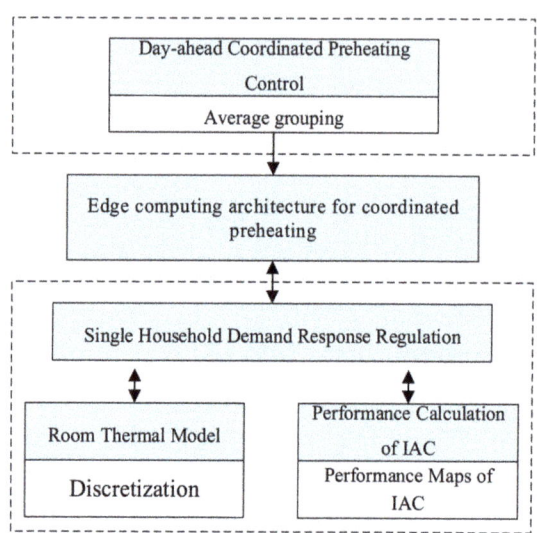

Figure 1. Research architecture.

Table 1. The parameters of the room thermal model.

Parameter	Value	Parameter	Value
Thickness of solid brick/mm	240	Thermal conductivity of solid brick/(W/mK)	0.86
Gypsum thickness/mm	15	Gypsum thermal conductivity/(W/mK)	0.386
Window thermal conductivity/(W/m^{2}K)	5.2	Convection–radiation transfer coefficient of outer surface of exterior wall/(W/m^{2}K)	24
Convection–radiation transfer coefficient of inner surface of exterior wall/(W/m^{2}K)	8.4	Solar heat gain coefficient SHGC	0.7
Equivalent heat capacity of indoor internal mass per unit residential area/(kJ/Km^2)	150	Equivalent heat capacity of external wall per unit area/(kJ/Km^2)	376
Convective heat conduction coefficient of air and indoor mass Uam/Acon/(W/m2K)	10	Absorptance of surface for solar radiation	0.8
Lighting heat gain/W	720	Electric appliance heat gain/W	780
Human body thermal radiation gain/W	300	The convective split for solar heat gain	0.6
The convective split for lighting heat gain	0.6	The convective split for electric appliance heat gain	0.8
The convective split for human body heat gain	0.5	The radiative split for solar heat gain	0.4
The radiative split for lighting heat gain	0.4	The radiative split for electric appliance heat gain	0.2
The radiative split for human body heat gain	0.5		

In this paper, a certain type of apartment in southern China is used as the standard type. The geometric parameters are: length 13.6 m, width 8.6 m, and height 2.6 m. The residence has a north/south external wall, with one external wall on the east or west side, totaling three external walls. The window-to-wall ratio of each orientation is 0.25 in the north direction; 0.35 in the south direction; and 0.2 in other directions.

3. Residential Electric Heating System Model

3.1. Room Thermal Model

In order to link the indoor temperature change with the user energy consumption scheduling, a residential ETP model is established, which consists of the outdoor part, the exterior wall and the indoor part, as seen in Figure 2. There are three main modes of heat transfer in the model: heat conduction, heat convection and heat radiation. The outer wall is equivalent to a thermal resistance and two thermal capacitances, which are denoted as R and C, respectively. For the external surface of the wall, it obtains the heat gain $Q_{solar,w}$ from solar radiation, obtains heat from the inside of the wall, and transfers heat to the outdoor air by convection. Equation (1) depicts the change rate of the exterior wall temperature T_{we}, which is related to heat transfer. Similarly, Equation (2) depicts the change rate of the internal wall surface temperature T_{wi}, which is influenced by both indoor air convection and wall conduction heat transfer. Equation (3) gives the energy balance equation of indoor air. Indoor air has a certain heat storage capacity, and it exchanges heat with internal mass, the air outside walls and windows. In the model, indoor air also absorbs heat gain $Q_{solar,a}$ from solar radiation, internal heat gain $Q_{gain,a}$ and heat Q_{AC} generated by IAC, but will leak heat Q_{lk} due to the gap at the junction. Equation (4) gives the energy balance equation of internal mass. The internal mass such as room partitions and furniture have large thermal inertia, and their equivalent heat capacity C_m can maintain the slow change of mass temperature T_m. In addition to the heat exchange with indoor air, the internal mass absorbs the solar radiation heat gain and internal heat gain, which are denoted as $Q_{solar,m}$ and $Q_{gain,m}$. Equations (5)–(7) are the expressions of solar heat gain absorbed by exterior walls, internal mass and indoor air, respectively. Equation (8) is the heat loss. Equations (9) and (10) represent the absorption of internal heat gain by mass and air respectively. Equation (11) represents three sources of internal heat gain: household appliances, lighting and human body heat radiation.

$$C_w \frac{dT_{we}}{dt} = \frac{T_o - T_{we}}{R_{wo}} + \frac{T_{wi} - T_{we}}{R_w} + Q_{solar,w} \tag{1}$$

$$C_w \frac{dT_{wi}}{dt} = \frac{T_{we} - T_{wi}}{R_w} + \frac{T_a - T_{wi}}{R_{wi}} \tag{2}$$

$$C_a \frac{dT_a}{dt} = \frac{T_m - T_a}{R_{am}} + \frac{T_{wi} - T_a}{R_{wa}} + \frac{T_o - T_a}{R_{win}} + Q_{solar,a} + Q_{gain,a} + Q_{AC} - Q_{lk} \tag{3}$$

$$C_m \frac{dT_m}{dt} = \frac{T_a - T_m}{R_{am}} + Q_{solar,m} + Q_{gain,m} \tag{4}$$

$$Q_{solar,w} = k_{solar} A_w I_{solar} \tag{5}$$

$$Q_{solar,m} = f_{solar,m} \times SHGC \times A_{win} I_{solar} \tag{6}$$

$$Q_{solar,a} = f_{solar,a} \times SHGC \times A_{win} I_{solar} \tag{7}$$

$$Q_{leak} = \rho C_p V_{room} \times ACH \times (T_a - T_o)/3600 \tag{8}$$

$$Q_{gain,m} = f_{gain,m} Q_{gain} \tag{9}$$

$$Q_{gain,a} = f_{gain,a} Q_{gain} \tag{10}$$

$$Q_{gain} = Q_{equip} + Q_{lamp} + Q_{occup} \tag{11}$$

where k_{solar} denotes absorptance of surface for solar radiation; A_w denotes geometric area of the exterior wall; I_{solar} denotes solar radiation; $SHGC$ represents solar heat gain coefficient; $f_{solar,m}$ and $f_{solar,a}$ represent the radiative/convective split for the solar heat gain respectively; A_w denotes geometric area of the window; ρ is the air density; C_p denotes the specific heat of air at constant pressure. V_{room} represents the room volume; ACH denotes the air exchange per hour; $f_{gain,m}$ and $f_{gain,a}$ represent the radiative/convective split for the internal heat gain respectively. The thermal dynamic model of the room can be added to the convex optimization problem as constraint conditions, which can be easily solved with advanced optimization techniques. The discretization expression is shown in (12)–(15).

$$T_{we}(t) = \left(1 - \frac{\Delta t}{C_w R_{wo}} - \frac{\Delta t}{C_w R_w}\right) T_{we}(t-1) + \frac{\Delta t}{C_w R_{wo}} T_o(t-1) + \frac{\Delta t}{C_w R_w} T_{wi}(t-1) + \frac{\Delta t}{C_w} Q_{solar,w}(t-1) \quad (12)$$

$$T_{wi}(t) = \left(1 - \frac{\Delta t}{C_w R_w} - \frac{\Delta t}{C_w R_{wi}}\right) T_{wi}(t-1) + \frac{\Delta t}{C_w R_w} T_{we}(t-1) + \frac{\Delta t}{C_w R_{wi}} T_a(t-1) \quad (13)$$

$$T_a(t) = \left(1 - \frac{\Delta t}{C_a R_{am}} - \frac{\Delta t}{C_a R_{wa}} - \frac{\Delta t}{C_a R_{win}}\right) T_a(t-1) + \frac{\Delta t}{C_a R_{am}} T_m(t-1) + \frac{\Delta t}{C_a R_{wa}} T_{wi}(t-1) \\ + \frac{\Delta t}{C_a R_{win}} T_o(t-1) + \frac{\Delta t}{C_a}(Q_{solar,a} + Q_{gain,a} + Q_{AC} - Q_{lk}) \quad (14)$$

$$T_m(t) = \left(1 - \frac{\Delta t}{C_m R_{am}}\right) T_m(t-1) + \frac{\Delta t}{C_m R_{am}} T_a(t-1) + \frac{\Delta t}{C_m}(Q_{solar,m} + Q_{gain,m}) \quad (15)$$

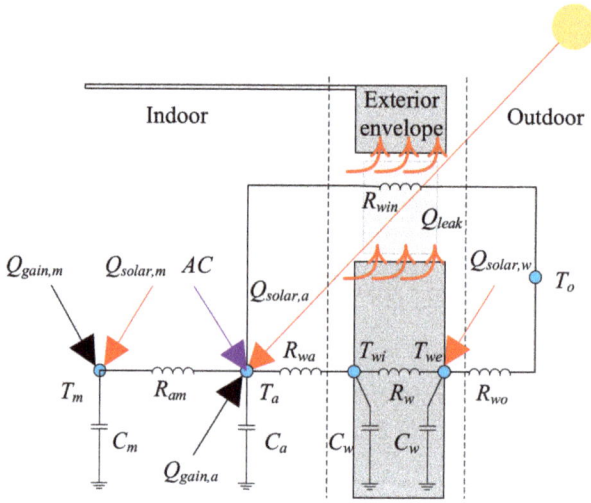

Figure 2. ETP model of residences.

3.2. Performance Maps of IAC

3.2.1. Performance Calculation of IAC

Steady-state IAC models can be used to study the cycling characteristics of IAC under different environment and control conditions. Specifically, the performance of IAC includes power consumption P_{AC} and coefficient of performance COP, which are affected by compressor speed N_{comp}, outdoor temperature T_o and indoor temperature T_a. Figure 3 shows the main components of the air conditioning heating system and the cycle process, which meets the conservation of energy and mass. Starting from point A, the cycle is analyzed. The specific enthalpy h_A of refrigerant steam is calculated from the evaporation temperature T_{evap} and the degree of superheat SH, and the steam changes into high-temperature and high-pressure steam through the compressor module (including the accumulator, suction pipe, compressor and exhaust pipe, etc.). Then the compressor power W_{comp} and the mass flow rate m_{comp} can be obtained. The high-temperature and high-pressure steam releases heat to the indoor air through the condenser and cools to subcooled liquid. The degree of subcooling SC, is calculated by the condenser model. The refrigerant liquid with medium temperature and high pressure is depressurized by the thermal expansion valve, and the mass flow rate m_{exp} is calculated according to the expansion valve model. The low-temperature and low-pressure liquid absorbs heat and turns into a gas through the evaporator, and the output specific enthalpy h_{evap} is calculated according to the evaporator model. There are three independent variables in the above cycle, namely $[D_{evap}, D_{cond}, SH]$, where D_{evap} is the difference between the inlet temperature at the air side of the evaporator and the evaporation temperature, and D_{cond} is the difference between the condensation temperature and the inlet temperature at the air side of the condenser. Accordingly, the cycle should also meet three balance constraints, namely, specific enthalpy constraints at the start and end of the cycle, condenser subcooling constraints and mass flow balance constraints. The constraints are represented by residuals $[\Delta 1, \Delta 2, \Delta 3]$ respectively. When the steady state model converges, the residual values are all zero.

$$P_{AC} = W_{comp} + W_{cond,fan} + W_{evap,fan} \tag{16}$$

$$COP = \frac{Q_{cond} + W_{cond,fan}}{P_{AC}} \tag{17}$$

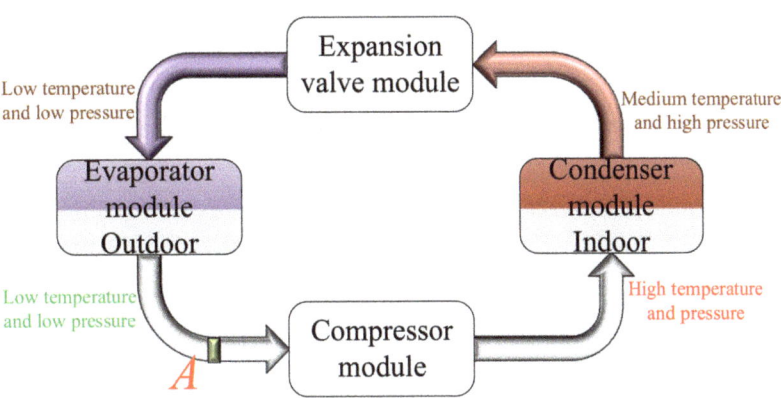

Figure 3. Schematic diagram of AC heating cycle.

Figure 4 shows the numerical calculation flow of the system, which is mainly divided into four parts: initial value calculation, independent variable update, pressure drop correction and performance parameter calculation. The operating condition parameters include the indoor air temperature T_a, the outdoor temperature T_{out} and the rotor speed N_{comp} of the inverter compressor. The model first solves the independent variables in the minimal model as the initial value of the iteration. Secondly, the Broyden algorithm is used to solve the independent variable iteratively to make the residual close to zero, and the pressure drop at the high pressure side ΔP_{h+}, and low pressure side Δp_l are calculated when converged. The pressure drop is introduced into the compressor model, and the pressure drop and independent variables are updated iteratively under this condition. The iteration stops when the pressure drop difference is less than the threshold. Finally, the performance parameters are calculated. As shown in Equations (16) and (17), the power consumption of IAC is mainly related to the compressor, condenser fan and evaporator fan. At the same time, the electric energy consumed by the condenser fan is eventually converted into the internal energy of the indoor air. Therefore, in addition to the heat release of the condenser, the numerator of COP also contains the internal energy.

3.2.2. 3D Storage of the Mapping

The calculation of the steady state model is time-consuming, while the demand response problem needs to ensure the accuracy and speed of the calculation. To this end, this paper calculates the performance parameters of IAC at each operating point in advance, and uses the 3D map to store the mapping relationships.

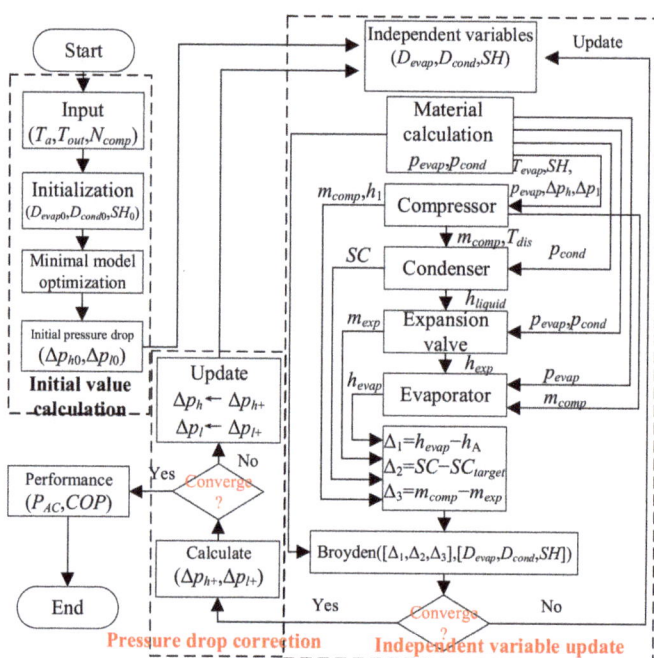

Figure 4. Flow chart of air conditioning system simulation calculation.

As shown in Figure 5, the power consumption of IAC increases with the rotational speed, while COP increases first and then decreases with the speed. As shown in Figure 5b, the COP corresponding to 30 Hz is lower than that of 40 Hz. At the same frequency, the closer T_a and T_{out} are, the higher the COP. According to Figure 5a,b, there is the following mapping: $Q_{AC} = f(P_{AC}, T_a, T_{out})$. With T_a and T_{out} determined, Q_{AC} can be expressed as a piecewise linear function of P_{AC}, denoted as f_{PQ}. It can be easily added to mixed integer linear problems as constraints. T_{out} is obtained from day-ahead weather forecasts, while T_a is related to the temperature set by the household.

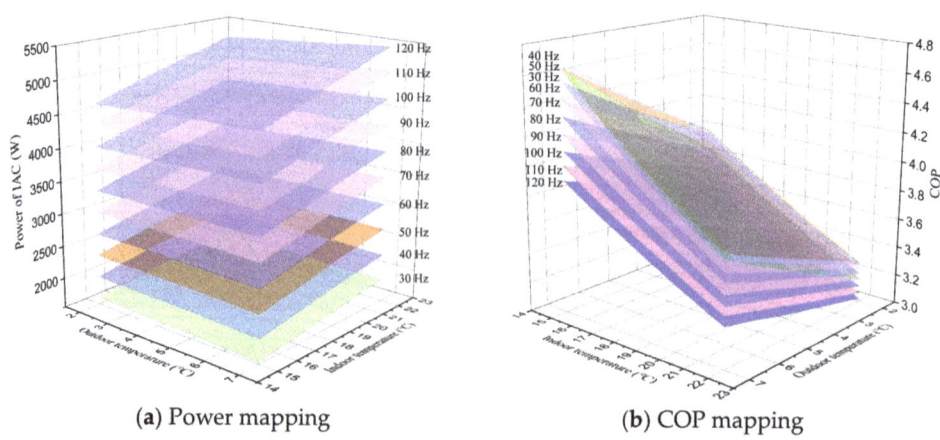

(a) Power mapping (b) COP mapping

Figure 5. Performance maps of IAC.

4. Day-Ahead Coordinated Preheating Control

The preheating control utilizes the thermal inertia of the room to transfer heat loads to improve the economy of the heating system. Specifically, houses are preheated with cheap electricity during the high generation period of renewable energy or the low load period of the system, so as to reduce the energy consumption of marginal units during the high load period of the system. At present, most of the southern areas of China are in the mode of intermittent air conditioning heating, and the heat load is flexible. This paper firstly proposes the heat load scheduling of a single house to minimize the operating cost of IACs while meeting the user's temperature requirements. We then propose an efficient grouping coordinated control framework considering that the large-scale heat load regulation will change the marginal cost of the system.

4.1. Single Household Demand Response Regulation

The objective of single household demand response regulation (SDR) is to reduce user bills within the scheduling period while meeting the user's temperature demands. The optimization problem is formulated as Equations (16)–(18).

$$\min_{P_{AC}} \sum_{t \in \mathcal{T}} P_{AC}(t) \Delta t \cdot pr_t + \rho_e e_t \tag{18}$$

$$s.t.\ Eq.(12) \sim Eq.(15)$$

$$Q_{AC}(t) = f_{PQ}(P_{AC}(t), T_a^*(t), T_{out}(t)) \tag{19}$$

$$T_{lb,t} + e_t \leq T_a(t) \leq T_{ub,t} + e_t \tag{20}$$

Equation (18) is the objective function, where T is the scheduling interval; Δt is the scheduling time step; pr_t is the electricity price at time t; $e_t \geq 0$ is a slack variable, used to avoid failure to solve optimization problems under tight constraints, and ρ_e is the corresponding penalty factor. Equations (12)–(15) are the discretized equivalent heat balance constraints, and each time step corresponds to four heat balance equations. Equation (19) is the COP mapping relationship of IACs, which is discretized into a piecewise linear constraint of $P_{AC} \rightarrow Q_{AC}$. The addGenConstrPWL function in Gurobi makes it easy to add to constraints. Equation (20) represents the user's demand constraints at different times, that is, the room temperature should be within the upper and lower limits set by the user. The upper and lower temperature limits at time t are respectively denoted as $T_{lb,t}$ and $T_{ub,t}$. There are different settings for working and sleeping. In addition, considering the threshold of IAC in actual operation, P_{AC} is set as a semi continuous variable, that is, it should meet $P_{AC}(t) = 0$ or $P_{lb,t} \leq P_{AC}(t) \leq P_{ub,t}$.

4.2. Grouping Coordinated Preheating Framework

Although the above optimization model can reduce user bills, if all users schedule based on the same price curve, the marginal cost of power generation will change, which cannot guarantee the reduction in total power generation cost. In order to solve this problem, this paper proposes a grouping coordinated preheating framework with edge computing and central coordination.

As shown in Figure 6, each household acts as an edge computing unit to solve the demand response problems (18)–(20) according to the received marginal electricity price. The central controller is the key to coordinate all groups, which contains three functional modules: user grouping, marginal cost calculation and specifying scheduling group. According to the central controller, the user adjusts the heat load planning based on the updated marginal cost. And in each round of adjustment, only a small number of users participate in scheduling so that the marginal cost will not be significantly changed.

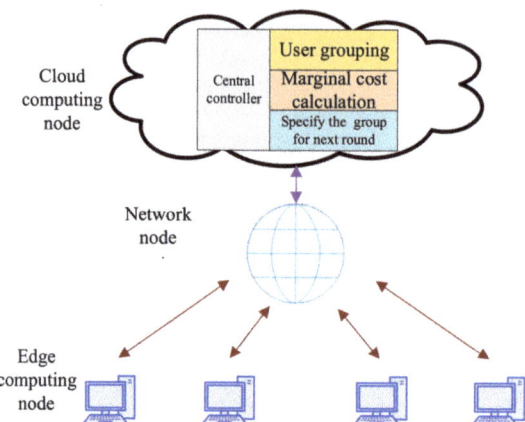

Figure 6. Edge computing architecture for coordinated preheating.

The interactive process of coordinated preheating is shown in Figure 7, which consists of the original load upload and several scheduling rounds. First, users of all groups upload the original load curve shown in ①. Next, Group # 1 ~ Group # G participates in each scheduling round in turn. Taking Group # 1 as an example, each round incorporates three steps: (1) The central controller sends the marginal generation cost curve to Group # 1 as process ②. (2) All local controllers in Group # 1 receive marginal cost curve based on Equations (18)–(20) to update the heat load curve. (3) The local controller uploads the new load curve, namely process ③. After the central controller updates the marginal cost curve, Group # 2 conducts the next round of adjustment. The iteration will continue until the groups balance or the set number of rounds are met. The process is summarized as Algorithm 1.

Algorithm 1 Coordinated Preheating Control Algorithm

Grouping: Group #1 ~ Group #G
Local controller input: Day-ahead travel planning and temperature range of each user
Central controller input: Generation cost function
1: Each user sets the day-ahead travel plan and acceptable temperature range, then the local controller calculates and uploads the heat load curve based on the on-off control.
2: The central controller calculates the marginal generation cost curve λ_n based on the aggregated load curve and passes it to Group #1.
3: **for** iteration = 1 to N **do**
4:    After receiving the marginal cost curve, all users within selected group get their own optimized heat load curve based on Equations (18)–(20). The local controller uploads the curve.
5:    The central controller updates the load curve, calculates the new marginal generation cost curve and passes it on to the next group.
6: **end for**

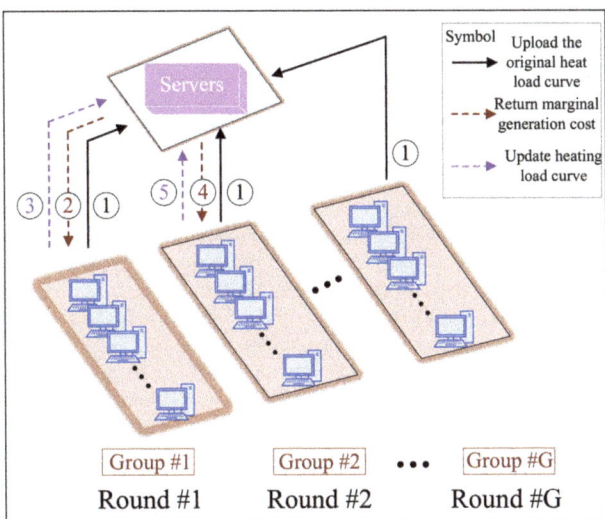

Figure 7. Interactive process of coordinated preheating.

5. Simulation Results

This paper will evaluate the effect of coordinated preheating from indicators such as energy consumption cost, user bills, renewable energy consumption, and iteration rounds.

The demo is a regional power system including 20,000 households, in which the installed capacity of wind power is 30 MW. All users are divided into ten groups to ensure that the adjustment of a single group has less impact on the system. $T_{lb,t}$ and $T_{ub,t}$ in Equation (20) are related to the user status, which includes awake at home, out of home and sleeping. The user status is divided by the time points shown in Table 2, where t_{up} denotes wake-up time, t_{leave} denotes departure time, t_{return} denotes home time, t_{down} denotes bedtime. In addition, the relationship between these time points is as follows: $t_{up} = t_{leave} - d_{mor}$, $t_{down} = t_{return} + d_{eve}$, in which d_{mor} and d_{eve} are the awake time spent at home in the morning/evening respectively. This paper assumes that t_{leave} and t_{return} follow truncated normal distribution, and d_{mor} and d_{eve} follow uniform distribution, as shown in Table 3. The value of the time step is an integer between 0 and 96, and the time step is the scheduling interval, which is 15 min.

Table 2. User status division.

Status	Time Ranges	Temperature Range
Status 1: Awake at home	$[t_{up}, t_{leave}]$, $[t_{return}, t_{down}]$	[21 °C, 25 °C]
Status 2: Out of home	$[t_{leave}, t_{return}]$	/
State 3: Sleeping	$[t_{start}, t_{up}]$, $[t_{down}, t_{end}]$	[18 °C, 24 °C]

Table 3. Distribution of parameters.

	Distribution	Range
Time to leave home/(15 min)	$t_{leave} \sim \mathcal{N}(31, 3^2)$	[25, 37]
Time to get home/(15 min)	$t_{return} \sim \mathcal{N}(76, 3^2)$	[70, 82]
Activity duration at home in the morning/(15 min)	$d_{mor} \sim \mathcal{U}(2, 9)$	/
Activity duration at home at night/(15 min)	$d_{eve} \sim \mathcal{U}(12, 21)$	/

The generation cost function required by the central controller can be in any form, including expressions, step function curve and even calculation programs. It is worth mentioning that the central controller only needs to perform a limited number of marginal cost calculations based on the generation cost during iteration, which can be extended to any user scale and network topology. As in [24–26], this paper sets the cost function in the form of quadratic function, that is, $C_{total} = ax^2 + bx + c$. The coefficient of the quadratic term is 5 ¥/MW² h, the coefficient of the primary term is 200 ¥/MWh, and the constant term is 800 ¥.

5.1. Coordinated Preheating Results and Comparison

The simulation will be implemented within the time range of 04:00~24:00. In order to comprehensively analyze the performance of the coordinated preheating algorithm, this paper compares several cases, and the results are shown in Figure 8. The gray area in the figure represents the base load, and the yellow area represents the wind power output. The base load is not adjustable. The black solid line is the baseline heating schedule, in which all users set heating schedules by on-off mode according to the set temperature. The solid red line is the result of coordinated preheating. It can be seen that the curve is smoother than the baseline because part of the peak load is shifted to the valley. And the red solid line is almost all above the yellow area, so the wind power consumption is greatly promoted. Another two cases were analyzed to demonstrate the superiority of the coordinated preheating strategy. In case 1, all households schedule their heat load curves based on the baseline electricity price to reduce the energy bill on the premise of meeting the temperature demand. The baseline electricity price curve is the marginal generation cost curve in the baseline case. It can be seen that in case 1, a forward and higher peak is formed due to the neglect of marginal cost change, which leads to excessive preheating and heat loss instead. Therefore, concentrated scheduling in the absence of coordination will lead to preheating failure and increase the cost of power generation. To further illustrate the necessity of coordination, case 2 is used for comparison. In case 2, all households schedule their heat load curves based on the marginal cost curve of the coordinated preheating case, which can be regarded as the optimal electricity price curve. However, it also failed to reduce the peak, indicating that the scheduling based on a single price curve is not feasible and the coordination is necessary.

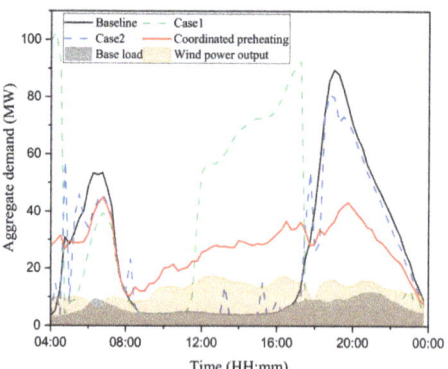

Figure 8. Total load under different conditions.

A total of 30 adjustment rounds were conducted in the simulation, namely Round # 1~Round # 30. In Round #1, Group #1 performs rescheduling based on the electricity price given by the central controller. Similarly, Group #10 performs rescheduling in Round #10. In Round # 11, it returns to Group # 1. The change of each evaluation index with the number of rounds, is shown in Figure 9. Figure 9a shows the change of total generation cost, which is reduced by 36.8% due to coordinated preheating. Figure 9b shows that the total electricity bill of users has decreased by 48.3%. In addition, it can be seen from Figure 9c that wind power curtailment is significantly reduced, and almost all wind power can be consumed. It is worth noting that the total user bill decreases more than the total generation cost, which is attributed to the smoother load curve after preheating adjustment. The two values are quite close when reaching the group equilibrium, which indicates that the proposed algorithm achieves the optimal adjustment.

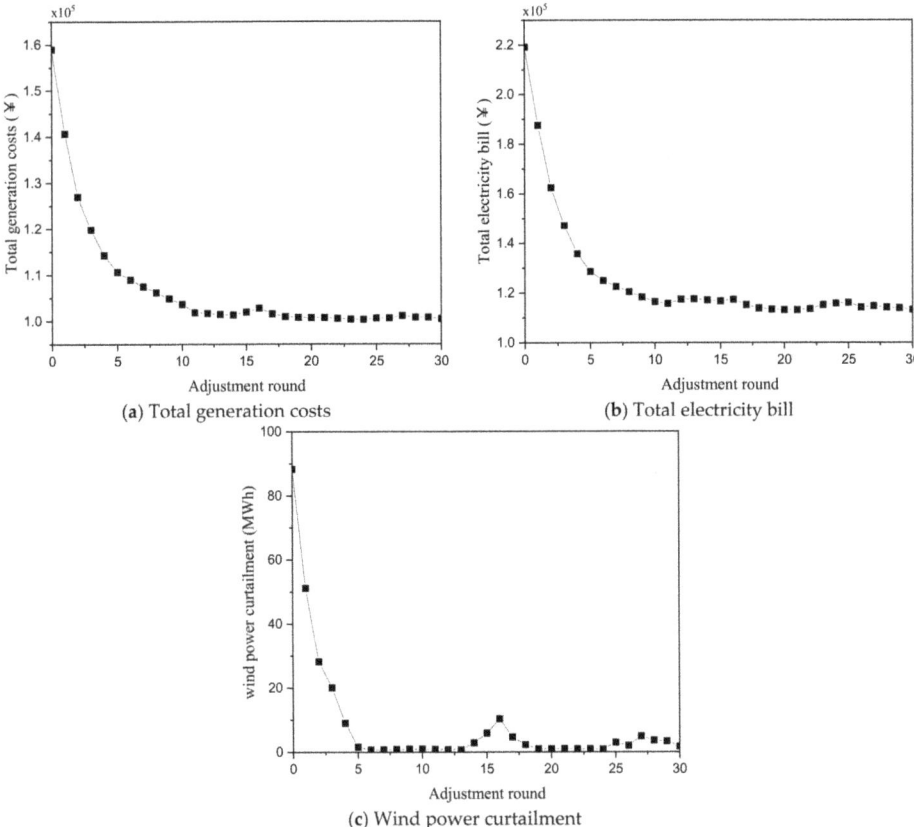

Figure 9. Changes of Evaluation Indexes with the Adjustment Rounds.

In addition, it can be found that all indicators have reached a balance in ten rounds. At this time, all groups performed a round of adjustment, and subsequent optimization produced only slight fluctuations. In the simulation, limited by the computing power of a single computer, users in the group need to schedule one by one, so it takes a long time. However, in practical applications, with the help of users' edge computing capabilities, users in the group solve optimization problems in parallel, and the adjustment time for each round is within 10 s. The whole coordinated preheating process can reach equilibrium within 2 min, which is suitable for day-ahead scheduling. In addition, the time-consuming is related to the number of groups, but not related to the user scale due to the group-by-group coordination method, which is extensible for super-large-scale coordination. On the contrary, the one-by-one coordination method requires 20,000 rounds of adjustment in this example, whose calculation time is linearly related to the user size, so it is not scalable.

Table 4 presents the performance comparison under different models, and two conclusions can be drawn: (1) A coordination mechanism is necessary. Preheating without coordination will lead to the failure of adjustment. Its net peak and generation cost are even higher than those without preheating; (2) A grouping mechanism is necessary, which can achieve scale-independence, while coordination one by one is unacceptable in terms of computing time.

Table 4. Comparison of different preheating effects.

	Generation Cost (10,000 ¥)	User Bill (10,000 ¥)	Net Peak (MW)	Wind Power Curtailment (MW)	Required Rounds
No preheating	15.90	21.92	73.33	88.34	—
Preheating without coordination	22.42	33.16	92.89	50.32	1
Coordinated preheating group by group	10.04	11.33	38.40	1.03	20
Coordinated preheating one by one	—	—	—	—	20,000 (Unacceptable)

In order to analyze the impact of each round of adjustment on the system, the important process curves are given in Figure 10, and their meanings have been marked in the figure. It can be seen from Figure 10a that both the morning and evening peaks of the total load decrease with the adjustment. The peak decreases more obviously in the first few rounds, and gradually approaches the balance in the later rounds. It is worth noting that compared with the morning peak, the evening peak has dropped more significantly. This is because the preheating period before the evening peak is long and there is no specific heat demand (the user has not yet arrived home), so the adjustable range is large. The preheating period before the morning peak has strict temperature restrictions (to ensure a proper sleep temperature), and the adjustment range is limited. In addition, since the earlier the preheating is, the greater the heat loss will be, the preheating period will not be too long. The gray part in Figure 10b represents the wind power curtailment area, and it can be seen that the curve gradually leaves the area with adjustment. Figure 10c shows the change process of the marginal cost curve, which becomes smoother with adjustment.

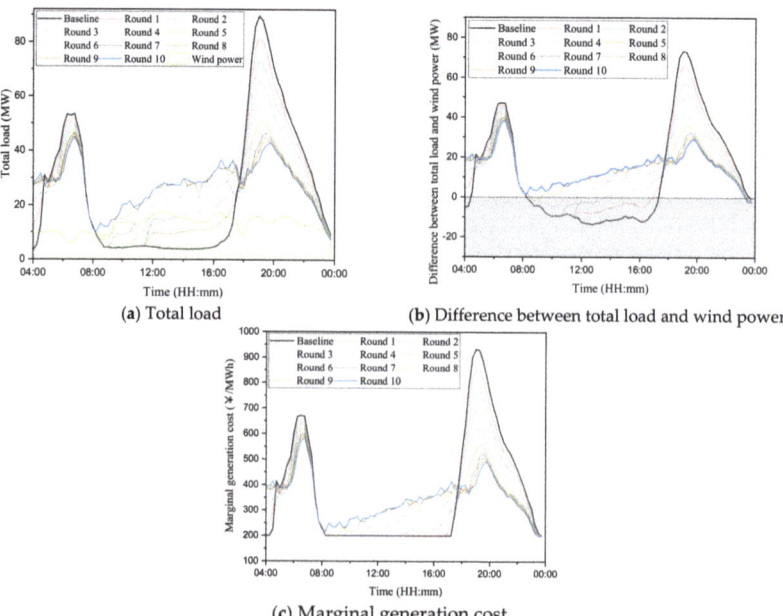

Figure 10. Process display of each round.

5.2. Analysis of Temperature and Energy Consumption in the Coordinated Preheating Process

To further explore the details of the preheating process, Figure 11 shows the temperature and power curves of users participating in Round #1, including the adjusted and original curves. In the temperature chart, the blue dotted line represents the temperature limit when awaking at home; The red dotted line represents the temperature limit when sleeping at home; The green area represents the sleeping period set by the user; The grey area refers to the period of awaking at home. Therefore, the indoor temperature curve must be in the orange and blue areas to meet user demands. The original curve adopts an on–off adjustment mode to ensure that the temperature does not exceed the limit.

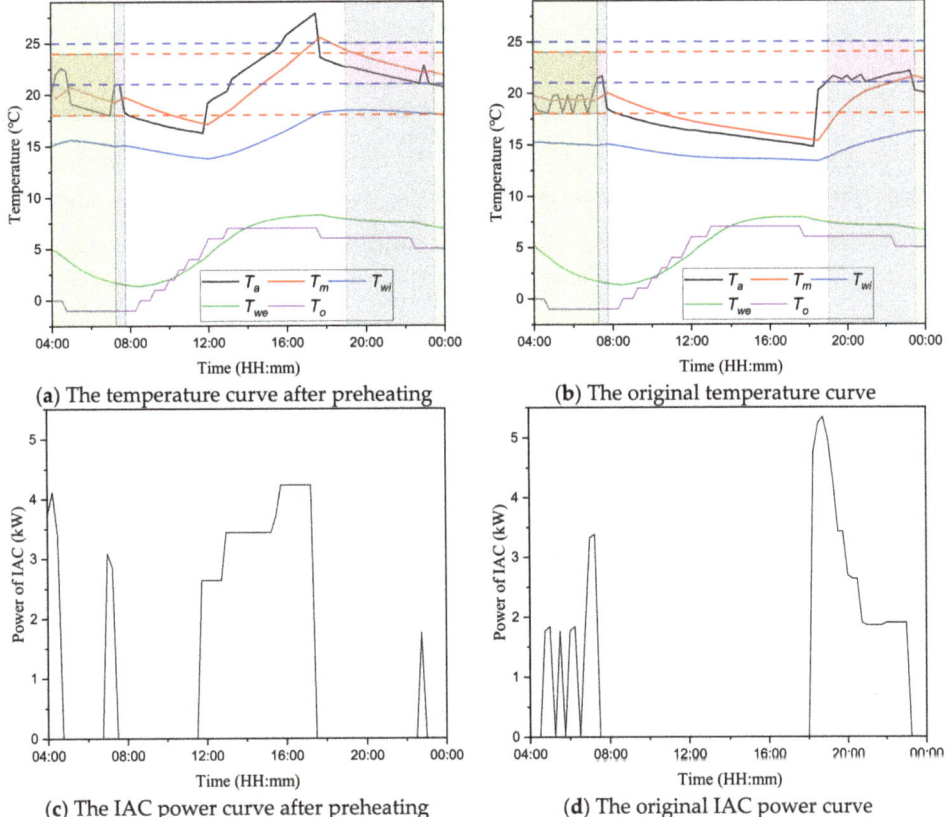

Figure 11. Comparison between optimization curves and original curves of a family participating in the Round #1.

Figure 11b is the original temperature curve, and Figure 11d is the corresponding IAC power curve. It can be seen that the black solid line representing the indoor air temperature is located at the bottom of the orange and blue areas, which both meet the user's temperature demands and save energy. From 04:00 to 07:15, the user was sleeping, so the IAC adjusts the power to ensure that the temperature does not exceed the lower limit. The power is increased around 07:00 to ensure a comfortable temperature when the user wakes up. After the user leaves home, the IAC stops running, so T_a and T_m starts to drop. T_m decreases more slowly than T_a because the heat capacity of internal mass is larger than that of air. At this time, the internal mass transfers heat to the air. As the temperature difference between T_a and T_{wi} gradually decreases, the heat loss becomes smaller, so the temperature drops faster first and then slower. The IAC power starts to rise at about 18:15 to ensure a comfortable temperature when users get home. As the air heat capacity is small, T_a rises rapidly. And due to the large temperature difference between the air and internal mass, the air begins to transfer a lot of heat to internal mass until temperature T_m is close to temperature T_a. In addition, as the temperature demand decreases during sleep, although the IAC power drops to zero at 23:15, the temperature still meets the demands when the user sleeps (around 23:30).

Figure 11a,c are the temperature and power curves after preheating adjustment. The marginal cost curve received by the user is the baseline curve in Figure 10c. It can be seen that the marginal cost is lower in the period of 04:00~04:30 and the period out of home, so the two periods are selected for preheating. Compared with the original power curve, the power of the IAC is higher during the preheating period, and the room temperature also increases while keeping within the set range. It is worth noting that, in order to reduce heat loss, the preheating power increases gradually in the second half of the time out of home. When users get home, the IAC is shut down for a long time, while the indoor temperature is still in a comfortable range, thus reducing the energy consumption during the high marginal cost period.

6. Conclusions

In this paper, a grouping coordinated preheating method based on a demand response model is proposed for large-scale electric heating load, and the effectiveness of the proposed method is verified by simulation. The main conclusions are as follows:

(1) The proposed framework can effectively coordinate the preheating scheduling of users, take into account the changes in the marginal cost of the system, so as to solve the problem of regulation failure under large-scale preheating, namely, new peaks and increasing system costs, which promotes the consumption of renewable energy. The effectiveness of the coordination is proved by the results in Figures 8–10.

(2) The room thermal parameter model and the performance map of IAC are integrated into the demand response model. The model can take into account the travel planning of users and ensure that the indoor temperature does not exceed the limit while reducing the electricity bill. A kind of edge computing and central regulation framework is applied to the coordination mechanism, in which customized heat consumption plan can be formulated for each user according to their temperature preferences and travel planning. The effectiveness is proved by the results in Figure 11.

(3) The parallel computing structure within the adjustment group under the coordination framework ensures the computing efficiency to meet the scheduling time requirements. The proposed framework can be extended to larger scale systems. The effectiveness is proved by the results in Table 4.

Author Contributions: Conceptualization, G.G. and Y.G.; methodology, G.G. and Y.G.; investigation, G.G. and Y.G.; writing—original draft preparation, G.G. and Y.G.; writing—review and editing, G.G. and Y.G.; supervision, G.G. and Y.G. All authors have read and agreed to the published version of the manuscript.

Funding: This research received no external funding.

Institutional Review Board Statement: Not applicable.

Informed Consent Statement: Not applicable.

Data Availability Statement: All data included in this study are available upon request by contact with the corresponding author.

Conflicts of Interest: The authors declare no conflict of interest.

References

1. CCTV NEWS. Power Load Reaches a New High in the Cold Wave Attacking. Available online: https://baijiahao.baidu.com/s?id=1688289589720589745&wfr=spider&for=pc (accessed on 1 September 2022).
2. Polaris Power Grid. Texas Blackout. Available online: https://news.bjx.com.cn/html/20210220/1136852.shtml (accessed on 1 September 2022).
3. Jia, Y. Influences of Occupant Ventilation-Behavior during Off-Periods on Heating Energy Consumption and Indoor Thermal Environment in Intermittently Heated Buildings. Ph.D. Thesis, Donghua University, Shanghai, China, 2021.
4. Jinan Daily. The Correct Way of 'Coal to Electricity' for Heating—Air Source Heat Pump. Available online: http://www.jnrdyxgs.com/jnrd/page?id=298c7f0b-4ec4-11e9-8081-bb132fb4f7bb&type=last (accessed on 1 September 2022).
5. Eurovent. EU Strategy for Energy System Integration (GEN-1138.00). Available online: https://eurovent.eu/?q=articles/eu-strategy-energy-system-integration-gen-113800 (accessed on 1 September 2022).
6. National Energy Administration. Answers to Reporters' Questions about Energy Supply This Winter and Next Spring. Available online: http://www.nea.gov.cn/2021-09/29/c_1310217739.htm (accessed on 1 September 2022).
7. Golmohamadi, H.; Larsen, K.G.; Jensen, P.G.; Hasrat, I.R. Optimization of power-to-heat flexibility for residential buildings in response to day-ahead electricity price. *Energy Build.* **2021**, *232*, 110665. [CrossRef]
8. Duman, A.C.; Erden, H.S.; Gönül, Ö.; Güler, Ö. A home energy management system with an integrated smart thermostat for demand response in smart grids. *Sustain. Cities Soc.* **2021**, *65*, 102639. [CrossRef]
9. Hong, Y.Y.; Lin, J.K.; Wu, C.P.; Chuang, C.C. Multi-objective air-conditioning control considering fuzzy parameters using immune clonal selection programming. *IEEE Trans. Smart Grid* **2012**, *3*, 1603–1610. [CrossRef]
10. Wang, D.; Meng, K.; Gao, X.; Qiu, J.; Lai, L.L.; Dong, Z.Y. Coordinated dispatch of virtual energy storage systems in LV grids for voltage regulation. *IEEE Trans. Ind. Inform.* **2017**, *14*, 2452–2462. [CrossRef]
11. Qi, N.; Cheng, L.; Xu, H.; Wu, K.; Li, X.; Wang, Y.; Liu, R. Smart meter data-driven evaluation of operational demand response potential of residential air conditioning loads. *Appl. Energy* **2020**, *279*, 115708. [CrossRef]
12. Cheng, L.M.; Bao, Y.Q. A day-ahead scheduling of large-scale thermostatically controlled loads model considering second-order equivalent thermal parameters model. *IEEE Access* **2020**, *8*, 102321–102334. [CrossRef]
13. Wang, D.; Meng, K.; Gao, X.; Coates, C.; Dong, Z. Optimal air-conditioning load control in distribution network with intermittent renewables. *J. Mod. Power Syst. Clean Energy* **2017**, *5*, 55–65. [CrossRef]
14. Hu, M.; Xiao, F.; Wang, L. Investigation of demand response potentials of residential air conditioners in smart grids using grey-box room thermal model. *Appl. Energy* **2017**, *207*, 324–335. [CrossRef]
15. Hu, M.; Xiao, F. Price-responsive model-based optimal demand response control of inverter air conditioners using genetic algorithm. *Appl. Energy* **2018**, *219*, 151–164. [CrossRef]
16. Hu, M.; Xiao, F.; Jørgensen, J.B.; Wang, S. Frequency control of air conditioners in response to real-time dynamic electricity prices in smart grids. *Appl. Energy* **2019**, *242*, 92–106. [CrossRef]
17. Pau, M.; Cremer, J.L.; Ponci, F.; Monti, A. Day-ahead scheduling of electric heat pumps for peak shaving in distribution grids. In *Smart Cities, Green Technologies, and Intelligent Transport Systems*; Springer: Cham, Switzerland, 2017; pp. 27–51.
18. Zhang, X.; Dong, Z.; Huang, W.; Zhang, N.; Kang, C.; Strbac, G. A novel preheating coordination approach in electrified heat systems. *IEEE Trans. Power Syst.* **2021**, *37*, 3092–3103. [CrossRef]
19. Nguyen, H.T.; Al-Sumaiti, A.S.; Turitsyn, K.; Li, Q.; El Moursi, M.S. Further optimized scheduling of micro grids via dispatching virtual electricity storage offered by deferrable power-driven demands. *IEEE Trans. Power Syst.* **2020**, *35*, 3494–3505. [CrossRef]
20. Shao, S.; Shi, W.; Chen, H.; Li, X.; Yan, Q. Simulation on variable frequency air conditioner. In Proceedings of the 10th Annual Conference of the Chinese Society of Engineering Thermophysics, Beijing, China, 2001; pp. 140–144.
21. Zhou, R.; Zhang, T.; Catano, J.; Wen, J.T.; Michna, G.J.; Peles, Y.; Jensen, M.K. The steady-state modeling and optimization of a refrigeration system for high heat flux removal. *Appl. Therm. Eng.* **2010**, *30*, 2347–2356. [CrossRef]
22. Zakula, T. Heat Pump Simulation Model and Optimal Variable-Speed Control for a Wide Range of Cooling Conditions. Master's Thesis, Massachusetts Institute of Technology, Cambridge, MA, USA, 2010.
23. Hu, M.; Xiao, F.; Cheung, H. Identification of simplified energy performance models of variable-speed air conditioners using likelihood ratio test method. *Sci. Technol. Built Environ.* **2020**, *26*, 75–88. [CrossRef]
24. Hua, H.; Qin, Y.; Hao, C.; Cao, J. Optimal energy management strategies for energy Internet via deep reinforcement learning approach. *Appl. Energy* **2019**, *239*, 598–609. [CrossRef]
25. Du, Y.; Li, F. Intelligent multi-microgrid energy management based on deep neural network and model-free reinforcement learning. *IEEE Trans. Smart Grid* **2019**, *11*, 1066–1076. [CrossRef]
26. Xu, Q.; Zhao, T.; Xu, Y.; Xu, Z.; Wang, P.; Blaabjerg, F. A distributed and robust energy management system for networked hybrid AC/DC microgrids. *IEEE Trans. Smart Grid* **2019**, *11*, 3496–3508. [CrossRef]

Article

Local Energy Market-Consumer Digital Twin Coordination for Optimal Energy Price Discovery under Thermal Comfort Constraints

Nikos Andriopoulos [1,*], Konstantinos Plakas [1], Christos Mountzouris [1], John Gialelis [1], Alexios Birbas [1], Stylianos Karatzas [2] and Alex Papalexopoulos [3]

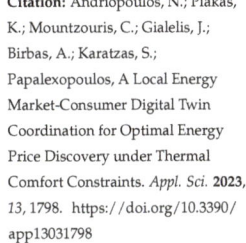

Citation: Andriopoulos, N.; Plakas, K.; Mountzouris, C.; Gialelis, J.; Birbas, A.; Karatzas, S.; Papalexopoulos, A Local Energy Market-Consumer Digital Twin Coordination for Optimal Energy Price Discovery under Thermal Comfort Constraints. *Appl. Sci.* **2023**, *13*, 1798. https://doi.org/10.3390/app13031798

Academic Editor: Chun Sing Lai

Received: 30 December 2022
Revised: 26 January 2023
Accepted: 28 January 2023
Published: 30 January 2023

Copyright: © 2023 by the authors. Licensee MDPI, Basel, Switzerland. This article is an open access article distributed under the terms and conditions of the Creative Commons Attribution (CC BY) license (https://creativecommons.org/licenses/by/4.0/).

[1] Applied Electronics Laboratory, Department of Electrical and Computer Engineering, University of Patras, Rion, 26 504 Patras, Greece; kplakas@ece.upatras.gr (K.P.); mountzou@ece.upatras.gr (C.M.); gialelis@ece.upatras.gr (J.G.); birbas@ece.upatras.gr (A.B.)
[2] Department of Civil Engineering, University of Patras, Rion, 26 504 Patras, Greece; stylianos.karatzas@upatras.gr
[3] ECCO International, Inc., 268 Bush Street, Suite 3633, San Francisco, CA 94104, USA; alexp@eccointl.com
* Correspondence: nandriopoulos@ece.upatras.gr

Abstract: The upward trend of adopting Distributed Energy Resources (DER) reshapes the energy landscape and supports the transition towards a sustainable, carbon-free electricity system. The integration of Internet of Things (IoT) in Demand Response (DR) enables the transformation of energy flexibility, originated by electricity consumers/prosumers, into a valuable DER asset, thus placing them at the center of the electricity market. In this paper, it is shown how Local Energy Markets (LEM) act as a catalyst by providing a digital platform where the prosumers' energy needs and offerings can be efficiently settled locally while minimizing the grid interaction. This paper showcases that the IoT technology, which enables control and coordination of numerous devices, further unleashes the flexibility potential of the distribution grid, offered as an energy service both to the LEM participants as well as the external grid. This is achieved by orchestrating the IoT devices through a Consumer Digital Twin (CDT), which facilitates the optimal adjustment of this flexibility according to the consumers' thermal comfort level constraints and preferences. An integrated LEM-CDT platform is introduced, which comprises an optimal energy scheduler, accounts for the Renewable Energy System (RES) uncertainty, errors in load forecasting, Day-Ahead Market (DAM) feed in/out tariff, and a fair price settling mechanism while considering user preferences. The results prove that IoT-enabled consumers' participation in the energy markets through LEM is flexible, cost-efficient, and adaptive to the consumers' comfort level while promoting both energy transition goals and social welfare. In particular, the paper showcases that the proposed algorithm increases the profits of LEM participants, lowers the corresponding operating costs, addresses efficiently the stochasticity of both energy demand and generation, and requires minimal computational resources.

Keywords: local energy markets; consumer digital twin; transactive energy; thermal comfort; DER

1. Introduction

In recent years, technological advancements and policy directives in the European Union (EU) [1] and the United States of America [2] have led to a significant increase in the number of Distributed Energy Resources (DER) that are primarily connected to the energy grid [3]. Therefore, the traditional energy consumers have been transformed into prosumers, i.e., active entities of the energy market that simultaneously consume, produce and share energy, depending on the regulatory framework, the weather and the operating conditions [4]. Prosumers may own multiple energy assets, primarily small-scale DERs for energy generation, and batteries or Electric Vehicles (EVs) for energy storage [5]. These trends reshape the conventional and centralized power system and eventually disrupt the existing energy system. In this emerging landscape, the power system must undergo structural changes to adapt and leverage the benefits of Internet of Things (IoT) technologies, digitalization, and decarbonization policies.

The edge grid is a structural component of the power system; thus, the concepts of Transactive Energy System (TES) and Local Energy Market (LEM) are considered novel solutions that enable energy exchange between prosumers, increase the power system's efficiency, and reliability, and support the coordination between DERs [6]. Financial and engineering advancements are embraced by TES and LEM, and they form an integral part of the hierarchical energy marketplace, which includes wholesale, retail, and local markets. Inevitably, the solutions of TES and LEM offer incentives for participation for all stakeholders and maximize the social welfare in the energy market.

The exponential growth of computational power and IoT has created a strong technological infrastructure for collecting, transmitting, and processing data in a reliable, efficient, and cost-effective manner. In this context, the concept of Digital Twin (DT), i.e., virtual representations of physical entities that encompass the most discriminative characteristics of the corresponding entity, has attracted the attention of researchers and the industry. Bidirectional and automatic data flow between the physical and virtual entities are used by DTs to produce predictive analytics, perform actions and support informed decisions. In the power sector, DTs are considered promising solutions for sustainability, demand side management, control of energy assets, reliable energy distribution and monitoring of energy grid operations [7].

As flexibility can be obtained either at the residential or community level, this work focuses on residential flexibility, which is then procured to the external grid through LEM. Controlling Heating, Ventilation and Air Conditioning (HVAC) systems based on consumers' thermal comfort tolerance, scheduling household appliances and EV charging are the typical ways to obtain residential flexibility. Based on the consumer's preferences, designated flexible loads are activated at optimal intervals during the desired time window to minimize energy costs. Moreover, by allowing the consumer to specify the subjective intensity of importance for each flexible load, prioritized Demand Response (DR) scheduling can be implemented.

As residential flexibility is becoming critical for power systems and is now at the forefront of the energy market, LEMs are acting as a catalyst in procuring residential flexibility and empowering small RES owners [8]. LEMs simplify and accelerate this process, enabling energy consumers at the edge of the grid to evolve from passive entities to active integral energy market actors. Despite the fact that LEMs expedite consumers' participation in the energy market, prosumers' involvement will not be materialized as long as their market engagement is conducted in a complicated way, limiting the interest on LEM and leading to potential depreciation and eventually failure. Apparently, a seamless and consumer-friendly way of market engagement is a decisive factor for LEM's success. To this end, the Consumer Digital Twin (CDT) serves as a powerful information tool that provides automated data streams on consumers' key characteristics and personalized preferences to minimize their active involvement in operations, increase the efficiency of operations and enhance the consumer-centricity of the LEM.

In this work, a LEM-CDT structure is introduced in order to bridge the gap between local flexibility potential and the consumers' preferences. Towards this end, the benefits of both concepts are leveraged. On one hand, the local flexibility is procured efficiently to the wholesale energy markets while guaranteeing monetary and social benefits for LEM's participants and at the same time the preferences of LEM members are not only respected but also, and most importantly, incorporated in the optimal LEM scheduling. By combining these concepts, the aim is to harness optimally the local flexibility capacity at the distribution grid and attract more participants at LEM initiatives in order to create local sustainable energy communities. To achieve this, CDT is an essential tool since it provides the necessary information to the LEM operator in order to consider the particularities and preferences of each participant placing them at the center of the electricity market. This user-centric approach is one of the main novelties of this paper, offering an efficient solution for LEM operators to model their energy scheduling accurately and attract more participants.

The envisaged LEM–CDT structure is shown in Figure 1. Flexibility stems from the prosumers' side, either in the form of production from various DERs or through DR actions. As shown, prosumers are at the center of the energy market, contributing to the energy mix with any means they own, such as DERs, EVs, or even via DR schema. Supplying flexibility has both technical advantages (e.g., quick response time) and widens the pool of potential LEM members; thus, consumers who do not own DER assets can also be members and contribute to the overall supply through their load flexibility. In addition, participation via DR schema promotes democratization within the community, as all members contribute to the aggregation generation and benefit from their participation. The stochasticity of both energy demand and generation is taken into account to ensure a constant, reliable, and cost-effective energy supply, thereby mitigating the cost of remedial action.

Figure 1. The consumer-centric structure of LEM

This paper is divided into five sections: In Section 2, a comprehensive literature review for both LEMs and CDTs is provided. In Section 3, the proposed solution for the integration of CDT and LEM is presented. In Section 4, the results from experiments are displayed revealing the benefits of integrating CDT into LEM. Finally, in Section 5, the conclusions of the work are provided.

2. State of the Art
2.1. Local Energy Market

Recent research by Honarmand et al. [9] and Doumen et al. [10] emphasized the emergence of the LEM as a solution that prioritizes consumers in integrating DERs into distribution networks effectively. The LEM approach allows the efficient management of DERs, thereby increasing their utilization and overall impact on the distribution grid.

LEMs facilitate prosumers' unfettered access to the electricity market, leading to increased flexibility in the grid and creating revenue streams for small-scale prosumers while supporting balancing, congestion management, and ancillary services [11]. An LEM comprises small-scale energy deployments and energy assets located in a small or a wide geographic area. They are connected to the distribution grid in a decentralized structure where participants cooperate with the available resources (DERs, DR, EVs) on a community level, as depicted in Figure 2. LEM's main objective is to encourage market participation by providing monetary incentives to prosumers to trade energy with one another with minimum or no intermediate (e.g., energy aggregators) [12]. In this market, prosumers can share the benefits of local flexibility within the community, promoting the deployment of distributed renewable generation and DR [13]. To promote prosumers' participation, the LEM achieves load balancing at a lower price compared to the external grid. Consumers (buyers) can reduce energy costs by buying energy at a lower price. In comparison, producers (sellers) can increase their profit by offering energy at a higher price compared to the external grid. Moreover, from a social perspective, it allows participants to be active members of their communities by supporting and enabling them to consume renewable energy and benefit from its distributed generation.

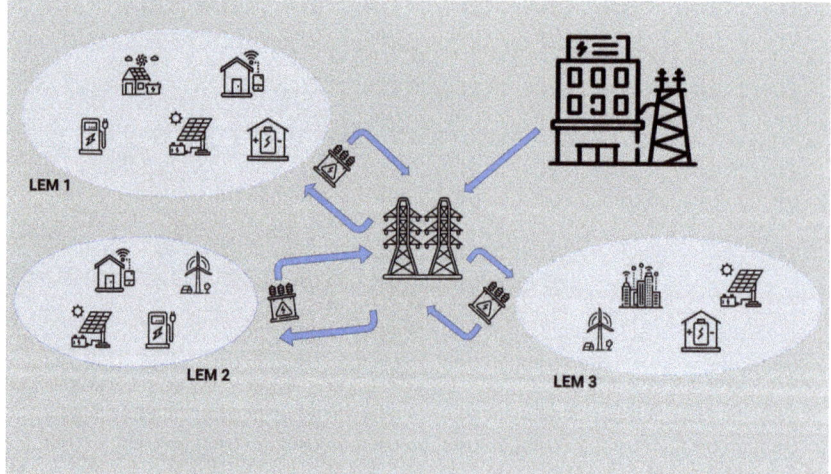

Figure 2. The future decentralized electricity system.

The pricing mechanisms of an LEM constitute one of its most important elements; many algorithms for the LEM clearing price have been proposed. Specifically, Tushar et al. [14] investigated the feasibility of social cooperation among prosumers participating in a peer-to-peer (P2P) energy trading market by utilizing a canonical coalition game approach. The results indicated that the proposed scheme can increase the prosumers' willingness to participate in P2P energy trading schemes. Lee et al. [15] proposed a direct electricity trading market, in which the electricity pricing scheme achieves a fair allocation of profits between consumers and small-scale energy suppliers by using the asymptotic Shapley value function. Tsaousoglou et al. [16] presented a TES where an auction mechanism is implemented with non-convex prosumer models and resource constraints. Long et al. [17] presented three examples of a P2P market structure, namely bill sharing, mid-market rate, and auction-based pricing, to validate the effectiveness of the proposed markets. These market structures were applied on a residential community microgrid with a PV system. Mengelkamp et al. [18] proposed a blockchain-based microgrid energy market without central coordination and evaluated the 'Brooklyn Microgrid Project' as a case study. Finally, Paudel et al. [19] introduced a game-theoretic approach for P2P energy trading among prosumers, where consumers can adjust their consumption according to market conditions.

Since LEMs are not fully restricted in electricity production, Brolin et al. [20] presented a corresponding multi-energy structure that utilizes its full flexibility potential, while Hayes et al. [21] facilitated the efficient flexibility procurement by an aggregator. In the proposed platform, an aggregator can communicate directly with the participants and determine costs and rewards between them so the benefits for both the aggregator and the electrical systems are mutually increased. Lyu et al. [22] proposed a comprehensive energy-sharing framework for smart buildings considering multiple dynamic components covering heating, ventilation, air conditioning, battery energy storage systems, and EVs. Bachoumis et al. [23] investigated the provided ancillary services to the external grid and particularly the fast frequency response service. Finally, Huo et al. [24] considered the uncertainty of PV production by employing the chance-constraints optimization method for the operation of an energy hub.

The current LEMs lack consumer-centrism, failing to take into account the individual preferences of participants. This proposal aims to address this knowledge gap by designing a consumer-centric LEM that respects the priorities of LEM participants, as expressed through personalized preferences for the flexible residential loads and the individual's indoor thermal comfort level. By prioritizing the preferences of consumers, the proposed LEM will increase the effectiveness of DERs integration in distribution networks, providing more efficient and cost-effective energy solutions for the participants.

2.2. Consumer Digital Twin

In electricity markets, DTs are promising solutions for sustainability, control of energy assets, demand-side management, control of energy assets, reliable energy distribution, and monitoring of energy grid operations [25]. Danilczyk et al. [26] proposed a DT to address security issues and detect potential failures in a microgrid, such as instabilities and failures in power distribution, in a timely manner. Darbali-Zamora et al. [27] introduced a real-time DT that optimizes DER operations for distribution voltage regulation and increases the awareness of power system dynamics. Podvalny et al. [28] proposed a scalable and evolutionary DT framework to simulate the behavior of a power system under critical events by employing a neural network as a decision support infrastructure. Wu et al. [29] introduced a DT of grid batteries to diagnose faults in time and control their usage to extend their lifetime, while Jain et al. [30] proposed a virtual replica of solar PVs to promptly detect operational faults and evaluate their power generation performance. Atalay et al. [31] proposed a DT that performs simulations over the virtual copy of the physical grid to detect possible power supply interruptions. Dembski et al. [32] introduced an urban DT representing a real community to enable the execution of scenarios over the virtual twin and provide customized energy services to prosumers. Bazmohammadi et al. [33] mentioned the enhancement of microgrid operations through DTs. Nguyen-Huu et al. [34] and Han et al. [35] utilized DTs as an orchestration mechanism to coordinate the operation of LEM and DER, respectively. Aghazadeh Ardebili et al. [36] used DTs as a tool to predict energy production in power systems with a high volume of RES. Zhou et al. [37] highlighted the benefits of DTs for providing flexibility in industrial power systems.

In the energy sector, the employment of DTs has gained significant attention as a means to optimize the management of DERs. However, while DTs have proven to be versatile tools in representing and simulating physical entities, their application to human entities remains a challenge. The complexity of human behavior, influenced by factors such as mental activities, ethics, and social interactions, makes it difficult to model human behavior deterministically [38]. As a result, human DTs tend to only include key attributes and selected characteristics to represent the corresponding human entity from a specific socioeconomic perspective. Despite this limitation, the development of human-oriented DTs holds potential for further advancements in the energy sector, particularly in the areas of demand response and local energy markets.

The proposed CDT is a human-oriented, simplified virtual replica that represents the entity of an electricity consumer within the context of an energy market. It incorporates the most informative and distinguishing characteristics, attributes, and behaviors of its physical counterpart, while ensuring synchronous and bidirectional data flow between the physical and virtual entities. To achieve this, raw data from both physical sources (smart meters and wearable devices) and digital sources (REST API services) are collected and processed to extract knowledge and facilitate informed decision-making.

The functionalities performed by the CDT include developing dynamic constructs of prosumer energy behaviors, while also identifying consumer preferences with respect to energy usage, thermal comfort tolerance and openness to engaging in flexibility and DR actions, and the assessment of prosumer's indoor thermal comfort level according to the ASHRAE-55 standard [39] using environmental and physiological parameters captured by a wrist-worn device. Thermal comfort expresses the personal thermal satisfaction associated with indoor thermal environmental conditions and adheres to an ideal thermal condition and the appropriate tolerance limits within which the consumer feels comfortable. Continuous and automatic consumer thermal comfort assessment in conjunction with consumer preferences is critical since it enables pertinent and optimal demand side management while preserving the desirable thermal tolerance limits making the consumer predictable energy-wise.

By utilizing time series of predicted weather data, CDT produces forecasts of the consumer's energy demand and projected energy production from owned RES, which are further optimized based on the user's preferences. With this in mind, the proposed CDT is a core element that facilitates the deployment of human-centric DR optimization strategies, it enables personalized and non-intrusive control functions of energy assets without compromising the consumer's desired thermal comfort tolerance, and it provides consumer flexibility to aggregators. Additionally, it ensures the improvement of short and mid-term demand forecasting by using real data streams from the consumer's energy assets to address the stochasticity of the distribution grid and minimize DR strategy overruns.

The CDT consists of a front-end, a back-end, and a database intending to act as a web-based tool that records and processes the user's preferences to produce priority vectors through multi-criteria decision analysis, and user's environmental and physiological parameters through ML methods to assess indoor thermal comfort. Additionally, CDT completes analytics and enables interoperability to cater to aggregator platforms with critical flexibility information in real-time.

The overall contribution of this paper can be summarized as follows:

- An energy market design that enables the unfettered participation of small-scale, local DERs and residential flexible loads in electricity markets, allowing the exchange of energy without any external involvement and eliminating the requirement for ownership of energy assets;
- A consumer-centric LEM that respects the priorities of LEM participants, as expressed through personalized preferences for the flexible residential loads and the individual's indoor thermal comfort level;
- The maximization of potential benefit for LEM participants by integrating CDT information into the LEM marketplace. The CDT provides information regarding the consumer's energy demand and production, the energy consumption of electric appliances within the household, the consumer's personalized preferences, and the consumer's indoor thermal comfort level;
- An optimal energy scheduling considering the stochastic nature of both the generation assets and the local demand by employing the chance-constraints method.

3. Framework Implementation

In this section, the CDT and LEM-implemented models are described and their integration is presented. The main goal of the proposed framework is to integrate the benefits of CDT into the LEM architecture. This will enable every member of an LEM to be an active prosumer and empower its position in the energy market through the LEM. In that context,

CDT offers the opportunity of seamless participation in the local market and at the same time the incorporation of consumers' preferences into the market outcome.

3.1. Consumer Digital Twin

Consumer preferences constitute a set of decision criteria that reflect the consumer's prioritized demand response potential within an LEM. As such, the consumer defines the subjective intensity of importance for each residential flexible load, i.e., HVAC system, heat pump, EV charger, battery storage unit, dishwasher, and washing machine, and the desired operation time window of each load, to generate a priority vector as input to the LEM. Additionally, the boundaries of the consumer's thermal comfort tolerance are defined, adhering to Fanger's 7-point thermal sensation scale [40], as depicted in Figure 3. In this scale, each state of thermal sensation corresponds to a numerical value between -3 and $+3$, where -3 and $+3$ denote the cold and hot thermal states, respectively, and 0 denotes the neutral state.

The hierarchical ordering of consumer preferences needs to be arranged through a formal methodological approach. Therefore, the Analytical Hierarchy Process (AHP), which is a decision-making framework that allocates weights to a set of N decision criteria and produces a priority vector imposing pairwise comparisons between them, is employed. A 9-pointed balanced importance scale is utilized, in which the consumer defines whether a criterion is superior or inferior to the compared one in terms of verbal appreciation. More specifically, the scale's values come under the discrete set of $\Delta_1 = \{1, 3, 5, 7, 9\}$ and indicate equal, moderate, strong, very strong, and extreme importance, respectively, whereas intermediate values of the discrete set $\Delta_2 = \{2, 4, 6, 8\}$ are omitted as they represent a compromise between the compared criteria. To a superior criterion, the corresponding numerical value of the verbal response is assigned as priority value a_{ij}, while the reciprocal a_{ij}^{-1} one is assigned to the inferior one. The priority values are allocated to a squared decision matrix $A_{N \times N}$ to derive the normalized priority vector for the set of decision criteria (preferences). To evaluate whether the obtained weights from the AHP method are plausible, the consistency ratio metric (CR) is applied to the normalized priority vector. This metric employs the random consistency index, whose value results from a predefined set of constant values with respect to the number of decision criteria. The results of the method are considered sufficiently consistent and acceptable if the CR index is less than 0.1. If this condition is not met, the stakeholder should revise the intensity of importance for each pairwise comparison, and the process of the method is iterated.

CDT provides two matrices as input to LEM; the load flexibility matrix, denoted by **LF**, whose elements represent the relative importance of each flexible load as determined by the AHP method along with the operating time window of each flexible load as defined by the consumer, and the thermal flexibility matrix, denoted by **TF**, whose elements include the consumer's current thermal comfort level and thermal comfort tolerance deviation.

The **LF** matrix allocates to each row the weight of the corresponding flexible load and the operating time window intervals, respectively, so that the first column vector $lf_{:,1}$ of **LF** matrix represents the priority vector determined by the AHP method. For instance, the vector $lf_{n,:} = [0.2, 2, 7]$ represents a weight of 0.2 assigned to the corresponding load and a desired load's operating time window between 2:00 a.m. and 7:00 a.m., whereas the vector $lf_{:,1} = [0.2, 0.1, 0.1, 0.3, 0.1, 0.2]$ indicates weights of 0.2, 0.1, 0.1, 0.3, 0.1, 0.2 to the HVAC system, the heat pump, the EV charger, the battery storage unit, the dishwasher and the washing machine, respectively. At the same time, CDT provides data streams with the consumer's current thermal comfort level and the thermal comfort tolerance deviation from the desired range in 15-min intervals. For instance, the **TF** = $[-1.1, +0.9, +2.1]$ represents a consumer with thermal comfort tolerance desired range from -2 (cool) to $+1$ (slightly warm) and a current thermal comfort level of -1.1.

To insert and update the subjective intensity of importance for each residential flexible load, as well as the desired load's operation time window and thermal comfort tolerance boundaries, CDT offers a user-friendly graphical interface. This interface allows consumers to determine their thermal comfort tolerance over the thermal comfort scale, as presented in Figure 3 and the intensity of importance for each load through pairwise comparisons and the desired operation's time window intervals over a 9-pointed importance scale, as presented in Figure 4.

Figure 3. The implemented thermal comfort scale.

Figure 4. The implemented 9-pointed importance scales.

3.2. Local Energy Market

The LEM is a digital platform that facilitates transactions between a number of energy actors, i.e., consumers and prosumers, at a local level. A community, comprising at least two participants engaging in energy trading, can be characterized as LEM. The LEM optimizes the Day-Ahead (DA) scheduling to minimize its operating costs. The price at which the transaction is cleared within LEM can be determined using different approaches. In this section, two applied pricing algorithms are presented and compared along with the market design.

3.2.1. Market Design

LEM allows small-scale DER owners to actively engage in energy trading among themselves and to participate in the wholesale and retail energy markets. The fundamental feature of the LEM is the lower market clearing price compared to the external grid, which provides an incentive for prosumers and consumers to participate in such a cooperative market mechanism. Another important feature of the proposed LEM market is the clear definition of the market architecture and the pricing rules. It is evident that the rules of the design should be disseminated and explained in detail to the LEM participants so that each participant knows in advance the operation of the market. In our case, a two-sided market is considered, meaning that several buyers hold items for sale and several buyers consider buying these items. The key concept in such a market is that every participant (either buyer or seller) has a different valuation and risk profile of the held items, products and services. An efficient market maximizes the total profit obtained both from the buyers' and sellers' sides. To achieve an efficient local market, the total profit must be maximized for both groups.

3.2.2. Pricing Algorithms

For the validation of the proposed coordination scheme between CDT and LEM, the first step is the local price clearing at which the transactions take place in the market. Regarding the pricing algorithms, two different approaches are considered. In both algorithms, the price resolution is hourly, similar to the price signals of the external day-ahead market (DAM). Therefore, there is a different arbitrage (difference between LEM price and external price) for each particular hour. The two algorithms are summarized below:

- Peer-to-peer (P2P) pricing algorithm: The first approach is a direct P2P pricing mechanism. In a P2P transaction, the buyer and the seller transact directly with each other in terms of the delivery of the good or service and the exchange of payment. Specifically, after the initial submission of the bids from both sides, the order book of the pairs of transactions is created. Multiple price levels are initially created since different energy levels are offered at different prices. Prices are ranked for sellers from the lowest to the highest and vice versa for buyers. If the lowest price for consumers is lower than the highest price for sellers, the transaction can be executed. Otherwise, there is a case of supply deficit; in that case, the minimum of the supply-demand pair is cleared within the internal market and the rest is supplied by the grid. The clearing price P_{lem} (the average value of the two prices) creates one universal price inside the LEM. A universal (same) price is desirable since it is easier to evaluate the efficiency of the market. The clearing price results from:

$$P_{lem} = \frac{P_{prod_{low}} + P_{cons_{high}}}{2} \quad (1)$$

- CDT-LEM pricing algorithm: In this work, the price is calculated based on the LEM's production and consumption values. In other words, the participants are not directly participating in the market in the form of bids. This pricing mechanism has three main advantages. First, it minimizes the participants' involvement so that the LEM is accessible to more potential members by lowering the entry barriers. Secondly, the elimination of a bidding process strengthens the resiliency of LEM against market manipulation concerns. Finally, the solution's applicability is straightforward since all the necessary data are directly taken from smart meters or IoT devices. The process of determining the clearing price is analyzed in more detail in the following section.

In both pricing algorithms, the internal market is cleared at a price lower than the selling price of the retail external grid and higher than the buying price of the external grid. The internal energy is practically exempt from transfer losses and any other monetary burdens (such as transfer may incur due to the transition of energy through a large-scale grid) and thus the internal price can be lower. The time horizon that transactions take place depends on the market design (Day-ahead market, Real-time market, etc.), while the granularity of the algorithm's solution is determined mainly by how often the system updates its information and control signals.

3.2.3. LEM and CDT Integration

The potential impact of CDT on LEM services is significant due to the integration of consumers' energy flexibility information with supplementary parameters from third-party resources, such as the energy retailer's price. As shown in Figure 5, bidirectional and automatic data flows between the LEM and the CDT platform are enabled by the communication layer, so thus individual parameters that affect the consumer's energy flexibility potential are seamlessly elicited by CDT for each LEM participant, specifically, the participant's energy demand and production, the prioritized DR scheduling based on the participant's preferences and the participant's thermal comfort level. The LEM platform refers to the energy market and the assets that locally generate electricity while CDT involves information regarding user comfort. The information exchange between the platforms aims at the maximization of users' benefits. To this end, CDT serves as a strong knowledge base for enhancing the efficiency of LEM operations and price discovery mechanism and optimizing energy management services.

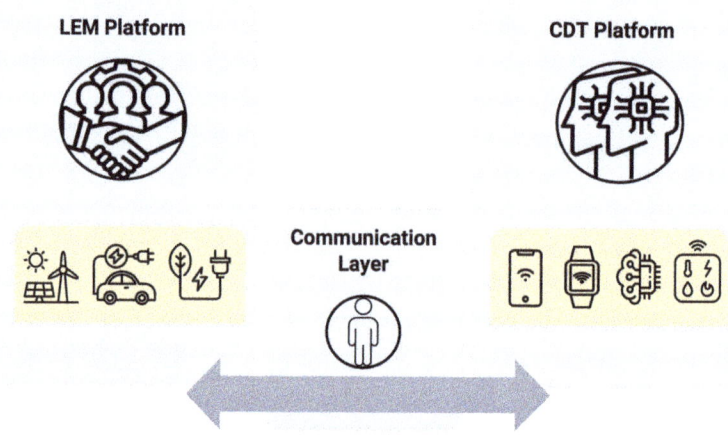

Figure 5. Integration of LEM and CDT.

The information exchange between LEM and CDT is critical for the integration of the two platforms and provides the following benefits:

- Information about the LEM participant's priorities and preferences. CDT employs a multi-criteria methodological framework, as described in Section 3.1, from which the importance level of a family of energy criteria is determined. Thus, the consumer-centrism of LEM is enhanced and LEM operations are personalized.
- Information about LEM participants energy consumption and production. The consumption levels of the consumer are elicited from the IoT devices installed within the domicile, i.e., smart meters and sub-meters, on the desired time scale. The energy production levels from RES are retrieved from the smart sensors installed on household rooftop solar. In addition, CDT analyzes historical consumption data to forecast future energy demand and employs a forecast model for energy production from solar PVs based on the predictive analytics of outdoor weather conditions. For the implementation of LEM, energy demand and generation data are essential for both price discovery and energy scheduling. The forecasted values of energy consumption and production allow LEM to plan its operation in a more efficient way.
- Optimization of Energy Management System (EMS) services. CDT can optimize the EMS services since it contributes valuable information about the behavior of an LEM participant. More specifically, it provides data related to the individual's energy consumption, which is utilized to discover energy consumption patterns and classify consumers into groups. Thus, LEM can be operated in a coordinated manner to initially connect users with similar behavioral patterns, and then seek alternative solutions. In addition, the LEM operator can classify LEM participants based on their flexibility potential, as retrieved from the CDT information. In a nutshell, the LEM-CDT integration can offer a highly scalable and easy-to-implement solution that enables LEM to harness the available flexibility in its ecosystem in an optimal and efficient manner.

4. Results and Validation

Use Case Description

In this work, the CDT-LEM pricing is compared with P2P pricing [18]. The evaluation is conducted in terms of consumers' payment, prosumers' profit, and computational performance of both algorithms. To assess the applicability of LEM, both pricing algorithms are examined in a real-world test case: a residential community in central Germany. In particular, the total number of participants in the community is 27 residential members, 12 of them have installed PV systems on their rooftops, while the rest of them are solely energy consumers. The former group contributes to the LEM through their flexible loads, in our case their HVAC systems, via a DR scheme. This specific flexible load was chosen since it is the most energy-consuming appliance in households and, in our case, it is possible to control HVACs' operation through smart controllers. Moreover, the time period is a typical single day during summer, characterized by substantial PV production and high demand for HVAC load.

The pricing algorithm and the optimal scheduling are calculated on the LEM operator's cloud platform. Additional critical services, such as load and generation forecasting, interaction with the wholesale energy market, and integration of meteorological data, are also deployed in LEM's cloud. Regarding the CDT implementation, each participant is equipped with a wrist-worn wearable device, which assesses and transmits the thermal comfort level. In addition, energy preferences have been provided by the consumer and processed by CDT, as described in Section 3.1. Based on this information, participants are classified into three classes. To simulate the participation of the consumers, the different classes of flexibility capacity stemming from each participant are considered to follow the normal distribution. The classes are derived based on the values provided by the flexibility and thermal comfort matrices. The first class contains the non-flexible consumers whose load profile cannot be altered. The second class contains low-flexibility participants, while the third class contains fully flexible participants with no constraints of adjusting their load. In all three classes, the thermal comfort limits of each user are not violated. These classes are formulated as parameters in the optimization problem and determine the allowed shiftable demand of each participant. In that context, CDT extends the flexibility capabilities of LEM by creating a more stable and fair cooperative energy scheme, since even the members who do not own any energy production assets can contribute to the energy community. Accordingly, all participants contribute to LEM for mutual benefit. Finally, CDT offers a seamless way of performing load shifting based on participants' preferences. The main advantage of the CDT-LEM pricing algorithm is that active engagement of participants is not required. Consumers are encouraged to participate, as the price results are always within the range of feed-in and feed-out tariffs. The inputs of the algorithm are consumption and generation forecasts, along with feed-in and feed-out tariffs, and the output is the resulting LEM clearing price. The LEM price curve is a representation of the local generation and demand, and is made available to all participants. It is depicted as a three-dimensional surface, as illustrated in Figure 6. The curve is generated for each hour, although it can be created for any other desired time horizon.

On the contrary, in the case of P2P pricing, the price is calculated based on each participant's bidding strategy, which is prone to market manipulation, especially in the case of a single participant with significant market power (e.g., higher installed PV capacity). Hence, the adoption of the P2P algorithm requires a sufficient regulatory framework. Even so, both algorithms lead to lower energy costs, where consumers and prosumers interact with the external grid via feed-in and feed-out tariffs respectively.

LEM's main goal is to optimize the DA scheduling by minimizing its operating costs. Batteries' charging and discharging are among the decision variables of the optimization problem. The minimization of LEM's operating cost is calculated based on the forecasted values of local generation and local demand in a one-day horizon. The forecasting error, especially in low-scale energy deployments [41], leads to deviations in the output of the optimization problem. Hence, the DAM schedule differs from the optimal deterministic solution, leading to remedial actions that bear down higher costs on LEM participants. Incorporating stochastic variables in the optimization problem is the key to ensuring a constant, reliable power supply and achieving a cost-efficient LEM operation.

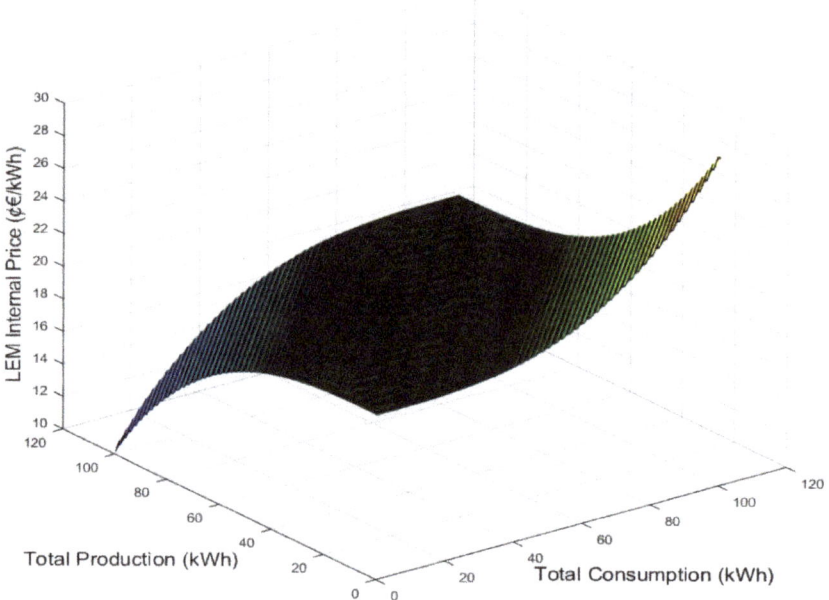

Figure 6. LEM pricing curves with feed-out and feed-in tariffs as limits.

The stochasticity of DERs generation and local consumption affects the LEM scheduling [42]. It is crucial to design a market that effectively deals with stochasticity, as the cost for remedial actions will be significant in the case of high uncertainty levels [43]. A wide variety of stochastic optimization methods have been employed to cope with uncertainty in power systems operation, namely scenario-based approaches, robust optimization, and chance-constrained optimization [44]. The chance-constraints method allows certain unexpected events to violate specific constraints considering the overall constraint satisfaction is satisfied with a predefined level of probability. Chance constraints are transformed into deterministic equivalents, and a standard solution method is then employed to solve the problem. The uncertainty of DERs and demand is incorporated into the optimal scheduling process by using statistical moments of the parameters' distribution, which are derived from historical data such as mean and standard deviation [45]. The ability to accommodate a wide range of distributions eliminates the need for discretization of a probability space for scenario sampling [46] or the derivation of a finite uncertainty set [47].

In this work, a chance-constraints approach for the DA scheduling is employed, which explicitly incorporates the stochasticity of DERs generation and local demand and analyzes the effect of uncertainty level on LEM's social welfare. The rationale behind our choice was that compared to the other two stochastic optimization methods, the chance-constrained leads to less conservative results [48]. Another important factor is that the level of uncertainty can be tuned via a confidence interval [49]. Finally, by incorporating the stochastic optimization problem into a chance constraints formulation, the convexity of the optimization problem is maintained. In this paper, stochasticity is accounted for on both the supply (i.e., solar PV) and demand sides. The granularity of the optimal scheduling of the LEM is equal to 1 h and has an interval horizon of 24 h, similar to the wholesale DAM market architecture. The objective function of the optimization problem is formulated as follows:

$$\min C_{total} = \sum_{i=1}^{24}(\pi_{int}^i \times P_{int}^i + \pi_{buy}^i \times P_{in}^i - \pi_{sell}^i \times P_{out}^i) \qquad (2)$$

where i, l, k, and n indicate the hour, flexible consumers, PV owners and battery owners, respectively. C_{total}^i denotes the total LEM cost, π_{int}^i is the LEM price, and π_{buy}^i and π_{sell}^i are the respective feed in and feed out grid tariffs. P_{in}^i and P_{out}^i denote power import and export from and to the grid.

Equations (3)–(6) outline the energy conservation laws and constitute the constraints of the optimization problem:

$$P_{int}^i = P_{gen}^i + P_{BBdis}^i - P_{load}^i + FI \times SC_{l,i}^+ - FI \times SC_{l,i}^- - P_{BBch}^i \qquad (3)$$

$$P_{gen}^i = \sum_{1}^{k}(P_{gen}^{i,k}) \qquad (4)$$

$$P_{load}^i = \sum_{1}^{l}(P_{load}^{i,l}) \qquad (5)$$

$$P_{int}^i = P_{gen}^i + P_{BBdis}^i - P_{load}^i + FI \times SC_{l,i}^+ - FI \times SC_{l,i}^- \qquad (6)$$

where FI indicates the flexibility index derived from the flexibility matrix and thermal comfort matrix, $SC_{l,i}^+$ and $SC_{l,i}^-$ indicate the amount of power consumption for the specific i-hour, shifted by each household at each time-step and the amount of power consumption that has previously been shifted and is now consumed, respectively. Additionally, P_{gen}^i indicates the total LEM production, P_{BBdis}^i and P_{BBch}^i indicate the battery's discharging and power, respectively, and P_{load}^i indicates the LEM's load profile. The notations of k, l and n indicate the number of producers, consumers and storage owners, respectively:

$$0 \leq P_{in}^i \leq P_{in}^{max} \qquad (7)$$

$$0 \leq P_{out}^i \leq P_{out}^{max} \qquad (8)$$

The allowed bounds of energy exchange between the LEM and the grid are defined by constraints (7) and (8):

$$e_{BB}^{n,i} = e_{BB}^{n,i-1} + \eta_{ch}^n P_{BBch}^{n,i} \times \Delta t - (1/\eta_{dis}^n) \times P_{BBdis}^{n,i} \times \Delta t \qquad (9)$$

$$0 \leq P_{BBch}^{n,i} \leq P_{BBch,max}^n \qquad (10)$$

$$0 \leq P_{BBdis}^{n,i} \leq P_{BBdis,max}^n \qquad (11)$$

$$0 \leq e_{BB}^{n,i} \leq e_{BB,max}^n \qquad (12)$$

$$e_{BB}^{n,1} = e_{BB}^{n,T} = (1/2)e_{BB,max}^n \qquad (13)$$

Equations (9)–(13) denote the storage energy state update, in which $e_{BB}^{n,i}$ denotes the battery's energy state and η_{ch}^n and η_{dis}^n show charging and discharging efficiency levels, respectively. Lastly, $P_{BBch,max}^n$ and $P_{BBdis,max}^n$ denote the maximum and minimum storage charging rate while $e_{BB,max}^n$ and SD_t define the maximum storage capacity and shifted demand, respectively,

$$FI \times SC_{l,i}^{+} \leq FI \times SD_t + SC_{l,i}^{-} \tag{14}$$

$$SC_{l,i}^{-} = SC_{k,i-1}^{+} \tag{15}$$

$$SC_{l,I}^{+} = 0 \tag{16}$$

Demand can be shifted by one hour at a fixed rate determined for each tenant based on their preferences. Constraint (14) assures that the amount of the shifted demand does not exceed the maximum residential load, by restricting the amount of shifted demand to the residential load for the specific time step plus the shifted demand in the previous time step. Equation (15) guarantees that the already shifted demand is either consumed in the current time-step or shifted further. The shifted demand at the end of the time horizon for the optimization problem in Equation (16) should equal zero, ensuring that the demand will not be shifted to the next day's optimization problem:

$$Pr\{P_{gen}^i \leq \tilde{P}_{gen}^i\} \geq \alpha' \tag{17}$$

$$Pr\{P_{gen}^i \geq \tilde{P}_{gen}^i\} \leq \beta' \tag{18}$$

$$\alpha' + \beta' \leq 1 \tag{19}$$

Equations (17)–(19) denote the probability of the actual production to be in a specific range, where $\tilde{P}_{gen}^i$ is equal to the forecasted generation value plus the forecast error ϵ_i. Finally, α' and β' denote the probabilities of upper and lower bounds, respectively.

The proposed method in [50] is applied in order to transform the probabilistic chance constraints into deterministic values that can be used as input to the optimization procedure.

The maximum PV forecasting error ϵ_i is considered equal to 20% following the normal distribution $N(0, \sigma^2)$, with a 99.7% confidence interval achieved in $[-3\sigma, +3\sigma]$ range. At time interval i, σ is equal to $0.1 P_{Fgen}^i$, which is the forecasted PV generation. Moreover, ϵ_i is constrained between maximum installed capacity P_{gen}^{max} and the forecasted value P_{Fgen}^i; therefore, the error distribution ϵ_i belongs within the range $[-P_{Fgen}^i, P_{gen}^{max} - P_{FPV}^i]$. The ϵ_i follows the conditional probability distribution given by:

$$\Phi_i(x) \sim N(0, 0.01(P_{FPV}^i)^2) \tag{20}$$

where $\Phi_i(x)$ is the conditional probability distribution:

$$\Phi_i'(x) = \frac{\Phi_i(x) - \Phi_i(-P_{Fgen}^i)}{\Phi_i(P_{gen}^{max} - P_{Fgen}^i) - \Phi_i(-P_{Fgen}^i)} \tag{21}$$

$$\Phi_i^{-1'}(x) = \Phi_i^{-1}[x\Phi_i(P_{gen}^{max} - P_{Fgen}^i) + (1-x)\Phi_i(-P_{Fgen}^i)] \tag{22}$$

If (22) is solved and $\Phi_i^{-1'}(x)$ can be found, then (17) and (18) can be transformed into the following equation:

$$F_{\beta'}^{-1}\{\tilde{P}_{gen}^i\} \leq P_{gen}^i \leq F_{1-\alpha'}^{-1}\{\tilde{P}_{gen}^i\} \tag{23}$$

where $F^{-1}\{\tilde{P}_{gen}^i\}$ is the inverse forecasted PV production distribution.

To address the stochastic nature of demand, a similar approach as the one described above for the generation is followed. Across the literature, there are two approaches for modeling load uncertainty as a way to ensure that demand will not be shifted with equality constraints containing stochastic parameters in the optimization problem. The first method converts the equality constraints into inequality ones, whereas the second method eliminates variables [51]. However, by following the second one, the final variable values will remain uncertain; this is because the aforementioned variables depend on stochastic parameters. In our case, those variables are the charging/discharging level of the batteries and the demand shift, while the parameter is the day-ahead load. Moreover, if the variables that will be eliminated contain stochastic parameters as coefficients, eliminating such variables leads to nonlinear optimization problems. On that account, this approach is not recommended because the eliminated variables (i.e., the charging/discharging level of the batteries and the demand shift) depend on the stochastic parameters, which is the day-ahead demand. Moreover, it is evident that the selection of variables determines the subsequent optimization problem. In other words, the choice of different eliminating variables leads to different optimization problems.

Based on the above, the first approach is followed and the equality constraints (4) and (5) are converted into inequality constraints (25) and (26), respectively:

$$k = P_{gen}^i + P_{BBdis}^i - P_{load}^i - P_{BBch}^i \tag{24}$$

$$k - d \leq P_{int}^i \leq k + d \tag{25}$$

$$\sum_{1}^{l}(P_{load}^{i,l}) - d \leq P_{load}^i \leq \sum_{1}^{l}(P_{load}^{i,l}) + d \tag{26}$$

where d is a small parameter to ensure that the above inequalities are tight at optimality.

5. Results

In Figure 7, the hourly prices for both pricing algorithms are presented. The price levels of both algorithms are within the bounds defined by the external grid (feed-in and feed-out tariffs); therefore, consumer participation in an LEM framework is beneficial under both pricing algorithms. In particular, the prices of our proposed algorithm are lower than those of the P2P approach during hours of high PV production. This is to be expected since, in the CDT-LEM pricing, no direct bids are submitted by the participants; the price behavior follows the pattern of residual load. Therefore, during time intervals with excess PV production, the community energy demand is lower and the prosumer is not adequately compensated. On the other hand, the prices of the proposed algorithm are significantly higher during the night hours of the day (7:00 p.m.–6:00 a.m.), as shown in Figure 7.

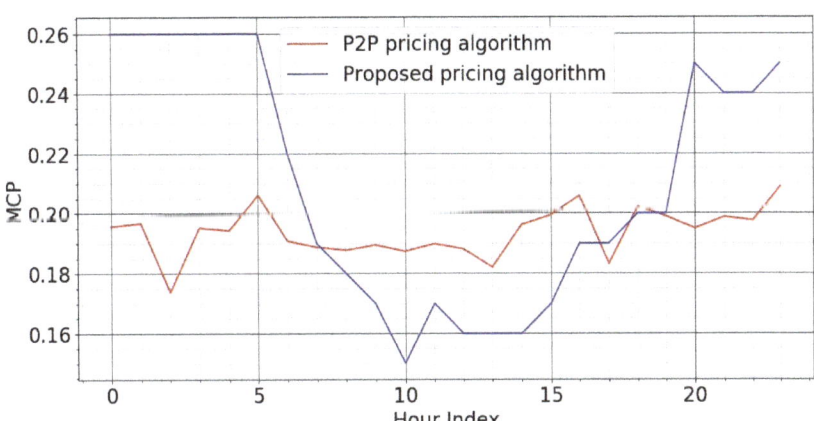

Figure 7. Hourly prices with the two (2) different pricing algorithms.

In Table 1, it is evident that, without an LEM, the procurement daily costs are significantly higher. Specifically, employing the proposed algorithm without CDT, the cost is reduced by 20.7%, while the cost reduction with CDT is 27.8%. It is noticeable that, while P2P pricing also leads to cost reduction, the percentage drop is smaller compared to the respective one from our proposed algorithm. This is due to the different way, in which the LEM operates; under the P2P algorithm, there is a higher trade of energy with the external grid, while the proposed algorithm seeks to optimize the energy within the LEM framework.

Table 1. Total Daily Cost in Euros.

Market	No LEM (e)	P2P w/o CDT (e)	Proposed Algorithm w/o CDT (e)	P2P w/t CDT (€)	Proposed Algorithm w/t CDT (e)
Total Cost	20.091 €	18.792 €	17.490 €	16.351 €	15.612 €

In Table 2, consumer payments are the lowest under the proposed method with CDT. The producers' profit is also the highest under the same schema since the traded energy is higher with the proposed algorithm than with the P2P approach, which leads to profit maximization.

Table 2. Total payment and profit in euros.

	No LEM (e)	P2P Algorithm (e)	Proposed Algorithm (e)
Payment	48.555 €	37.087 €	33.767 €
Profit	36.670 €	51.096 €	58.196 €

In Table 3, the results for both pricing algorithms are displayed, with and without the utilization of the CDT. The results focus on two time periods, between 2:00 a.m. and 8:00 p.m. and between 2:00 p.m. and 8:00 p.m. These time periods are selected because the pricing algorithms generate different state-of-charge values during these time periods. In particular, under P2P pricing, the excess energy (during midday) is sold to the external grid, leading the batteries to reach their lowest accepted levels (20% state of charge) regardless of the CDT. On the other hand, with CDT-LEM pricing, this energy is used to charge the batteries, which is why there are no abrupt peaks. This fact leads to a more "self-sufficient" LEM since the interaction with the external grid is lower than in the P2P case. In the CDT-LEM pricing mechanism, the LEM prioritizes the local energy needs and then the energy trading with the external grid via the wholesale markets. Apparently, our algorithm leads to battery charging when there is a higher local generation while the P2P algorithm sells the excess capacity to the wholesale market.

Figures 8 and 9 show how the probability of constraint violation affects the total operational LEM costs. As the probability decreases, the total expected cost increases due to the fact that the LEM operation becomes more "conservative" in order to respect the optimization constraints. Furthermore, the CDT-LEM pricing coupled with the CDT results in the lowest costs, regardless of the probability. Clearly, there is a trade-off between higher operating costs and low-risk scheduling, and the associated decision rests with the LEM operator and the specific pricing mechanism.

In Figure 10, the energy exchange levels with the external grid are presented. Positive values represent the sale of energy to the grid and negative values represent the purchase of energy from the grid. As mentioned earlier and shown in Figure 10, the energy exchange is actually higher in the P2P approach, while our algorithm leads to lower dependence on the external grid. This is due to the fact that LEM first resolves its local imbalances and then interacts with the external grid. The hourly intervals with the highest grid interactions are around midday since the production level within LEM is high during these hours leading to a large energy surplus.

Finally, in Table 4, the scalability and the computational performance of the two proposed algorithms are examined for an increasing number of LEM participants. Apparently, our proposed algorithm outperforms the P2P pricing, as it is multiple times faster, particularly as the number of participants increases. This is due to the fact that the P2P algorithm requires a computationally demanding matching algorithm between the energy supply offers and energy demand bids. This procedure is conducted through an optimization problem in order to achieve optimal matching. Hence, as the number of participants increases, the optimization problem is more complex with a higher computational burden. On the other hand, the CDT-LEM pricing computational needs are minimal, since it results from a heuristic process that requires only the generation/demand values and the external market tariffs.

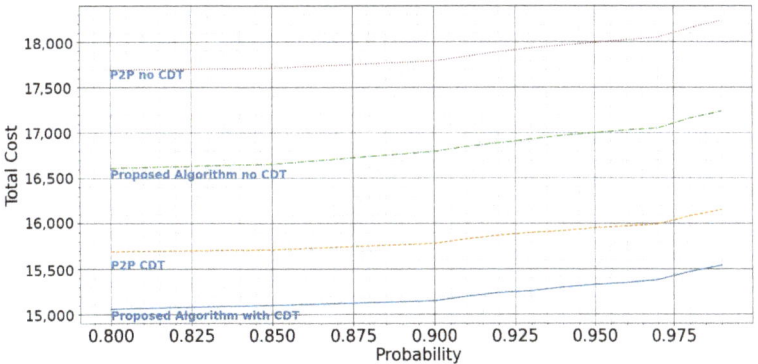

Figure 8. Total cost vs. probability of generation stochasticity violation

Table 3. Batteries state of charge.

Hour	P2P w/o CDT (%)	Proposed Algorithm w/o CDT (%)	P2P w/t CDT (%)	Proposed Algorithm w/t CDT (%)
2 a.m.	22.0	23.4	27.9	27.0
3 a.m.	21.5	33.0	30.0	37.3
4 a.m.	48.2	59.1	38.0	74.0
5 a.m.	73.0	100.0	74.3	74.0
6 a.m.	100.0	100.0	100.0	100.0
7 a.m.	25.0	75.0	20.0	73.9
8 a.m.	22.5	75.1	20.0	73.5
2 p.m.	22.0	24.0	20.0	67.0
3 p.m.	20.0	25.7	20.0	66.2
4 p.m.	20.0	75.2	20.0	65.3
5 p.m.	76.7	74.0	40.0	72.0
6 p.m.	76.7	75.2	42.3	69.2
7 p.m.	76.7	75.0	36.1	65.0
8 p.m.	25.0	26.0	20.0	20.0

Table 4. Computational performance of the P2P and proposed algorithm.

Algorithm	N = 10	N = 20	N = 50	N = 100
P2P	0.1 s	0.12 s	0.47 s	3.2 s
Proposed	0.004 s	0.0035 s	0.0065 s	0.017 s

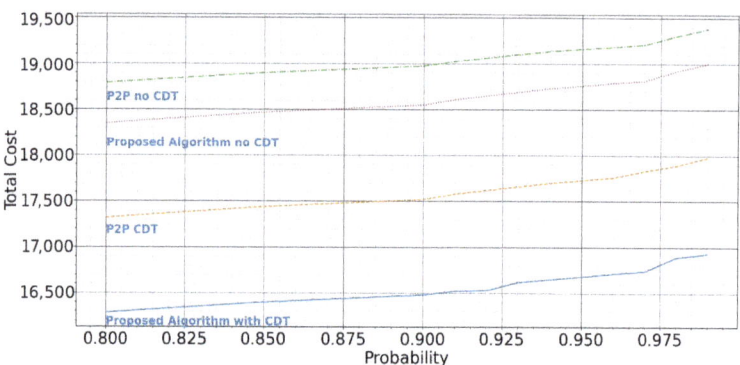

Figure 9. Total cost vs. probability of demand stochasticity violation.

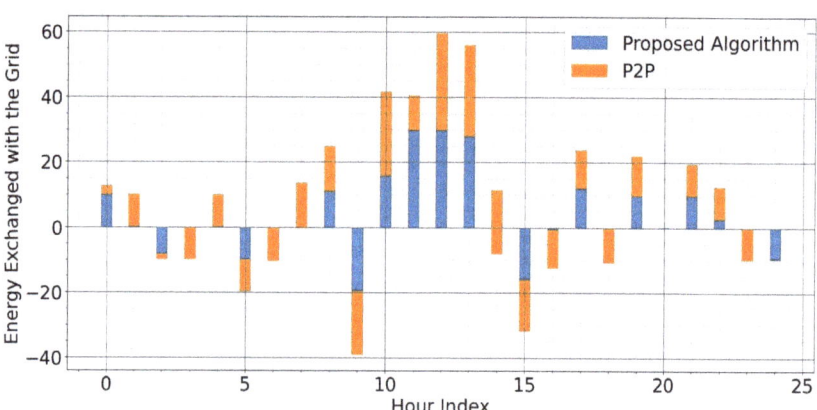

Figure 10. Energy exchange with the external grid.

Visualization of Results through CDT

In the proposed integration, CDT additionally serves as a visualization tool for consumers to track key indicators of their participation in LEM, as shown in Figure 11, promoting energy flexibility as a financial incentive. Specifically, a dashboard displays information on weekly energy production and demand with a daily resolution, financial profit from LEM participation, and specific parameters that affect thermal comfort.

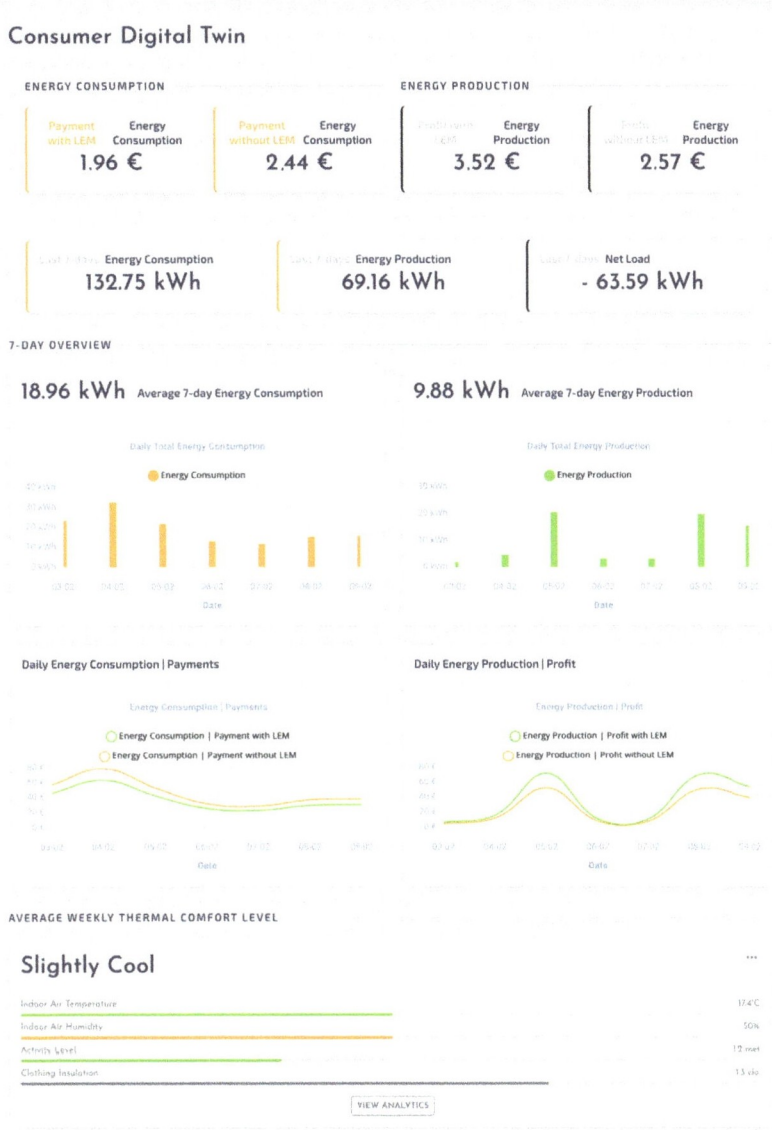

Figure 11. Consumer digital twin dashboard.

6. Conclusions

This paper presents an integrated framework of LEM-CDT that maximizes the flexibility potential of the participants and improves the market's operations efficiency. A detailed market design with different pricing mechanisms for handling LEM transactions is also introduced. The results of the proposed framework are summarized as follows:

- Consumer preferences regarding thermal comfort and residential loads are proved to be valuable inputs for optimizing LEM operations.
- The attainment of optimal energy exchange with the external grid and the maximization of social welfare is accomplished.

- The enhancement of consumer-centricity and ease of implementation, as the participants are not required to actively submit energy selling or buying offers through a bidding process.
- The democratization of LEM through CDT-enabled automated participation broadens the potential participant base, positioning it as a more environmentally-friendly, attractive, and consumer-centric alternative to traditional energy markets.
- The generation and demand stochasticity is modeled by a chance-constrained scheduling optimization algorithm that ensures lower balancing needs with minimal requirements for market participation or remedial actions, despite the higher costs.

The proposed solution encounters challenges with regard to regulatory compliance and ensuring the confidentiality of participants' data. Despite a current dearth of clear guidelines for LEMs design and operations, efforts are being undertaken to rectify this. In order to establish trust and attract new members, LEMs must prioritize creating a reliable environment. Additionally, the integration of a large number of IoT devices accentuates the necessity for robust cybersecurity measures to safeguard against potential breaches.

The proposed model holds the potential for further enhancement in two directions. One area of improvement would be the integration of a more comprehensive collection of consumer preferences, with the aim of augmenting its consumer-centric orientation and further refining the consumer priorities for LEM operation. Another avenue for research would be to incorporate the distribution network constraints within LEM, in order to prevent voltage and line congestion incidents. Additionally, the proposed integration of LEM-CDT should be assessed in a larger-scale study utilizing a larger volume of data.

Author Contributions: Conceptualization, N.A., K.P., C.M. and J.G.; Methodology, N.A., K.P., C.M., J.G. and A.B.; Software, N.A. and C.M.; Validation, N.A., K.P., C.M. and J.G.; Writing—original draft preparation, N.A., K.P., C.M. and J.G.; Writing—review and editing, A.B., S.K. and A.P.; Visualization, N.A., K.P., C.M. and J.G.; Funding acquisition, J.G. and A.B. All authors have read and agreed to the published version of the manuscript.

Funding: This work has been supported by TwinERGY project, No 957736, "Intelligent Interconnection of Prosumers in Positive Energy Communities with Twins of Things for Digital Energy Markets-TwinERGY", H2020-LC-SC3-2020-EC-ES-SCCRIA.

Institutional Review Board Statement: Not applicable.

Informed Consent Statement: Not applicable.

Data Availability Statement: Not applicable.

Conflicts of Interest: The authors declare that the research was conducted in the absence of any commercial or financial relationships that could be construed as a potential conflict of interest.

Abbreviations

The following abbreviations are used in this manuscript:

AHP	Analytic Hierarchy Process
CDT	Consumer Digital Twin
DA	Day-Ahead
DAM	Day-Ahead Market
DES	Distributed Energy Resources
DR	Demand Response
DT	Digital Twin
EVs	Electric Vehicles
HVAC	Controlling Heating, Ventilation and Air Conditioning
LEM	Local Energy Market
P2P	Peer-to-Peer
RES	Renewable Energy Resources
TES	Transactive Energy Systems

References

1. Meeus, L.; Nouicer, A. *The EU Clean Energy Package*; European University Institute: Fiesole, Italy, 2018.
2. Cartwright, E.D. FERC Order 2222 Gives Boost to DERs. *Clim. Energy* **2020**, *37*, 22. [CrossRef]
3. Eurostat. Renewable energy statistics. *Stat. Focus.* **2020**, *56*, 1–8.
4. Usef, S. *The Framework Explained*; USEF: Lexington, KY, USA, 2015.
5. Xu, Z. *The eLectricity Market Design for Decentralized Flexibility Sources*; Springer: Berlin/Heidelberg, Germany, 2019.
6. Chen, T.; Pourbabak, H.; Su, W. Electricity market reform. In *The Energy Internet*; Elsevier: Amsterdam, The Netherlands, 2019; pp. 97–121.
7. Singh, M.; Fuenmayor, E.; Hinchy, E.P.; Qiao, Y.; Murray, N.; Devine, D. Digital Twin: Origin to Future. *Appl. Syst. Innov.* **2021**, *4*, 36. [CrossRef]
8. Lee, S.; Whaley, D.; Saman, W. Electricity Demand Profile of Australian Low Energy Houses. *Energy Procedia* **2014**, *62*, 91–100. [CrossRef]
9. Honarmand, M.E.; Hosseinnezhad, V.; Hayes, B.; Siano, P. Local Energy Trading in Future Distribution Systems. *Energies* **2021**, *14*, 3110. [CrossRef]
10. Doumen, S.C.; Nguyen, P.; Kok, K. The State of the Art in Local Energy Markets: A Comparative Review. In Proceedings of the 2021 IEEE Madrid PowerTech, Virtual Event, 28 June–2 July 2021; pp. 1–6.
11. Herenčić, L.; Ilak, P.; Rajšl, I. Effects of local electricity trading on power flows and voltage levels for different elasticities and prices. *Energies* **2019**, *12*, 4708. [CrossRef]
12. Rassa, A.; van Leeuwen, C.; Spaans, R.; Kok, K. Developing local energy markets: A holistic system approach. *IEEE Power Energy Mag.* **2019**, *17*, 59–70. [CrossRef]
13. Khorasany, M.; Mishra, Y.; Ledwich, G. A decentralized bilateral energy trading system for peer-to-peer electricity markets. *IEEE Trans. Ind. Electron.* **2019**, *67*, 4646–4657. [CrossRef]
14. Tushar, W.; Saha, T.K.; Yuen, C.; Liddell, P.; Bean, R.; Poor, H.V. Peer-to-peer energy trading with sustainable user participation: A game theoretic approach. *IEEE Access* **2018**, *6*, 62932–62943. [CrossRef]
15. Lee, W.; Xiang, L.; Schober, R.; Wong, V.W. Direct electricity trading in smart grid: A coalitional game analysis. *IEEE J. Sel. Areas Commun.* **2014**, *32*, 1398–1411. [CrossRef]
16. Tsaousoglou, G.; Pinson, P.; Paterakis, N.G. Transactive energy for flexible prosumers using algorithmic game theory. *IEEE Trans. Sustain. Energy* **2021**, *12*, 1571–1581. [CrossRef]
17. Long, C.; Wu, J.; Zhang, C.; Thomas, L.; Cheng, M.; Jenkins, N. Peer-to-peer energy trading in a community microgrid. In Proceedings of the 2017 IEEE Power & Energy Society General Meeting, Chicago, IL, USA, 16–20 July 2017; pp. 1–5.
18. Mengelkamp, E.; Gärttner, J.; Rock, K.; Kessler, S.; Orsini, L.; Weinhardt, C. Designing microgrid energy markets: A case study: The Brooklyn Microgrid. *Appl. Energy* **2018**, *210*, 870–880. [CrossRef]
19. Paudel, A.; Chaudhari, K.; Long, C.; Gooi, H.B. Peer-to-peer energy trading in a prosumer-based community microgrid: A game-theoretic model. *IEEE Trans. Ind. Electron.* **2018**, *66*, 6087–6097. [CrossRef]
20. Brolin, M.; Pihl, H. Design of a local energy market with multiple energy carriers. *Int. J. Electr. Power Energy Syst.* **2020**, *118*, 105739. [CrossRef]
21. Hayes, B.P.; Thakur, S.; Breslin, J.G. Co-simulation of electricity distribution networks and peer to peer energy trading platforms. *Int. J. Electr. Power Energy Syst.* **2020**, *115*, 105419. [CrossRef]
22. Lyu, C.; Jia, Y.; Xu, Z. Fully decentralized peer-to-peer energy sharing framework for smart buildings with local battery system and aggregated electric vehicles. *Appl. Energy* **2021**, *299*, 117243. [CrossRef]
23. Bachoumis, A.; Andriopoulos, N.; Plakas, K.; Magklaras, A.; Alefragis, P.; Goulas, G.; Birbas, A.; Papalexopoulos, A. Cloud-Edge Interoperability for Demand Response-Enabled Fast Frequency Response Service Provision. *IEEE Trans. Cloud Comput.* **2021**, *10*, 123–133. [CrossRef]
24. Huo, D.; Gu, C.; Greenwood, D.; Wang, Z.; Zhao, P.; Li, J. Chance-constrained optimization for integrated local energy systems operation considering correlated wind generation. *Int. J. Electr. Power Energy Syst.* **2021**, *132*, 107153. [CrossRef]
25. Zhang, X.; Shen, J.; Saini, P.K.; Lovati, M.; Han, M.; Huang, P.; Huang, Z. Digital twin for accelerating sustainability in positive energy district: A review of simulation tools and applications. *Front. Sustain. Cities* **2021**, *3*, 35. [CrossRef]
26. Danilczyk, W.; Sun, Y.; He, H. ANGEL: An Intelligent Digital Twin Framework for Microgrid Security. In Proceedings of the 2019 North American Power Symposium (NAPS), IEEE, Wichita, KS, USA, 13–15 October 2019. [CrossRef]
27. Darbali-Zamora, R.; Johnson, J.; Summers, A.; Jones, C.B.; Hansen, C.; Showalter, C. State Estimation-Based Distributed Energy Resource Optimization for Distribution Voltage Regulation in Telemetry-Sparse Environments Using a Real-Time Digital Twin. *Energies* **2021**, *14*, 774. [CrossRef]
28. Podvalny, S.L.; Vasiljev, E.M. Digital twin for smart electricity distribution networks. *IOP Conf. Ser. Mater. Sci. Eng.* **2021**, *1035*, 012047. [CrossRef]
29. Wu, B.; Widanage, W.D.; Yang, S.; Liu, X. Battery digital twins: Perspectives on the fusion of models, data and artificial intelligence for smart battery management systems. *Energy AI* **2020**, *1*, 100016. [CrossRef]
30. Jain, P.; Poon, J.; Singh, J.P.; Spanos, C.; Sanders, S.R.; Panda, S.K. A Digital Twin Approach for Fault Diagnosis in Distributed Photovoltaic Systems. *IEEE Trans. Power Electron.* **2020**, *35*, 940–956. [CrossRef]

31. Atalay, M.; Angin, P. A Digital Twins Approach to Smart Grid Security Testing and Standardization. In Proceedings of the 2020 IEEE International Workshop on Metrology for Industry 4.0. & IoT, Roma, Italy, 3–5 June 2020. [CrossRef]
32. Dembski, F.; Wössner, U.; Letzgus, M.; Ruddat, M.; Yamu, C. Urban Digital Twins for Smart Cities and Citizens: The Case Study of Herrenberg, Germany. *Sustainability* **2020**, *12*, 2307. [CrossRef]
33. Bazmohammadi, N.; Madary, A.; Vasquez, J.C.; Mohammadi, H.B.; Khan, B.; Wu, Y.; Guerrero, J.M. Microgrid Digital Twins: Concepts, Applications, and Future Trends. *IEEE Access* **2022**, *10*, 2284–2302. [CrossRef]
34. Nguyen-Huu, T.A.; Tran, T.T.; Tran, M.Q.; Nguyen, P.H.; Slootweg, J. Operation Orchestration of Local Energy Communities through Digital Twin: A Review on suitable Modeling and Simulation Approaches. In Proceedings of the 2022 IEEE 7th International Energy Conference (ENERGYCON), Riga, Latvia, 9–12 May 2022; pp. 1–6. [CrossRef]
35. Han, J.; Hong, Q.; Syed, M.H.; Khan, M.A.U.; Yang, G.; Burt, G.; Booth, C. Cloud-Edge Hosted Digital Twins for Coordinated Control of Distributed Energy Resources. *IEEE Trans. Cloud Comput.* **2022**, 1–15. [CrossRef]
36. Aghazadeh Ardebili, A.; Longo, A.; Ficarella, A. Digital Twins bonds society with cyber-physical Energy Systems: A literature review. In Proceedings of the 2021 IEEE International Conferences on Internet of Things (iThings) and IEEE Green Computing & Communications (GreenCom) and IEEE Cyber, Physical & Social Computing (CPSCom) and IEEE Smart Data (SmartData) and IEEE Congress on Cybermatics (Cybermatics), Melbourne, Australia, 6 December 2021; pp. 284–289. [CrossRef]
37. Zhou, Y.; Su, P.; Wu, J.; Sun, W.; Xu, X.; Abeysekera, M. Digital Twins for Flexibility Service Provision from Industrial Energy Systems. In Proceedings of the 2021 IEEE 1st International Conference on Digital Twins and Parallel Intelligence (DTPI), Beijing, China, 15 July–15 August 2021; pp. 274–277. [CrossRef]
38. Shengli, W. Is Human Digital Twin possible? *Comput. Methods Programs Biomed. Update* **2021**, *1*, 100014. [CrossRef]
39. *ASHRAE Standard 55-2013*; Thermal Environmental Conditions for Human Occupancy. ANSI/ASHRAE: Peachtree Corners, GA, USA, 2013.
40. Fanger, P.O. Assessment of man's thermal comfort in practice. *Occup. Environ. Med.* **1973**, *30*, 313–324. [CrossRef]
41. Andriopoulos, N.; Magklaras, A.; Birbas, A.; Papalexopoulos, A.; Valouxis, C.; Daskalaki, S.; Birbas, M.; Housos, E.; Papaioannou, G.P. Short Term Electric Load Forecasting Based on Data Transformation and Statistical Machine Learning. *Appl. Sci.* **2021**, *11*, 158. [CrossRef]
42. Wong, S.; Fuller, J.D. Pricing energy and reserves using stochastic optimization in an alternative electricity market. *IEEE Trans. Power Syst.* **2007**, *22*, 631–638. [CrossRef]
43. Kazempour, J.; Pinson, P.; Hobbs, B.F. A stochastic market design with revenue adequacy and cost recovery by scenario: Benefits and costs. *IEEE Trans. Power Syst.* **2018**, *33*, 3531–3545. [CrossRef]
44. Dall'Anese, E.; Baker, K.; Summers, T. Chance-constrained AC optimal power flow for distribution systems with renewables. *IEEE Trans. Power Syst.* **2017**, *32*, 3427–3438. [CrossRef]
45. Fang, X.; Hodge, B.M.; Du, E.; Kang, C.; Li, F. Introducing uncertainty components in locational marginal prices for pricing wind power and load uncertainties. *IEEE Trans. Power Syst.* **2019**, *34*, 2013–2024. [CrossRef]
46. Mieth, R.; Dvorkin, Y. Data-driven distributionally robust optimal power flow for distribution systems. *IEEE Control Syst. Lett.* **2018**, *2*, 363–368. [CrossRef]
47. Mieth, R.; Dvorkin, Y. Distribution electricity pricing under uncertainty. *IEEE Trans. Power Syst.* **2019**, *35*, 2325–2338. [CrossRef]
48. Bienstock, D.; Chertkov, M.; Harnett, S. Chance-constrained optimal power flow: Risk-aware network control under uncertainty. *Siam Rev.* **2014**, *56*, 461–495. [CrossRef]
49. Wu, H.; Shahidehpour, M.; Li, Z.; Tian, W. Chance-constrained day-ahead scheduling in stochastic power system operation. *IEEE Trans. Power Syst.* **2014**, *29*, 1583–1591. [CrossRef]
50. Hu, Z. *Energy Storage for Power System Planning and Operation*; John Wiley & Sons: Hoboken, NJ, USA, 2020.
51. Ben-Tal, A.; El Ghaoui, L.; Nemirovski, A. *Robust Optimization (Princeton Series in Applied Mathematics)*; Princeton University Press: Princeton, NJ, USA, 2009.

Disclaimer/Publisher's Note: The statements, opinions and data contained in all publications are solely those of the individual author(s) and contributor(s) and not of MDPI and/or the editor(s). MDPI and/or the editor(s) disclaim responsibility for any injury to people or property resulting from any ideas, methods, instructions or products referred to in the content.

Article

Implementation Aspects of Smart Grids Cyber-Security Cross-Layered Framework for Critical Infrastructure Operation

Dennis Agnew [1], Nader Aljohani [1], Reynold Mathieu [1], Sharon Boamah [1], Keerthiraj Nagaraj [1], Janise McNair [1] and Arturo Bretas [1,2,*]

1 Department of Electrical & Computer Engineering, University of Florida, Gainesville, FL 32611, USA; dennisagnew@ufl.edu (D.A.); eng89nader@ufl.edu (N.A.); reynold.mathieu@ufl.edu (R.M.); sharonboamah@gmail.com (S.B.); k.nagaraj@ufl.edu (K.N.); mcnair@ece.ufl.edu (J.M.)
2 Distributed Systems Group, Pacific Northwest National Laboratory, Richland, WA 99354, USA
* Correspondence: arturo@ece.ufl.edu

Abstract: Communication networks in power systems are a major part of the smart grid paradigm. It enables and facilitates the automation of power grid operation as well as self-healing in contingencies. Such dependencies on communication networks, though, create a roam for cyber-threats. An adversary can launch an attack on the communication network, which in turn reflects on power grid operation. Attacks could be in the form of false data injection into system measurements, flooding the communication channels with unnecessary data, or intercepting messages. Using machine learning-based processing on data gathered from communication networks and the power grid is a promising solution for detecting cyber threats. In this paper, a co-simulation of cyber-security for cross-layer strategy is presented. The advantage of such a framework is the augmentation of valuable data that enhances the detection as well as identification of anomalies in the operation of the power grid. The framework is implemented on the IEEE 118-bus system. The system is constructed in Mininet to simulate a communication network and obtain data for analysis. A distributed three controller software-defined networking (SDN) framework is proposed that utilizes the Open Network Operating System (ONOS) cluster. According to the findings of our suggested architecture, it outperforms a single SDN controller framework by a factor of more than ten times the throughput. This provides for a higher flow of data throughout the network while decreasing congestion caused by a single controller's processing restrictions. Furthermore, our CECD-AS approach outperforms state-of-the-art physics and machine learning-based techniques in terms of attack classification. The performance of the framework is investigated under various types of communication attacks.

Keywords: cyber security; software-defined networking; network security; cyber-physical systems; cross-layered; power systems; machine learning

1. Introduction

Over the last decade, there has been a growing and significant demand for cyber-related smart grid (SG) security. Physical operation process dependability is the focus of current research on cyber-related power grid vulnerabilities, whereas the cyber-physical security of SGs is still evolving. The traditional power system is often safeguarded by isolated and uncoordinated equipment that offers ad-hoc solutions to each protection challenge. As these tools do not work together; they are vulnerable to dispersed attacks. The creation of cross-layer awareness of the smart grid system is a potential new technique in this field. The current study into the cyber-security of power grid operation is centered on a method known as State Estimation (SE). The fundamental purpose of SE is to offer a real-time grid monitoring technique by estimating system states utilizing measurements and static data on system topology [1]. False Data Injection Attack (FDI), which changes the measures utilized by SE, is the most prevalent cyber-attack in the literature. Multiple actors exploiting diverse security weaknesses in the physical and cyber domains of the cyber-physical system is a more realistic scenario for cyber-attacks. Denial-of-Service (DoS), Distributed Denial-of-Service (DDoS), and Man-in-the-Middle (MITM), False Data Injection (FDI) attacks might all be launched. These types of attacks impact data from several layers of the grid's physical structure. As a result, an integrated approach for identifying numerous attacks launched from different tiers inside the SG will improve grid security.

Machine Learning (ML) technology is presently being utilized to aid in the detection process or to acquire reliable findings through statistical data analysis. At various phases of the SE process, the work of the data-driven anomaly detection framework and the cross-layer viewpoint has progressed and been examined [2–6]. An integrated solution of a cyber-physical security framework based on a cross-layer approach that focuses on the detection of various cyber assaults is described in this study. The implementation is based on a real-world scenario. The power grid is modeled in Simulink and data from the grid is gathered every 4 s and delivered to the cloud for analysis. SimComponents [7] is used to simulate communication grid data, which is created and transferred to the cloud. In the analysis layer, SE and ML models are used to identify data threats. The SE exclusively uses measured data from the power grid, but the ML combines data from both the power and communication grids.

In addition, we propose an Open Network Operating System (ONOS) [8] three-controller distributed Software Defined Networking (SDN) architecture for the communication layer, which we tested and compared to the performance of one of the first SDN controllers, POX [9], which comes as standard with the Mininet emulation tool. SDN is a network architecture technology that allows networks to be intelligently and centrally managed, or programmed, using software applications. As its public release in 2009, software-defined networking (SDN) research has seen tremendous advancements and breakthroughs. SDN provides improved utilization, resource efficiency, network service flexibility, and lower maintenance costs as compared to conventional networks [10].

To the best of our knowledge, our suggested SDN framework is the first of its kind proposed in the literature. ONOS is the most widely used open-source SDN controller for next-generation SDN and Network function virtualization (NFV) applications. ONOS allows for network configuration as well as real-time control, removing the requirement for routing and switching control protocols to be executed inside the network fabric. By leveraging an SDN network topology, we can move the routing intelligence of the network to the ONOS controller to allow for better management, response, and visibility against cyber attacks against our network.

In a previous work [11], the authors have demonstrated that a distributed SDN framework may effectively and efficiently govern and assist in the protection of a smart grid. The authors were able to protect the smart grid from a denial of service attack by combining distributed SDN controller placement, intrusion detection systems (IDS), and state estimation. However, the framework uses a global SDN controller and a global security controller as the network and security masters, respectively, which introduce single points of failure. Furthermore, defense results against false data injection or man-in-the-middle (MiTM) attacks are not discussed. In another previous work [12,13], researchers suggest a distributed SDN framework to address scalability and reliability in smart grid systems. The authors of this study assume controller communication by using the BGP protocol to link two OpenDaylight controllers, allowing the controllers to share the workload of the network. This study, however, does not address smart grid security or protect against cyberattacks. Furthermore, the authors do not discuss or consider controller failure resistance in their framework.

In another previous work [4], The authors created the Ensemble CorrDet with Adaptive Statistics (ECD-AS) technique to analyze measurement data and packet contents. ECD-AS is another data-driven approach for detecting FDI attacks that takes into account the changing status of the SG. This method's drawback is that it solely employs measurement data, limiting its capacity to identify cyberattacks focused on the SG's communication network layer. However, the work in this research will make use of the analysis layer, which may also take into account data from the communication network that powers the SG, notably packet inter-arrival intervals, transmission delay (TD), and packet count (PC). Consideration of this form of data would broaden the model of an FDI cyberattack as well as uncover models of other types of cyberattacks that would be undetected by present methodologies in the literature. Previous work [3,14] demonstrated that ML may be utilized in the cyber domain to enhance bad data analysis by acting on the same data as the SE. This hybrid data-driven physics-model-based framework makes use of both temporal data through ML and the system's known topology through SE. However, this approach, like the other current research investigations, solely covers FDI attacks and relies on routine measurements of power systems. As a result, it overlooks the SG's cross-layer interdependence.

The remainder of the paper is organized as follows. Background information on the major components of the framework is presented in Section 2. Section 3 contains data flow information about the framework, and the implementation necessary to make it work. In Section 4, we present a case study. Lastly, in Section 5, we conclude the paper. The contribution of this work towards the state-of-the-art is two folds:

1. To the best of our knowledge, the first proposed three controllers distributed SDN architecture for the Smart Grid's communication layer.
2. Identification of cyber attacks against the power grid using a cross-layered framework.

2. Theoretical Background

In the following, the theoretical background of SE, ML, cyber-attack models, and communication networks performance metrics are presented. The equations utilized in this study are derived from our previous work [6].

2.1. State Estimation

In modern Energy Management Systems (EMS), the SE process is most important for situational awareness of power system operation and is used in many EMS applications, including the detection of bad data. The common approach to SE is using the classical Weighted Least Squares (WLS) method described in [15]. In this approach, the power grid is modeled as a set of non-linear equations based on the physics of the system:

$$\mathbf{z} = h(\mathbf{x}) + \mathbf{e} \tag{1}$$

where $\mathbf{z} \in \mathbb{R}^{1 \times d}$ is the measurement vector, $\mathbf{x} \in \mathbb{R}^{1 \times N}$ is the vector of state variables, $h : \mathbb{R}^{1 \times N} \to \mathbb{R}^{1 \times d}$ is a continuously non-linear differentiable function, and $\mathbf{e} \in \mathbb{R}^{1 \times d}$ is the measurement error vector. Each measurement error, e_i, is assumed to have zero mean, standard deviation σ_i and Gaussian probability distribution. d is the number of measurements and N is the number of states.

In the classical WLS approach, the best estimate of the state vector in (1) is found by minimizing the cost function $J(\mathbf{x})$:

$$J(\mathbf{x}) = \|\mathbf{z} - h(\mathbf{x})\|_{R^{-1}}^2 = [\mathbf{z} - h(\mathbf{x})]^T R^{-1} [\mathbf{z} - h(\mathbf{x})] \tag{2}$$

where R is the covariance matrix of the measurements. In [16], it is shown that the error can be decomposed into detectable and undetectable parts where the undetectable part is recovered through the Innovation Index. Hence, the composed measurement error CME is then used for Bad Data analysis, one of the main applications of SE [17], as

$$J_{CME}(\hat{\mathbf{x}}) = \sum_{i=1}^{d} \left[\frac{CME_i}{\sigma_i} \right]^2 > \chi_{d,p}^2 \tag{3}$$

where σ_i the measurement's standard deviation, p is the probability (typically $p = 0.95$) and d the degrees of freedom.

2.2. Machine Learning

ECD-AS is an adaptive data-driven anomaly detection framework presented in [4]. ECD-AS learns and adapts from the real-time SG data to distinguish any anomalous behavior from the normal behavior of the system. The ECD-AS detector learns a series of statistics (μ_m, Σ_m and τ_m), one for each bus m and then updates them with new incoming data samples to adapt them. ECD-AS uses mean (μ_m) and covariance matrix (Σ_m) of each bus established using normal samples data to calculate squared Mahalanobis distance (δ_m^{ECD-AS}) and uses it as a decision score for the bus m and detects any anomalies by comparing it to the adaptive threshold of bus m (τ_m). The squared Mahalanobis distance is calculated as

$$\delta_m^{ECD-AS}(\mathbf{z}_m) = (\mathbf{z}_m - \mu_m)^T \Sigma_m^{-1} (\mathbf{z}_m - \mu_m) \qquad (4)$$

where $\mathbf{z}_m$ is the measurement vector of bus m, μ_m is the mean and Σ_m^{-1} is the inverse covariance matrix of normal samples related to m^{th} bus. The mean, μ_m, and inverse covariance matrix, Σ_m^{-1}, for each bus are updated using the Woodbury Matrix Identity equations provided in [18]. The adaptive threshold (τ_m) of bus m is updated with a sliding window of size β over recent normal samples using the equation

$$\tau_m = \mu_{thr,m,-\beta} + \eta * \sigma_{thr,m,-\beta} \qquad (5)$$

where $\mu_{thr,m,-\beta}$ and $\sigma_{thr,m,-\beta}$ are the mean and standard deviation of the squared Mahalanobis distance values of normal samples in β most recent samples of selected measurements associated with m^{th} bus, and η is a hyper-parameter that decides how many standard deviations the threshold should be from the mean.

2.3. Software-Defined Networking

The concept and practice of SDN is enticing to those in the field of networking due to its visibility and ease of network device programmability. In recent years, SDN has taken shape. At Stanford University, the name SDN was coined to describe the concepts and techniques of Openflow [19]. SDN is divided into three planes:

1. **Application Plane:** It covers network management, policy implementation, and security services SDN applications.
2. **Control Plane:** This is a logically centralized control framework that runs the network operating system, operates the network operating system, and provides hardware abstractions to SDN applications. A flow in SDN is described as a set of instructions followed by a sequence of packets between the source and destination. Controllers install the flows into the flow tables of the forwarding devices.
3. **Data Plane:** A set of forwarding components used to move traffic flows in response to control plane instructions.

Figure 1 represents an overview example of a modern functional SDN architecture. Routers, switches, and access points comprise the infrastructure layer, as indicated in the diagram. The data plane is formed by this layer, which represents the physical network equipment in the network. Information is passed across planes of the SDN architecture through application programming interfaces (APIs). Southbound APIs like OpenFlow, ForCES, PCEP, NetConf, or IRS are used by the controller to communicate with the data plane. If there are multiple controllers, they interact via Westbound and Eastbound APIs like AlTO or Hyperflow. The application plane is the uppermost layer. The network operator can use functional applications for activities like energy efficiency, access control, mobility management, and security management at this layer. Northbound APIs such as FML, Procera, Frenetic, or RESTful are used by the application layer to communicate with the control layer. The network operator can use these APIs to relay the necessary modifications to the control layer, allowing the controller to make the appropriate adjustments in the infrastructure layer.

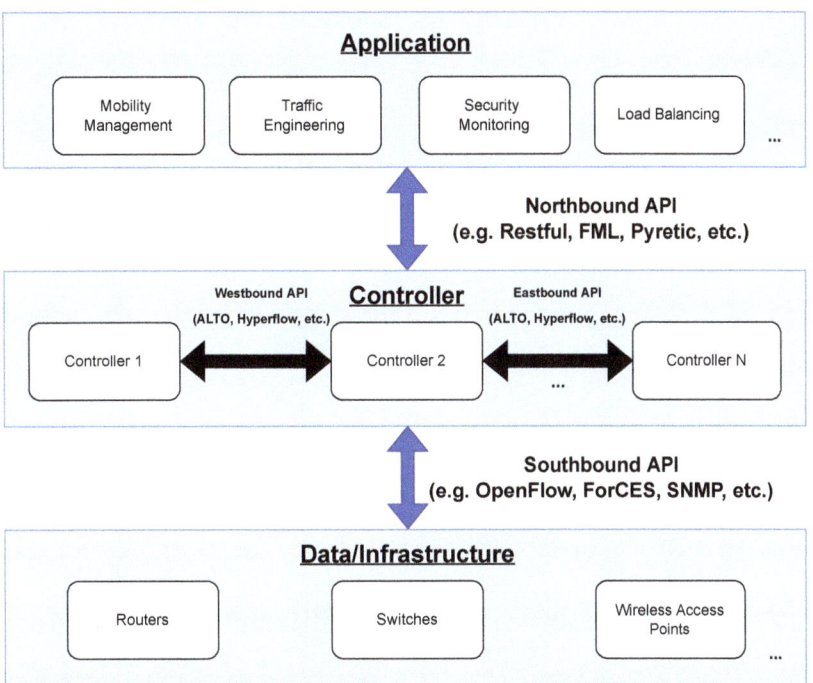

Figure 1. SDN Architecture.

To emulate the SDN framework, we use Mininet. Mininet [20] is an open-source networking software used to quickly prototype and emulate networks consisting of hosts, links, and switches on a single system. Mininet uses process-based virtualization and network namespaces to establish virtual networks, both of which are present in modern Linux kernels [21]. As hosts in Mininet are simulated as bash processes running in a network namespace, any code that would usually execute on a Linux server (e.g., web server or client software) operates as it normally would. Each Mininet "Host" have their own private network interface and will only be able to see the processes it is running. Software-based switches, such as Open vSwitch or the OpenFlow reference switch, are used in Mininet. Links are virtual ethernet pairs that reside in the Linux kernel and connect our simulated switches and hosts.

Without a controller to manage the network, no SDN network would be complete. The POX [22] controller is the default controller in Mininet and is an OpenFlow controller written in Python that is useful for quick prototyping. It was built from NOX [23], the first OpenFlow controller with only C++ language support. As the POX controller does not support multiple, distributed controllers, East/Westbound API communication, as shown in Figure 1, is not possible. It is, nonetheless, the go-to controller for quickly testing SDN frameworks in Mininet due to its simplicity of setup. Because of this, we use it as our standard of comparison for our proposed framework.

There are a variety of different controllers used in SDN literature research, such as Floodlight [24], OpenDaylight (ODL) [25], RYU [26], and Open Network Operating System (ONOS) [8]. Each has its own set of features and functions, as determined by the developers. As of its relative ease of use, multiple software applications, and network visibility in the form of a graphical user interface (GUI), we employ ONOS to construct a controller cluster to manage our SDN network to achieve our distributed three controller architecture. A detailed overview of our implementation of ONOS can be found in Section 4.

2.4. Network Performance Statistics

The cross-layered analysis framework is based on the IEEE 118-bus system, which uses TCP/IP protocols to imitate the Modbus RTU. This model, which resembles the Poisson traffic model [27], sends packets in groups of four every four seconds. Each bus represents the M/M/c queue [28], i.e., cc 1, in which packet arrival is Poisson and queue service time is exponential. The traffic intensity or utilization is represented by the following equation:

$$P_{util} = \frac{\lambda}{\mu} \quad (6)$$

The packets' arrival rate is denoted by λ, while the packets' service time is represented by μ. The time difference (Δt) between packet arrivals is known as the inter-arrival time (IAT). With parameter λ, it has an exponential distribution. For $t \geq 0$, the probability density function is defined as follows:

$$f(t) = \lambda e^{\lambda t}. \quad (7)$$

The average IAT is defined as

$$IAT = \frac{1}{\lambda} \quad (8)$$

The service time follows an exponential distribution with parameter μ. The probability density function is as follows:

$$g(s) = \mu e^{-\mu s}, \forall \geq 0 \quad (9)$$

where $\frac{1}{\mu}$ is the average service time of the system. Utilizing Little's theorem, the total waiting time is defined as transmission delays (TD), and represented as the following:

$$W = TD = \frac{1}{\mu - \lambda} \quad (10)$$

The normal distribution of network packet arrivals (i.e., non-attacked packets) into each system was decided by the probability of witnessing a number of packet arrivals in a period from [0, T]. This equation is used to model the traffic volume of the bus:

$$P(n\ arrivals\ in\ interval\ T) = \frac{(\lambda T)^n e^{-\lambda T}}{n!} \quad (11)$$

whereas T is the IAT, and the n represents the number of packets. The packet count (PC) is modeled as the following:

$$PC = \lambda T \quad (12)$$

2.5. Communication Layer

The SCADA network (Supervisory Control and Data Acquisition) is vulnerable to cyber-attacks. This section will detail how we implemented simulations for the DoS, MITM, and FDI attack scenarios.

2.5.1. Denial of Service Attack Simulation

A denial-of-service (DoS) attack is a type of cyber-attack in which the perpetrator tries to prevent intended users from accessing a node or network by temporarily or permanently disrupting the services of a host connected to the network. A DoS attack is regularly carried out in an attempt to overwhelm systems and prevent some or all real requests from being handled by flooding the targeted computer or resource with unnecessary requests or packets (e.g., TCP, UDP, SYN, etc.). Communication between nodes is disrupted during a DoS attack because a huge number of service requests are delivered to the target node, depleting all of the node's resources. A spike in network traffic to the victim (i.e., arrival rate) is created as a result of such an attack. As a result, large queues are formed, resulting in higher wait times and transmission delays [29]. Therefore, we record interarrival times (IAT) and transmission delays (TD) in our DoS attack datasets.

To simulate the effects of a DoS attack on our network traffic SimPy and component toolkit SimComponents [7] are used at the communication layer to simulate network traffic in the smart grid layer. SimPy is a discrete-event simulation framework based on Python. Active components such as packets, packet generators, packet sinks, switch ports, and port monitors may all be simulated using it. To define and replicate these components and their functionality, the SimComponents toolkit is employed. To achieve synchronization between the communication layer, smart grid layer, and machine learning model, we append the start time and index to the sample measurements. We generate data for one day's worth of measurements and create datasets to reflect DoS attack scenarios. DoS Algorithm 1 illustrates our pseudo-code and shows the overview of our attack data generation. Figure 2 shows a histogram of malicious traffic sink interarrival times for DoS attacks. When the victim node is attacked, an influx of packets is sent to it, causing the frequency of packets received to increase.

Algorithm 1 Denial of Service Attack

```
1:  for One day worth of measurements do
2:      Create arrays for the IAT & TD values
3:      Append Index & Current Time
4:      if An attack sample is detected then
5:          Set error equal to 1
6:          Create Attack Bus List
7:          Extend Attack Bus List with attack bus number
8:      else
9:          Set error equal to 0
10:     end if
11:     for Each smart grid measurement do:
12:         if error = 0 then
13:             Simulate normal traffic
14:             Append IAT & TD values to arrays
15:         else
16:             if a from bus is in the attack bus list then
17:                 Simulate malicious DoS traffic
18:                 Append IAT & TD values to arrays
19:             else
20:                 Simulate normal traffic
21:                 Append IAT & TD values to arrays
22:             end if
23:         end if
24:         Update IAT CSV File
25:         Update TD CSV File
26:     end for
27: end for
```

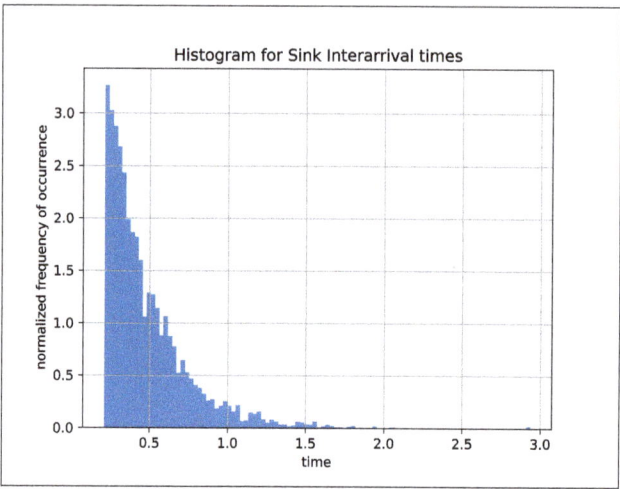

Figure 2. Histogram for sink interarrival times.

2.5.2. False Data Injection (FDI) Attack Simulation

False Data Injection Attacks in the communication layer occur when the adversary accesses the network layer and either manipulates the data within a packet or transmits wrong data packets. This subsequently affects the behavior of the packets in the network layer. Thus, FDI attacks have the capability to increase the inter-arrival time and transmission delay at the network level. Using these performance metrics, FDI attacks in the IEEE 118-bus system are emulated in the communication network based on the M/M/c queue, where $c \geq 1$. The SimComponents and SimPy libraries are used to emulate packet generation, transmission, and FDI attacks in the communication layer. The implementation of this cyber-attack is summarized in Algorithm 2. This scenario is demonstrated by generating the inter-arrival times and transmission delay of normal and malicious packet samples using the SimPy environment. The start time is appended to the sample measurements for synchronization between the network layer and the power system layer. The packet generator function is utilized to send normal or malicious packets with a fixed inter-arrival time distribution and packet size distribution for the sample buses. In addition, the switch port is simulated with exponential packet inter-arrival times and exponentially distributed packet sizes by setting the port rate and queue limits. The port rate of false data injections is set to be lower than the normal traffic, while the queue limit of false data injections is set to be higher than the normal traffic.

The implementation of FDI attacks in Algorithm 2 is done based on the presence of normal and attacked buses in the generated bus list for the IEEE 118-bus system. Considering the bus list, if a normal bus is detected, the switch port parameters for the port rate and queue limit are set to values of 10,000 and 100,000, respectively. The packet generator is used to generate the normal packets with exponential inter-arrival times and exponentially distributed packet sizes. Similarly, if a malicious bus is detected, the switch port parameters for the port rate and queue limit are assigned values of 30 and 10,000,000, respectively. Moreover, the packet generator generates malicious packets with exponential inter-arrival times and exponentially distributed packet sizes. The packet sink records the inter-arrival times and transmission delays and appends the values of each time measurement to a CSV file, which is utilized by the ML model in real-time. The experiment is conducted for 21,600 packet samples.

Figure 3a,b represent the statistics of the transmission delay and inter-arrival time of 21,600 samples taken for 691 measurements, respectively. The generated values for the packet transmission delay and inter-arrival time are exponentially distributed.

Algorithm 2 False Data Injection Attack

1: **Initialize** transmission delay (TD) array, inter-arrival time (IAT) array, switch function variables; port_rate_normal, queue_limit_normal, port_rate_malicious, queue_limit-malicious
2: **for** all samples **do**
3:     Append *Index* & *CurrentTime* to *IAT* array, *TD* array
4:     **if** *An attack sample is detected* **then**
5:         Create Attack Bus List packets sink
6:         Setup switch port using *port_rate_malicious*,
7: *queue_limit_malicious*
8:         Simulate malicious traffic
9:         Append *IAT* & *TD* values to arrays
10:    **else**
11:
12:         Create the packets generator and packets sink
13:         Setup switch port using *port_rate_normal*,
14: *queue_limit_normal*
15:         Simulate normal traffic
16:         Append *IAT* & *TD* values to arrays
17:    **end if**
18:    Create IAT CSV File
19:    Create TD CSV File
20: **end for**

	Transmission Delay	
	Mean	Standard deviation
3	0.008653868	0.001444722
4	0.008655773	0.001436691
5	0.008667217	0.001432317
6	0.008669691	0.001445522
7	0.008660648	0.001443733
8	0.008662527	0.001441175
9	0.008665332	0.001438561
10	0.008669188	0.001433087
11	0.008675082	0.001436771
12	0.008659845	0.001440388
13	0.008652211	0.001431122
14	0.008657773	0.001447691
15	0.008680384	0.001443478
16	0.008657822	0.001453899
17	0.008647711	0.001427985
18	0.00865814	0.001445786
19	0.008657252	0.001444495
20	0.008686628	0.001457041
21	0.008666389	0.001426464
22	0.008666574	0.001441871
23	0.008660409	0.001439144

(a)

	Inter-arrival time	
	Mean	Standard deviation
3	0.100271509	0.014779787
4	0.100263115	0.014686023
5	0.100163861	0.014539098
6	0.099953059	0.01448458
7	0.100008547	0.014555367
8	0.100251506	0.014507068
9	0.100109684	0.014572447
10	0.100181394	0.014572174
11	0.10028387	0.014634165
12	0.100071705	0.014600816
13	0.100169394	0.014579658
14	0.100411796	0.014717275
15	0.100143643	0.014695299
16	0.100210849	0.014674573
17	0.100409129	0.014786006
18	0.100204581	0.014641775
19	0.100069046	0.014662945
20	0.100266705	0.01472537
21	0.100010471	0.014621193
22	0.100311526	0.014675738
23	0.10035113	0.014826753

(b)

Figure 3. 21,600 samples taken for 691 measurements. (**a**) Statistics measurements for transmission delay. (**b**) Statistics measurements for Inter-arrival time.

Figure 4a,b are the plots for the transmission delay and inter-arrival time of 21,600 samples taken for the 690th normal traffic measurement. The values for the transmission delay and inter-arrival time fall within close ranges of each other for normal traffic measurement in the figures. Figure 5a,b illustrate the transmission delay and inter-arrival time of 21,600 samples for the 690th malicious traffic measurement. An instance of an FDI attack is demonstrated with a higher transmission delay at a maximum value of 8.30 in the 7712th sample, indicating malicious traffic. An occurrence of malicious traffic is represented in the 7718th sample, which shows an increased inter-arrival time with a maximum value of 3.83.

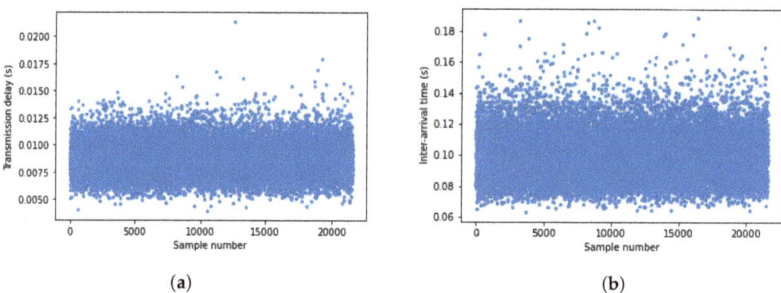

Figure 4. The 690th measurement for 21,600 samples with normal traffic. (**a**) Transmission delay for normal traffic. (**b**) Inter-arrival time for normal traffic.

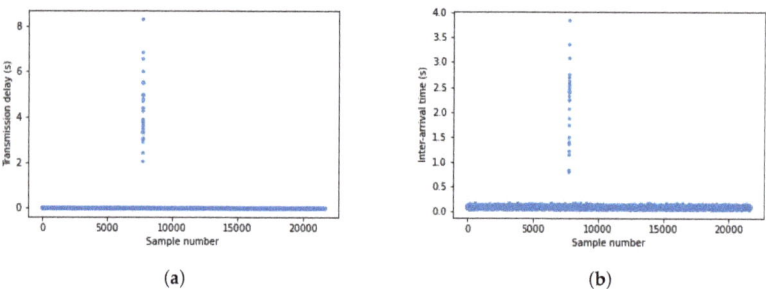

Figure 5. The 690th measurement for 21,600 samples with malicious traffic. (**a**) Transmission delay for malicious traffic. (**b**) Inter-arrival time for malicious traffic.

2.5.3. Man-in-the-Middle Attack Simulation

MITM attack is a type of attack during which malicious third parties position themselves between the communication of two other parties, or between a user and an application. The attack can be passive where the adversary eavesdrops and extract sensitive information. In this instance, communication confidentiality is compromised, and the adversary can remain undetected if a special network penetration test or analysis is not performed on a regular basis. The attack can also be active where the attackers hijack the router to which the victims are connected, or advertise false information using Address Resolution Protocol(ARP) messages. In doing so, they can gain access to the information being shared, and manipulate the data [30].

Our simulation considers only the active type of MITM attack because our code detects the difference in the number of packets transmitted and received between the victims. During an active MITM attack, the adversary flood the network with ARP messages to mislead the victims into thinking he is one of them. By doing so, there is a substantial increase in the number of packets transmitted that can be easily detected during the analysis of network statistics data. Moreover, active MITM attacks can also be detected with the analysis of network latency. Since the adversary is a pass-through between the two trusting parties, a delay is observed when the forwarding happens. We will only develop the concept of an increase in packet count in the following lines because this is the method used by our machine learning model to predict the behavior of the 118 bus system.

Algorithm 3 demonstrates how the principal purpose of a MITM cyber attack may be recognized and data collected. As we aforementioned, the traffic volume is affected by MITM attacks. This is why the method of detection is implemented by generating and testing packet counts. In this simulation, we alter the network traffic of randomly selected bus samples and as indicated in the algorithm and in the simulation result of Figure 6, we get alerts with the bus number and sample number for every bus that has an error.

Algorithm 3 Man-in-the-Middle Attack Detection

1: Create benchmark arrays from arPoisson.txt
2: Create array of sample data from matlab file
3: **for** the length of the victim list **do**
4:    **if** An attacked sample is detected **then**
5:       Set error equal to 1
6:       Extend Attacked Bus List with attack Bus number
7:       Alert of error with error number and bus number
8:    **else**
9:       Set error equal to 0
10:    **end if**
11:    **for** the length of victim list **do**:
12:       **if** error = 0 **then**
13:          Append packet count values to array from firsth
14:       **else**
15:          **if** a from bus is in the attack bus list at this index **then**
16:             Append packet count values to array from secondh
17:          **else**
18:
19:          **end if**
20:       **end if**
21:       Create packet count CSV File
22:    **end for**
23: **end for**

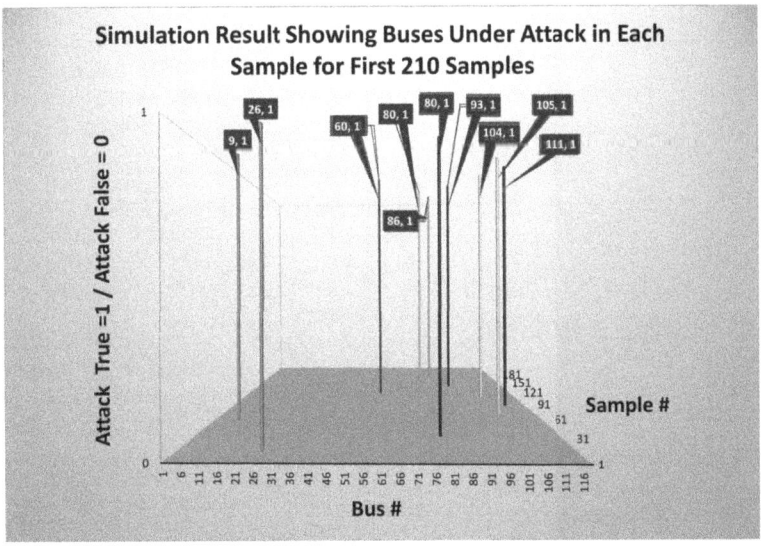

Figure 6. Simulation result showing buses under attack in each sample.

In conclusion, MITM attacks typically have a low impact or no impact at all on network performance but they are the first steps to achieve to jeopardize a cyber security system. During MITM attacks and depending on the type of MITM attack, a flux of packets could increase the network traffic. These kinds of attacks are amongst the most dangerous ones because they are very hard to detect and can easily open doors to other types of attacks like DoS or FDI.

2.6. Power Grid: FDI and Parameter Attacks

The power grid can be modeled through nodes and edges. Measurements collected from the grid are power flows into those edges (lines) and injections into the nodes. SE reads off those measurements and uses a model that is based on the connectivity of the nodes and the electrical characteristics of the lines (system database). Hence, SE is a monitoring tool to observe the healthiness of the power grid over time. Therefore, the attacks pertaining to the power grid could affect the collected measurements or the database used by SE. FDI attacks can be on measurements or databases [31]. Attack types result in different residual characteristics and patterns [17]. These will be used here for the correction. This work [32] was used towards parameter cyber-attack correction, while [33] is used for measurement FDI attack correction.

3. Framework

The cross-layered cyber security framework takes into account the cyber-physical domain of the concept of the smart grid. The physical domain is composed of a power grid and communication network while in the cyber domain, the collected and communicated data are analyzed. The top view of the cross-layer framework is illustrated in Figure 7. As shown in Figure 7 and reading the figure from left to right, the attacker is designed as an outside entity where the desired scenario. Data collected from the power grid and communication network are stored in files to be analyzed by the cyber-domain. The collected data then passes through three stages: detection, identification, and correction. In the detection stage, the data collected from the power grid as well as the communication network are combined and analyzed by Machine Learning techniques as shown in Figure 8. It is worth mentioning that the data stored in the "CSV" file are after the attack took place and their effects happened. For instance, for an FDI attack on measurement, the corresponding measurement in the "CSV" file named "Measurements" are altered based on the attack scenario constructed. The output of the detection stage is a flag that corresponds to the attack scenario initiated. In the identification stage, each flag will be passed to a corresponding routine for identifying the location of the attack within the network of interest. For instance, in an attack in the system database used by State Estimation (SE), the k-nearest neighbor (kNN) routine will be activated to identify which line (Edge/Arc) in the power network is being altered/attacked by the attack scenario.

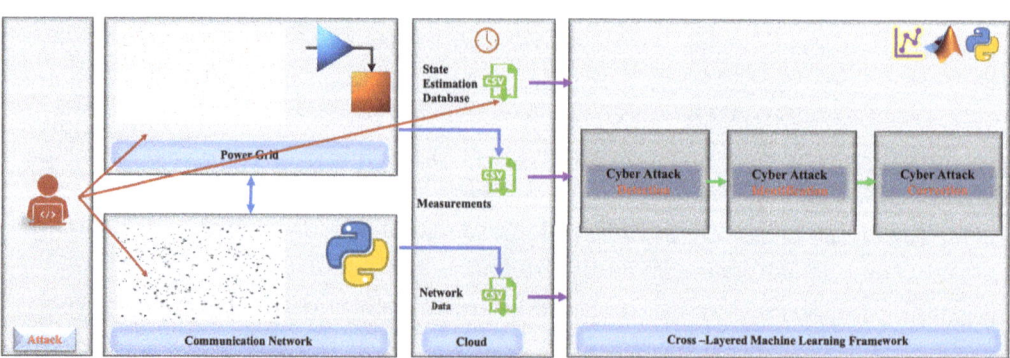

Figure 7. Proposed framework for cross-layer integration.

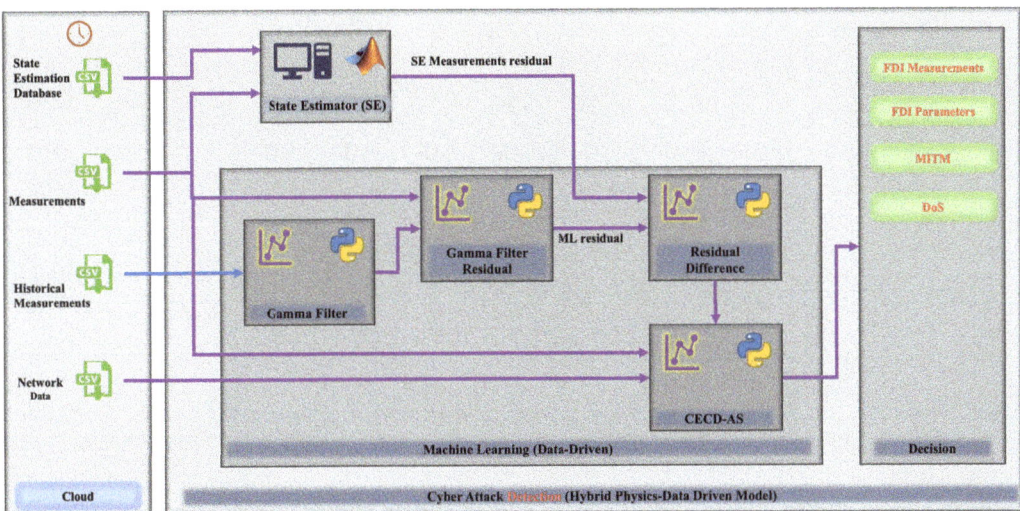

Figure 8. Detection framework.

3.1. Implementation of SDN

We use a distributed, three-controller SDN structure to serve as the communication layer for the smart grid. To avoid a single point of failure, we use three controllers, each of which has the same authority level in the network. This means that each controller has equal control over the network. There isn't a single point of failure in the network since there aren't any single controllers. We utilize ONOS as our SDN framework's control layer, which is primarily open source. It provides for network and configuration control in real-time.

ONOS also allows for quick redistribution of controller load, allowing each controller to function optimally for the network. We can monitor and regulate the flow of packets in our smart grid infrastructure from the ONOS cluster GUI. Furthermore, ONOS eliminates controller single points of failure by dynamically shifting the workload from a down controller to the remaining controllers if necessary. This is done by Atomix [34], a reactive java framework for building scalable fault-tolerant distributed systems. The Atomix cluster is responsible for ONOS cluster administration, service discovery, and data storage, as depicted in Figure 9. ONOS controllers may be quickly discovered and removed if down, thanks to the Atomix framework. First, we utilized Docker containers to build the Atomix and ONOS clusters in a local virtual machine (VM) installation on a local personal device, using ONOS version 2.3, Openflow version 1.3, and Atomix version 3.1.5. Then, using a python script with Mininet 2.3 APIs, we built 118 hosts and 45 Open vSwitches with OpenFlow 1.3, as illustrated in Figure 10. Open vSwitch [35] is an open-source distributed virtual multilayer switch solution and one of the most popular implementations of OpenFlow. This is our proposed SDN framework, which is modeled around the IEEE 118 bus system.

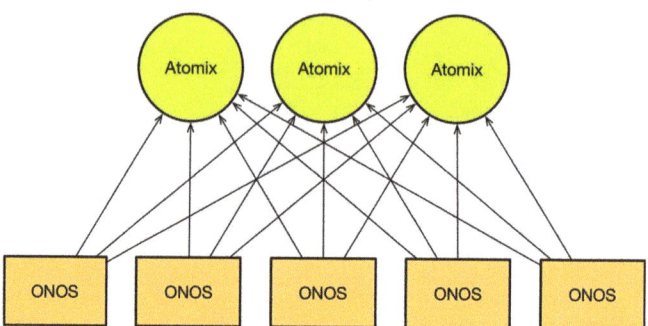

Figure 9. Atomix framework [34].

Figure 10. Proposed distributed three controller.

3.2. Details on the Apps: Cybersecurity Framework

The KNN (k Nearest Neighbors) algorithm is known as a non-parametric supervised learning classifier. This algorithm is typically used as a classification algorithm in Machine Learning. The favor of choosing this technique typically is the ease of interpretation as well as the low computation time. The main idea of the KNN algorithm is that most similar samples are clustered and grouped into the same class. Classification of a new sample in KNN starts by finding k nearest neighbors of the new sample in the training dataset, and then the new sample is classified to the major class in the k nearest neighbors.

The Cross-Layer Ensemble CorrDet with Adaptive Statistics (CECD-AS) algorithm proposed iteAllen2022Starke is utilized as the cyber threat detection technique in this paper and is used in conjunction with polling and synchronization steps to make it work for a real-time system. The CECD-AS algorithm combines data from measurement collection devices in SG and communication networks in real-time to detect any anomalous behavior caused due to cyber attacks such as FDI, DoS, or MITM. The power grid and communication layers generate measurement and network performance values every 4 s respectively and save them in corresponding CSV files. The data acquisition component in the ML layer is equipped with a polling module that polls for data every 2 s and checks if the CSV files are updated with any new data. If the CSV files are updated, then the data acquisition component collects newly added data samples along with index numbers and time stamps. The index numbers and timestamp values of data samples incoming from the communication layer and power grid layer are matched to synchronize and combine data as needed for the CECD-AS algorithm. The CECD-AS algorithm creates and updates statistical models for different regions of the SG using Woodbury Matrix Identities, anomaly scores using Mahalanobis distance measures, and updates adaptive thresholds for different regions based on the equations provided [4,6].

4. Case Study

We built a 691 × 10,000 matrix in the form of a CSV file to create our dataset. Each row represents a moment in time for the respective measurements, while each column indicates a measurement point in the grid. A sample of measurements for the full grid is considered one row of data. Mininet takes around 4 s to create one data point, or 45 min for a given network sample, or row of data. This creates a temporal constraint because our ML model requires 10,000 rows of data. We use SimComponent, as previously stated, to reduce the time spent obtaining network data. In comparison, SimComponent generates the data required for one data sample in roughly 0.80 s, which is substantially quicker than Mininet, allowing the completion of the necessary dataset to acquire the results for the CECD-AS method, as shown in Table 1.

Table 1. Performance results for FDI, DoS, and MITM attacks (FDI: False Data Injection attacks, DoS: Denial of Service attacks, MITM: Man In The Middle attacks).

Attack Type	Accuracy $\mu_{cv} \pm \sigma_{cv}$	Precision $\mu_{cv} \pm \sigma_{cv}$	Recall $\mu_{cv} \pm \sigma_{cv}$	F1-Score $\mu_{cv} \pm \sigma_{cv}$
MITM	92.48 ± 00.20	91.65 ± 00.29	86.41 ± 00.28	**88.91 ± 00.24**
FDI	99.95 ± 00.01	99.46 ± 00.34	99.87 ± 00.13	**99.61 ± 00.17**
DoS	99.88 ± 00.07	99.75 ± 00.09	99.80 ± 00.16	**99.78 ± 00.08**
FDI-DoS	99.63 ± 00.08	98.42 ± 00.26	99.95 ± 00.04	**99.20 ± 00.15**

CBench [36], a tool for benchmarking Openflow controllers, is used to test the SDN architecture. We can calculate the maximum throughput, or how much data was sent from a source at any particular moment, using CBench. CBench emulates the N amount of OpenFlow switches set by the researchers and connects it to the controller. Then, it emulates traffic and calculates and records the throughput. In the cluster, we test each ONOS controller individually to establish the maximum throughput for 45 open vswitches (15 switches for each controller). We ran this test again for the single POX controller for all 45 open vswitches at once and documented the results. Figure 11 shows the results of our controller benchmarking test. The ONOS cluster had an average throughput of 533.121 flows/ms, whereas the POX controller had a flow rate of 50.267 flows/ms, which is more than a tenfold improvement. Our tests revealed that a distributed controller design not only removes a single point of failure but also improves network and resource management throughput. By spreading the network workload, the ONOS cluster can endure the rigors of the smart grid better than a single controller.

For parameter attack in the lines of power grid used by State Estimator, each line is attacked to form classes for classification. In the system under study, we have 117 lines, which are transformed into 117 classes in the KNN algorithm. For identifying which line of the power grid is being attacked, the KNN algorithm is implemented. The dataset is split into training and testing. For the training dataset, each line is attacked with false data injection into its parameter. The classification accuracy of the presented KNN algorithm is illustrated in Figure 12.

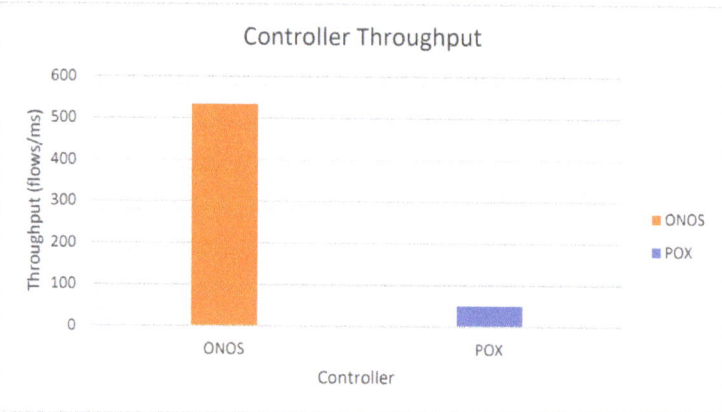

Figure 11. Throughput benchmark of controllers.

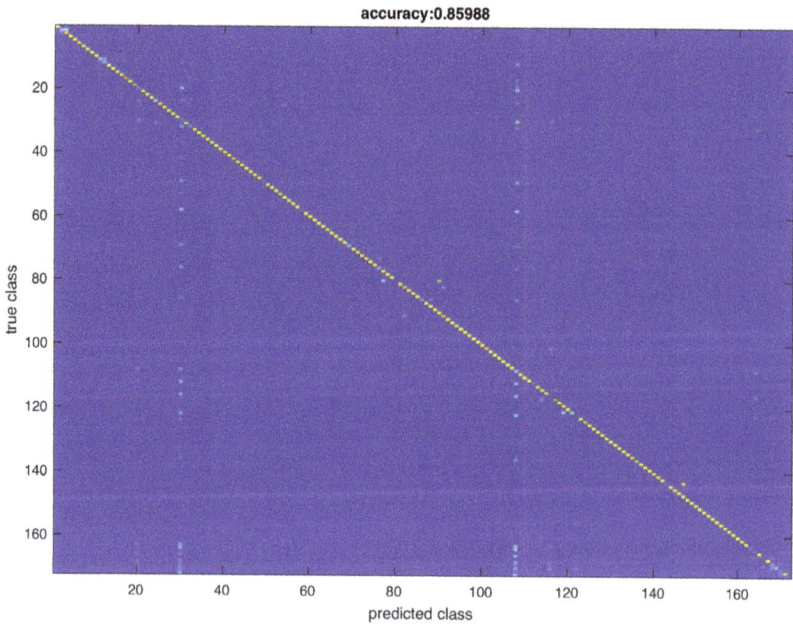

Figure 12. Prediction accuracy of KNN for attacks in power grid line parameter.

To validate the cyber threat detection framework, IEEE 118 bus system is selected as the power grid under study. The power grid is simulated such that every 4 s a set of measurements is sent to the cloud where data is stored and analyzed by SE and ML. Hence, the data in the cloud is updated every 4 s. The data from the smart grid layer and communication are combined in the cloud and then analyzed to detect cyber threats. The detection results of MITM, FDI, DoS, and simultaneous FDI-DoS attacks through the use of the real-time CECD-AS algorithm are presented in Table 1 in terms of accuracy, precision, recall, and F1-score values. In the paper [6] where CECD-AS is presented, its cyber threat detection performance is shown to outperform the state-of-the-art physics-based and machine learning-based techniques. The results in this paper are based on the data and experiments discussed [6]. Table 1 shows that the real-time CECD-AS algorithm performs extremely well for a variety of cyber threats discussed in this paper. The enhancement in the detection is due to the integration (combined) data from communication and power grid that added values to the ECD-AS algorithm. Hence, data from the two layers, i.e., power grid and communication grid, complement each other.

5. Conclusions and Future Work

The design of a Cross-Layerered framework for safeguarding the power grid's operation from physical component or communication network threats is described in this paper. To address power grid communication, we recommended adopting a distributed three-controller SDN architecture. We can manage our network with increased visibility, control, and responsiveness because of SDN. Moreover, using ONOS clusters eliminates the single point of failure that might occur when using a single controller. We benchmarked our proposed SDN framework against the conventional POX controller to demonstrate the network's increased performance and load management. SimComponents is used to quickly build and simulate DoS, Man-in-the-Middle (MiTM), and False Data Injection attacks (FDI) attacks. The state estimation is affected by FDI and DoS attacks, and all attacks have an impact on the communication network. To detect the corresponding attack, the state estimator and machine learning examine the consequences of all attacks. Simulink is used to represent the power grid, allowing for real-time simulation. SimComponent, a Python library, is used to simulate a communication network. To detect attacked samples, data from each layer is synced and evaluated using a real-time cross-layered machine learning technique. According to the results of our suggested architecture, a three-controller distributed arrangement outperforms a single controller by a factor of more than ten times the throughput. This allows for a greater flow of data throughout the network while reducing congestion caused by the processing constraints of a single controller. Moreover, our CECD-AS approach outperforms state-of-the-art physics and machine learning-based algorithms in attack classification.

In future work, we would like to extend this framework to include more types of cyber attacks, additional controllers, and P4 [37] and Stratum [38]-enabled switches. There are other cyber attacks we would like to defend against, such as ransomware, botnet, and host impersonation attacks. Furthermore, we would like to build upon our failure-resistant framework by adding additional standby controllers for each of the current controllers to increase protection against our model from unforeseen outages that may be experienced in the field. In addition, we would like to include P4 and Stratum-enabled switches/routers to allow for complete "white box" control of the forwarding devices. This would allow for the control of packet parsing at the forwarding device level to increase QoS in the network. The framework was developed in a real-time simulated environment, making it an ideal starting point for future research on data integrity cyber threats in smart grids.

Author Contributions: Conceptualization, D.A. and N.A.; methodology, D.A. and N.A.; software, D.A., N.A., R.M. and S.B.; validation, D.A., N.A., R.M. and S.B.; formal analysis, D.A., N.A., R.M. and S.B.; investigation, D.A., N.A., R.M. and S.B.; resources, A.B. and J.M.; data curation, A.B., K.N. and J.M.; writing—original draft preparation, D.A., N.A., R.M. and S.B.; writing—review and editing, A.B. and J.M.; visualization, A.B. and J.M.; supervision, A.B. and J.M.; project administration, A.B. and J.M.; funding acquisition, A.B. and J.M. All authors have read and agreed to the published version of the manuscript.

Funding: This work was supported by NSF grant ECCS-1809739.

Institutional Review Board Statement: Not applicable.

Informed Consent Statement: Not applicable.

Conflicts of Interest: The authors declare no conflict of interest.

References

1. Bretas, A.; Bretas, N.; London, J.; Carvalho, B. *Cyber-Physical Power Systems State Estimation*; Elsevier: Amsterdam, The Netherlands, 2021; Volume 1.
2. Trevizan, R.D.; Ruben, C.; Nagaraj, K.; Ibukun, L.L.; Starke, A.C.; Bretas, A.S.; McNair, J.; Zare, A. Data-driven Physics-based Solution for False Data Injection Diagnosis in Smart Grids. In Proceedings of the 2019 IEEE Power Energy Society General Meeting (PESGM), Atlanta, GA, USA, 4–8 August 2019; pp. 1–5. [CrossRef]
3. Ruben, C.; Dhulipala, S.; Nagaraj, K.; Zou, S.; Starke, A.; Bretas, A.; Zare, A.; McNair, J. Hybrid data-driven physics model-based framework for enhanced cyber-physical smart grid security. *IET Smart Grid* **2020**, *3*, 445–453. [CrossRef]
4. Nagaraj, K.; Zou, S.; Ruben, C.; Dhulipala, S.; Starke, A.; Bretas, A.; Zare, A.; McNair, J. Ensemble CorrDet with adaptive statistics for bad data detection. *IET Smart Grid* **2020**, *3*, 572–580. [CrossRef]
5. Nagaraj, K.; Aljohani, N.; Zou, S.; Ruben, C.; Bretas, A.; Zare, A.; McNair, J. State Estimator and Machine Learning Analysis of Residual Differences to Detect and Identify FDI and Parameter Errors in Smart Grids. In Proceedings of the 2020 52nd North American Power Symposium (NAPS), Tempe, AZ, USA, 11–13 April 2021; pp. 1–6.
6. Starke, A.; Nagaraj, K.; Ruben, C.; Aljohani, N.; Zou, S.; Bretas, A.; McNair, J.; Zare, A. Cross-layered distributed data-driven framework for enhanced smart grid cyber-physical security. *IET Smart Grid* **2022**. [CrossRef]
7. Van Rossum, G.; Drake, F. *Python 3 Reference Manual*; CreateSpace: Scotts Valley, CA, USA, 2009.
8. Berde, P.; Gerola, M.; Hart, J.; Higuchi, Y.; Kobayashi, M.; Koide, T.; Lantz, B.; O'Connor, B.; Radoslavov, P.; Snow, W.; et al. ONOS: Towards an open, distributed SDN OS. In Proceedings of the Third Workshop on Hot Topics in Software Defined Networking, Chicago, IL, USA, 22 August 2014; pp. 1–6.
9. Kaur, S.; Singh, J.; Ghumman, N.S. Network programmability using POX controller. In Proceedings of the ICCCS International Conference on Communication, Computing & Systems, Chennai, India, 20–21 February 2014; Volume 138, p. 70.
10. Sun, S.; Fu, X.; Luo, B.; Du, X. Detecting and mitigating ARP attacks in SDN-based cloud environment. In Proceedings of the IEEE INFOCOM 2020-IEEE Conference on Computer Communications Workshops (INFOCOM WKSHPS), Toronto, ON, Canada, 6–9 July 2020; pp. 659–664.
11. Ghosh, U.; Chatterjee, P.; Shetty, S. A security framework for SDN-enabled smart power grids. In Proceedings of the 2017 IEEE 37th International Conference on Distributed Computing Systems Workshops (ICDCSW), Atlanta, GA, USA, 5–8 June 2017; pp. 113–118.
12. Qureshi, K.N.; Hussain, R.; Jeon, G. A distributed software defined networking model to improve the scalability and quality of services for flexible green energy internet for smart grid systems. *Comput. Electr. Eng.* **2020**, *84*, 106634. [CrossRef]
13. Hussain, R.; Bashir, M.U. Model to Improve Scalability and Quality of Services in Software Define Networking. In Proceedings of the 2019 2nd International Conference on Communication, Computing and Digital systems (C-CODE), Islamabad, Pakistan, 6–7 March 2019; pp. 28–33.
14. Bretas, A.; Rossoni, A.; Trevizan, R.; Bretas, N. Distribution networks nontechnical power loss estimation: A hybrid data-driven physics model-based framework. *Electr. Power Syst. Res.* **2020**, *186*, 10639. [CrossRef]
15. Bretas, A.S.; Bretas, N.G.; Carvalho, B.E. Further contributions to smart grids cyber-physical security as a malicious data attack: Proof and properties of the parameter error spreading out to the measurements and a relaxed correction model. *Int. J. Electr. Power Energy Syst.* **2019**, *104*, 43–51. [CrossRef]
16. Bretas, N.G.; Bretas, A.S. The extension of the Gauss approach for the solution of an overdetermined set of algebraic non linear equations. *IEEE Trans. Circuits Syst. II Express Briefs* **2018**, *65*, 1269–1273. [CrossRef]
17. Bretas, N.G.; Bretas, A.S.; Martins, A.C.P. Convergence Property of the Measurement Gross Error Correction in Power System State Estimation, Using Geometrical Background. *IEEE Trans. Power Syst.* **2013**, *28*, 3729–3736. [CrossRef]
18. Alvey, B.; Zare, A.; Cook, M.; Ho, D.K.C. Adaptive coherence estimator (ACE) for explosive hazard detection using wideband electromagnetic induction (WEMI). *Proc. SPIE* **2016**, *9823*, 58–64. [CrossRef]
19. Kreutz, D.; Ramos, F.M.; Verissimo, P.E.; Rothenberg, C.E.; Azodolmolky, S.; Uhlig, S. Software-defined networking: A comprehensive survey. *Proc. IEEE* **2014**, *103*, 14–76. [CrossRef]
20. Kaur, K.; Singh, J.; Ghumman, N.S. Mininet as software defined networking testing platform. In Proceedings of the International Conference on Communication, Computing & Systems (ICCCS), Chennai, India, 20–21 February 2014; pp. 139–142.
21. Mininet/Mininet: Emulator for Rapid Prototyping of Software Defined Networks. Available online: https://github.com/mininet/mininet (accessed on 1 June 2022).
22. POX Controller Manual Current Documentation. Available online: https://noxrepo.github.io/pox-doc/html/ (accessed on 1 June 2022).
23. Gude, N.; Koponen, T.; Pettit, J.; Pfaff, B.; Casado, M.; McKeown, N.; Shenker, S. NOX: Towards an operating system for networks. *ACM SIGCOMM Comput. Commun. Rev.* **2008**, *38*, 105–110. [CrossRef]
24. Floodlight. Floodlight Sdn Openflow Controller. Available online: https://github.com/floodlight/floodlight (accessed on 1 June 2022).
25. OpenDaylight: A Linux Foundation Collaborative Project. 2022. Available online: https://www.opendaylight.org/ (accessed on 1 June 2022).
26. Faucetsdn. Ryu Component-Based Software Defined Networking Framework. Available online: https://ryu-sdn.org/ (accessed on 1 June 2022).

27. Jain, R.; Routhier, S. Packet trains–Measurements and a new model for computer network traffic. *IEEE J. Sel. Areas Commun.* **1986**, *4*, 986–995. [CrossRef]
28. Haviv, M. *Queues—A Course in Queueing Theory*; The Hebrew University of Jerusalem: Jerusalem, Israel, 2009; 219p.
29. Gao, J.; Chai, S.; Zhang, B.; Xia, Y. Research about DoS attack against ICPS. *Sensors* **2019**, *19*, 1542. [CrossRef] [PubMed]
30. Dowling, B.; Hale, B. Secure Messaging Authentication against Active Man-in-the-Middle Attacks. In Proceedings of the 2021 IEEE European Symposium on Security and Privacy (EuroS P), Vienna, Austria, 6–10 September 2021; pp. 54–70. [CrossRef]
31. Aljohani, N.; Bretas, A. A Bi-Level Model for Detecting and Correcting Parameter Cyber-Attacks in Power System State Estimation. *Appl. Sci.* **2021**, *11*, 6540. [CrossRef]
32. Zou, T.; Aljohani, N.; Nagaraj, K.; Zou, S.; Ruben, C.; Bretas, A.; Zare, A.; McNair, J. A Network Parameter Database False Data Injection Correction Physics-Based Model: A Machine Learning Synthetic Measurement-Based Approach. *Appl. Sci.* **2021**, *11*, 8074. [CrossRef]
33. Bretas, A.S.; Bretas, N.G.; Carvalho, B.; Baeyens, E.; Khargonekar, P.P. Smart grids cyber-physical security as a malicious data attack: An innovation approach. *Electr. Power Syst. Res.* **2017**, *149*, 210–219. [CrossRef]
34. Cluster Configuration in Owl (1.14). Available online: https://wiki.onosproject.org/pages/viewpage.action?pageId=28836788#:~:text=The%20Owl%20release%20(1.14)%20features,of%20a%20separate%20Atomix%20cluster (accessed on 1 June 2022).
35. Openvswitch. Openvswitch/OVS: Open Vswitch. Available online: https://www.openvswitch.org/ (accessed on 1 June 2022).
36. CBench: An Dedicated OpenFlow Controller Implementation for "Cbench" OpenFlow Controller Benchmark Suite. Available online: https://github.com/trema/cbench (accessed on 1 June 2022).
37. Programming Protocol-Independent Packet Processors (P4). 2022. Available online: https://opennetworking.org/p4/ (accessed on 1 June 2022).
38. Stratum—Enabling the Era of Next-Generation SDN. 2022. Available online: https://opennetworking.org/stratum/ (accessed on 1 June 2022).

Article

Fusing Local and Global Information for One-Step Multi-View Subspace Clustering

Yiqiang Duan [1], Haoliang Yuan [1,*], Chun Sing Lai [1,2,*] and Loi Lei Lai [1,*]

[1] Department of Electrical Engineering, School of Automation, Guangdong University of Technology, Guangzhou 510006, China; 2111904200@mail2.gdut.edu.cn
[2] Brunel Interdisciplinary Power Systems Research Centre, Department of Electronic and Electrical Engineering, Brunel University London, London UB8 3PH, UK
* Correspondence: haoliangyuan@gdut.edu.cn (H.Y.); chunsing.lai@brunel.ac.uk (C.S.L.); l.l.lai@gdut.edu.cn (L.L.L.)

Abstract: Multi-view subspace clustering has drawn significant attention in the pattern recognition and machine learning research community. However, most of the existing multi-view subspace clustering methods are still limited in two aspects. (1) The subspace representation yielded by the self-expression reconstruction model ignores the local structure information of the data. (2) The construction of subspace representation and clustering are used as two individual procedures, which ignores their interactions. To address these problems, we propose a novel multi-view subspace clustering method fusing local and global information for one-step multi-view clustering. Our contribution lies in three aspects. First, we merge the graph learning into the self-expression model to explore the local structure information for constructing the specific subspace representations of different views. Second, we consider the multi-view information fusion by integrating these specific subspace representations into one common subspace representation. Third, we combine the subspace representation learning, multi-view information fusion, and clustering into a joint optimization model to realize the one-step clustering. We also develop an effective optimization algorithm to solve the proposed method. Comprehensive experimental results on nine popular multi-view data sets confirm the effectiveness and superiority of the proposed method by comparing it with many state-of-the-art multi-view clustering methods.

Keywords: multi-view learning; subspace representation; graph learning; one-step clustering

1. Introduction

Clustering is a fundamental unsupervised learning problem that is widely used in the tasks of machine learning [1], computer vision [2], and data mining [3]. It attempts to help to understand the structure of unlabeled data by dividing the entire unlabeled samples into clusters, where the samples in the same cluster are not similar to samples in the other clusters [4–6].

With the continuous development of information technology, different features of the object can be easily acquired by different feature extractors, data sources or sensors. For example, an image can be depicted by the color, texture, and edge features. A news report is usually composed of text descriptions and pictures. In the field of autonomous driving, an obstacle can be captured by different types of sensors. These different features can be viewed as multi-view data. Since each view commonly contains view-specific information about the object, using only one view for clustering may yield poor results [7]. Therefore, it is reasonable and appropriate to fuse different views for clustering. It is known that multiple views come from the same object. Hence, multi-view data contain not only the consistency but also the diversity across views. How to reasonably utilize the consistency and diversity to find the underlying clustering structure of multi-view data has become an important research topic.

To deal with the multi-view data, a natural idea is to concatenate these different feature vectors into a new vector and then adopt some existing single-view clustering methods to group the multi-view data. Although this idea is intuitive and simple to deal with multi-view data, it ignores the consistency and complementary information across these views. To address this problem, lots of multi-view clustering methods have been developed to obtain the good clustering performance. For background reading, the reader can refer to the surveys on multi-view clustering [8–10]. In this paper, we mainly focus on the multi-view subspace clustering, which has received extensive attention due to its advanced clustering performance and good mathematical interpretability.

Multi-view subspace clustering attempts to construct an ideal subspace representation to describe the multiple linear subspace structure, and the clustering results are then obtained by utilizing the spectral clustering for this obtained subspace representation. The mechanism for computing the subspace representation is based on the self-expressive reconstruction model, where each sample is reconstructed by entire samples. Hence, subspace representation yielded by the self-expressive reconstruction model can exploit the global information but may ignore the local information of multi-view data. Nevertheless, exploring the local structure has been confirmed to improve the learning performance [11]. Moreover, most multi-view subspace clustering methods divide the learning subspace representation and clustering into two individual procedures, which ignores their communications.

To address the above-mentioned issues, in this paper, we propose a novel subspace clustering method fusing local and global information for one-step multi-view subspace clustering (LGOMSC). The proposed method combines the procedures of constructing subspace representation, multi-view information fusion, and clustering into a unified optimization framework. In this framework, as shown in Figure 1, to exploit the local and global information of multi-view data, we integrate graph learning into the self-expressive reconstruction model by adaptively exploring the local structure information for the construction of subspace representation. To capture latent consistency information across views, the proposed method adopts a multi-view information fusion to learn the common subspace representation from these specific subspace representations of different views. Meanwhile, in graph learning, a rank constraint is applied to the Laplacian matrix yielded by the common subspace representation to directly produce the clustering result. Therefore, the proposed method is a one-step multi-view subspace clustering method. The main contributions of the work are summarized as follows:

- A novel one-step multi-view subspace clustering method is proposed, which fuses the subspace representation (exploring local and global information), multi-view information fusion (constructing a common subspace representation by fusing different view-specific subspace representations), and clustering (imposing rank constraint on the Laplacian matrix from the common subspace representation) as a unified optimization framework to realize the end-to-end clustering.
- We develop an effective optimization algorithm to solve the proposed method. Comprehensive experiments on nine popular multi-view data sets confirm the effectiveness and superiority of the proposed method by comparing it with some state-of-the-art multi-view clustering methods.

The rest of this paper is organized as follows. In Section 2, we review the related works. In Section 3, we introduce the formulation of the proposed LGOMSC method. In Section 4, we provide the optimization algorithm to solve the proposed LGOMSC method, including the analysis of the convergence and computation complexity. In Section 5, we conduct the experiments on nine popular multi-view data sets and analyze the experimental results. Finally, we provide the conclusion in Section 6.

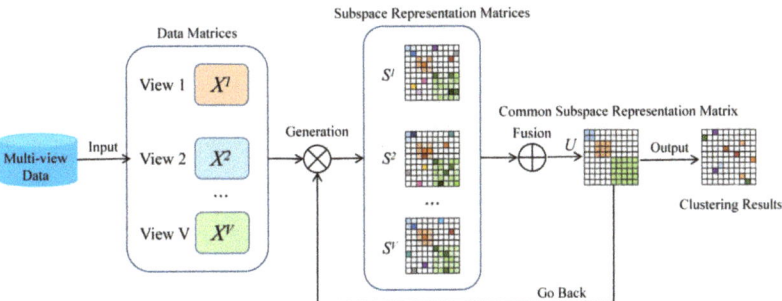

Figure 1. Framework of proposed LGOMSC.

2. Related Work

Multi-view clustering is a very powerful data analysis tools for unsupervised learning of data with heterogeneous features. In the past two decades, many multi-view clustering methods have been proposed to achieve robust clustering performance. In the following, we will briefly introduce several multi-view clustering methods from different perspectives.

Subspace-based methods have recently become the mainstay of multi-view clustering research, aiming to discover potential subspace structures across different views. For example, Gao et al. [12] propose a multi-view subspace clustering method that utilizes a common cluster structure to exploit the consistency information across multiple views. Cao et al. [13] propose a diversity-induced multi-view subspace clustering method that adopts the Hilbert–Schmidt independence criterion as the diversity term to explore the complementary information of multi-view data. Luo et al. [14] propose a multi-view subspace clustering method that simultaneously considers the consistency and specificity for learning the subspace representation. Wang et al. [15] propose a multi-view subspace clustering method that considers the complementarity of multi-view data by adopting a position-aware exclusivity term. Guo et al. [16] propose a rank consistency induced multi-view subspace clustering model that learns a consistent subspace structure. Brbić and Kopriva [17] propose a multi-view subspace clustering method that adopts an agreement term to ensure the consistency among these subspace representations. To capture the high-order correlations underlying multi-view data, the tensor technique is adopted to exploit the complementary information among different views. For example, Zhang et al. [18] propose a low-rank tensor constrained multi-view subspace clustering model that adopts a low-rank tensor constraint for the obtained subspace representations. Xie et al. [19] utilize the subspace representations of multiple views as a tensor data and then utilize the tensor-singular value decomposition on the rotated tensor to guarantee the consensus among different views. Zhang et al. [20] propose a tensorized multi-view subspace representation learning that adopts a low-rank constraint model for the subspace representation tensor. Yin et al. [21] propose a multi-view subspace clustering model by organizing the multi-view data as tensorial data, and the tensorial data can be represented by a t-linear combination with sparse and low-rank penalty. Recently, researchers considered partition-level multi-view information fusion and proposed a partition-based clustering model to construct joint optimization of multi-view subspace clustering. For example, Kang et al. [22] propose a unified multi-view subspace clustering model that implements the graph construction, the generation of basic partitions, and the fusion of consensus clustering in an interactive way. Lv et al. [23] propose a partition fusion-based multi-view subspace clustering method that utilizes the different partitions to find a shared partition. Zhang et al. [24] develop a consensus one-step multi-view subspace clustering method that fuses the subspace representation learning, partition learning, and clustering into a whole to iteratively optimize. Kang et al. [25] propose to integrate multi-view information in the partition space and obtain clustering results by assigning each partition with a respective rotation matrix. Furthermore, each view is assigned a weight to consider the differences in the clustering capacity of the views. The anchor-based model is proposed to fit for the large-scale multi-view data. Kang et al. [26] propose a large-scale multi-view subspace clustering method by integrating the anchor graphs from different views for spectral clustering. Wang et al. [27] propose a fast parameter-free multi-view subspace clustering by adaptively learning the anchors and graph structure. Sun et al. [28] propose to combine anchor learning and graph construction into a unified optimization framework, allowing the learned anchors to represent the actual latent data distribution more accurately, leading to a more discriminative clustering structure.

Matrix factorization-based methods refer to obtaining consistent latent representations through matrix factorization. Specifically, a given data matrix can be represented by the product of two or more low-dimensional matrices. Liu et al. [29] extended the traditional single-view non-negative matrix factorization algorithm to multi-view application scenarios and proposed a multi-view clustering algorithm based on non-negative matrix factorization. Guo et al. [30] propose to exploit group sparsity inducing norm in a matrix factorization framework to learn shared sparse subspace representations. Recently, Wang et al. [31] proposed a diversity non-negative matrix factorization multi-view clustering method by introducing a new diversity term to increase the diversity among multi-view representations and linearize the running time. Nie et al. [32] propose a new joint clustering method named Fast Multi-view Matrix Tri-Factorization to reduce the information loss in the matrix factorization process, while reducing the computational complexity and improving the operational efficiency. Liu et al. [33] propose a novel multi-view matrix factorization-based clustering method, which proposes to consider the higher-order relationships among features using an optimal graph regularization strategy and introduces the Hilbert–Schmidt independence criterion (HSIC) to fully explore the complementary information in different views. In addition, researchers have extended matrix factorization from the perspective of intact space learning [34]. For example, Zhang et al. [35] propose a latent multi-view subspace clustering that utilizes the latent representation for subspace clustering. Li et al. [36] propose a flexible multi-view representation learning that utilizes the kernel dependence measure to obtain a latent representation from different views for subspace clustering. Xie et al. [37] propose a multi-view subspace clustering method that fuses graph learning, latent representation, and clustering into a unified optimization framework.

Graph-based methods provides an effective way to solve the nonlinearly separated problems. For example, Tang et al. [38] propose a fusion process using linked matrix factorization to fuse the graph matrices corresponding to all views with multiple sources of information. Nie et al. [39] propose a multi-view graph clustering method based on the idea of manifold learning that can perform local structure learning and multi-view clustering at the same time and can also adaptively learn the weights corresponding to each view. Meanwhile, Nie et al. [40] propose an automatic weighting method to fuse a series of view-specific low-quality graphs into a high-quality unified graph, while extending the Laplacian rank approach to multi-view learning. Similarly, Zhan et al. [41] further design a notable clustering method based on twostep multiple graph fusion strategy. Recently, Zhan et al. [42] proposed a method to learn a consensus graph matrix by all views by minimizing disagreement between different views and constraining the rank of the Laplacian matrix. Wang et al. [43] propose another graph-based multi-view clustering method that automatically fuses multiple graph matrices to generate a unified graph matrix. The learned unified graph matrix can help the graph matrices of all views and gives the clustering indicator matrix. Recently, Zhao et al. [44] proposed to minimize the divergence between graphs using tensor Schatten p-norm regularization and integrate the tensor Schatten p-norm regularization and the manifold learning regularization into a unified framework to learn a shared common graph.

Although most of existing multi-view subspace clustering methods have achieved good clustering performance, they still have some limitations. First, the subspace representation generated by the self-expression reconstruction model usually ignores the local structure of the data set. Second, most multi-view subspace clustering methods usually divide the subspace representation learning process and the subsequent clustering task into two separate processes, ignoring the interactions between them. To address these issues, in this paper we propose an LGOMSC method that considers adding graph learning to explore local information adaptively for obtaining subspace representations. Moreover, LGOMSC performs multi-view information fusion directly on the subspace representation and introduces rank constraints on the Laplacian matrix of the common subspace representation matrix, which helps to naturally partition the data points into the desired number of clusters. Our approach integrates similarity learning, multi-view information fusion and clustering as a unified framework to achieve multi-view clustering in an end-to-end manner.

Duan et al. [45] propose a multi-view subspace clustering (MVSCLG) that also utilizes the local and global information to achieve the end-to-end clustering. The main differences between MVSCLG and LGOMSC include: (1) to explore the consistency between different views, MVSCLG adopts the spectral matrix fusion, but LGOMSC adopts the graph matrix fusion; (2) to achieve end-to-end clustering, MVSCLG adopts a rotation matrix to map the common spectral matrix to the final cluster label matrix, but LGOMSC adopts a rank constraint on the common Laplacian matrix to directly achieve clustering. Moreover, compared with MVSCLG, LGOMSC has some advantages. Firstly, MVSCLG involves many singular value decomposition procedures and contains many variables, which leads to longer running times and more memory usage than LGOMSC. Secondly, LGOMSC contains fewer hyperparameters than MVSCLG, which is more suitable for practical applications.

3. Proposed Method

3.1. Notations

For convenience, we list important mathematic notations that are used throughout the paper in Table 1. Matrices are represented in bold uppercase, while vectors are represented in bold lowercase.

Table 1. Notations and abbreviations.

Notation	Definition
$\mathbf{I}_n$	$n \times n$ Identity matrix
$\mathbf{1}$	All-ones column vector
n	Number of data sample
c	Number of clusters
V	Number of views
d_v	Feature dimension of the v-th view
$\mathbf{X}^v \in \mathbb{R}^{d_v \times n}$	Feature matrix of the v-th view
$x_{i,:}, x_j, x_{ij}$	Represented as the i − th row, j − th column, and ij − th element of matrix X, respectively
$\mathbf{X}^T$	The transpose of a matrix
$\mathbf{S}^v \in \mathbb{R}^{n \times n}$	Subspace representation matrix of the v-th view
$\mathbf{U} \in \mathbb{R}^{n \times n}$	Common subspace representation matrix
$\|\cdot\|_F$	The Frobenius norm
$Tr(\cdot)$	Trace operator of a matrix
$diag(\cdot)$	Vector of the diagonal elements of a matrix
$rank(\cdot)$	The rank of a matrix

3.2. Formulation

In this section, we provide the detailed modeling process of the proposed LGOMSC method. For a multi-view data set with V views, let $\mathbf{X}^1, \ldots, \mathbf{X}^V$ be the data matrices of the V views and $\mathbf{X}^v = \{x_1^v, \ldots, x_n^v\} \in \mathbb{R}^{d_v \times n}$ be the v-th view data, where d_v is the dimensionality of the v-th view, and n is the number of data points. Since each view contains view-specific information about the object, we respectively compute the view-specific subspace representation of each view to capture the diversity across views. The objective function of the self-expression model for the multi-view data can be formulated as:

$$\min_{S^v} \sum_{v=1}^{V} \|X^v - X^v S^v\|_F^2 + \lambda_1 \sum_{v=1}^{V} \|S^v\|_F^2$$
$$s.t. 0 \leq s_{ij}^v \leq 1, (S^v)^T \mathbf{1} = \mathbf{1}, diag(S^v) = 0 \quad (1)$$

where $S^v = \{s_1^v, s_2^v, \ldots, s_n^v\} \in \mathbb{R}^{n \times n}$ is the subspace representation matrix of the v-th view, s_{ij}^v is the j-th element of s_i^v, $\mathbf{1}$ denotes a column vector with all entries of one, $diag(\cdot)$ denotes a vector of the diagonal elements of a matrix.

Since Model (1) adopts the entire data set to linearly reconstruct each data sample, the subspace representation matrix S^v captures the global information of the v-th view data. However, this subspace representation obtained by Model (1) ignores the local structure to construct the subspace representations. In other words, two closed data samples should have similar subspace representations. Hence, to exploit the local information of multi-view data, we integrate the graph learning into Model (1) to compute the subspace representation. Hence, the objection function can be formulated as:

$$\min_{S^v} \sum_{v=1}^{V} \|X^v - X^v S^v\|_F^2 + \lambda_1 \sum_{v=1}^{V} \|S^v\|_F^2 + \sum_{v=1}^{V} \sum_{i=1}^{n} \sum_{j=1}^{n} \|x_i^v - x_j^v\|_2^2 s_{ij}^v$$
$$s.t. 0 \leq s_{ij}^v \leq 1, (S^v)^T \mathbf{1} = \mathbf{1}, diag(S^v) = 0 \quad (2)$$

Since multi-view data come from the same object, they should have latent consistency. To characterize this consistency, we adopt a multi-view information fusion term to obtain a common subspace representation matrix $U \in \mathbb{R}^{n \times n}$ from the subspace representation matrices $\{S^1, S^2 \ldots S^V\}$. This term can be represented as:

$$\min_{U} \sum_{v=1}^{V} \|S^v - U\|_F^2$$
$$s.t. U \geq 0, U^T \mathbf{1} = \mathbf{1} \quad (3)$$

Through minimizing Model (3), this common subspace representation matrix U can make these the subspace representation matrices $\{S^1, S^2 \ldots S^V\}$ to have latent consistency. Hence, we add this multi-view information fusion term into Model (2) as:

$$\min_{S^v, U, F} \sum_{v=1}^{V} \|X^v - X^v S^v\|_F^2 + \lambda_1 \sum_{v=1}^{V} \|S^v\|_F^2 + \sum_{v=1}^{V} \sum_{i=1}^{n} \sum_{j=1}^{n} \|x_i^v - x_j^v\|_2^2 s_{ij}^v$$
$$+ \sum_{v=1}^{V} \|S^v - U\|_F^2 \quad (4)$$
$$s.t. 0 \leq s_{ij}^v \leq 1, \mathbf{1}^T s_i^v = 1, diag(S^v) = 0, U \geq 0, U^T \mathbf{1} = \mathbf{1}$$

After obtaining the common subspace structure, we can get the affinity matrix $W = 1/2(U + U^T)$ and perform spectral clustering on such a subspace affinity matrix. However, the constructions of subspace representation and clustering are divided into two individual procedures, which ignore their interactions. To address this problem, we consider introducing a rank constraint [46] on the Laplacian matrix $L_U = D - W$, where the degree matrix D is defined as a diagonal matrix whose i-th diagonal element is $d_{ii} = \sum_{j=1}^{n} w_{ij}$. If the common subspace representation matrix U is non-negative, then the Laplacian matrix has the following theorem.

Theorem 1. *The number of connected components in the graph with U is equal to the multiplicity of zero eigenvalue of the Laplacian matrix L_U* [47].

According to Theorem 1, we consider making the number of zero eigenvalues of the Laplacian matrix L_U to be equal to the number of clustering clusters, i.e., $rank(L_U) = n - c$.

By adding the rank constraint $rank(L_U) = n - c$ into Model (4), the common subspace representation matrix U will have the ideal property. Therefore, we can directly obtain the cluster result from U without discretization.

However, it is difficult to directly solve the rank constraint $rank(L_U) = n - c$. It is known that $rank(L_U) = n - c$ is equivalent to $\sum_{i=1}^{c} \sigma_i(L_U) = 0$, where $\sigma_i(L_U)$ denotes the i-th smallest eigenvalues of L_U. Since L_U is positive semi-definite, $\sigma_i(L_U) \geq 0$. According to Ky Fan's Theorem [48], $\sum_{i=1}^{c} \sigma_i(L_U) = \min_{F^T F = I} Tr(F^T L_U F)$. Therefore, to hold $\sum_{i=1}^{c} \sigma_i(L_U) = 0$, the objective function of the proposed LGOMSC is formulated as:

$$\min_{S^v, U, F} \sum_{v=1}^{V} \|X^v - X^v S^v\|_F^2 + \lambda_1 \sum_{v=1}^{V} \|S^v\|_F^2 + \sum_{v=1}^{V} \sum_{i=1}^{n} \sum_{j=1}^{n} \|x_i^v - x_j^v\|_2^2 s_{ij}^v \\ + \sum_{v=1}^{V} \|S^v - U\|_F^2 + 2\lambda Tr(F^T L_U F) \quad (5)$$
$$s.t. \, 0 \leq s_{ij}^v \leq 1, \mathbf{1}^T s_i^v = 1, diag(S^v) = 0, U \geq 0, U^T \mathbf{1} = 1, F^T F = I_c$$

where $\lambda > 0$ is a parameter, $F = \{f_1, \ldots, f_c\} \in \mathbb{R}^{n \times c}$ is the embedding matrix, and $I_c \in \mathbb{R}^{c \times c}$ denotes the identity matrix.

In Model (5), when λ is large enough, the obtained common subspace representation U makes $\sum_{i=1}^{c} \sigma_i(L_U)$ zero. Hence, $rank(L_U) = n - c$ is satisfied. To effectively accelerate the optimization procedure, we determine the value of λ in a heuristic way. Moreover, in Model (5), we integrate subspace representation learning, multi-view information fusion, and clustering into a unified framework. The aim is to exploit the internal relationships of the three procedures to obtain a good clustering performance.

4. Optimization

There are several variables and constraints in the proposed method. To effectively solve these variables from LGOMSC, we developed an alternate optimization algorithm.

4.1. Update S^v

When U and F are fixed, the objective function about S^v becomes:

$$\min_{S^v} \|X^v - X^v S^v\|_F^2 + \lambda_1 \|S^v\|_F^2 + \sum_{i=1}^{n} \sum_{j=1}^{n} \|x_i^v - x_j^v\|_2^2 s_{ij}^v + \|S^v - U\|_F^2 \quad (6)$$
$$s.t. \, 0 \leq s_{ij}^v \leq 1, \mathbf{1}^T s_i^v = 1, diag(S^v) = 0$$

In this paper, we adopt a two-step approximation strategy [25] to optimize S^v. Firstly, we ignore the constraints in Model (6) to solve S^v as:

$$\min_{S^v} \|X^v - X^v S^v\|_F^2 + \lambda_1 \|S^v\|_F^2 + \sum_{i=1}^{n} \sum_{j=1}^{n} \|x_i^v - x_j^v\|_2^2 s_{ij}^v + \|S^v - U\|_F^2 \quad (7)$$

Through making the derivative of Model (7) of S^v as zero, we have:

$$\hat{S}^v = ((X^v)^T X^v + I_n + \lambda_1 I_n)^{-1} ((X^v)^T X^v + U - \frac{1}{2} B^v) \quad (8)$$

where $b_{ij}^v = \|x_i^v - x_j^v\|_2^2$ is the ij-th element of $B^v \in \mathbb{R}^{n \times n}$ and $I_n \in \mathbb{R}^{n \times n}$ denotes the identity matrix.

Secondly, through adding the constraints of S^v, the solution of S^v can be obtained by:

$$\min_{s_i^v} \|s_i^v - \hat{s}_i^v\|_2^2 \quad (9)$$
$$s.t. \, s_{ii}^v = 0, S_i^v \geq 0, \mathbf{1}^T S_i^v = 1$$

Model (9) is a constrained quadratic optimization problem, which can be effectively solved by the iterative algorithm in the work [49].

4.2. Update U

When S^v and F are fixed, the objective function about U is represented as:

$$\min_{U} \sum_{v=1}^{V} \sum_{i=1}^{n} \sum_{j=1}^{n} (s_{ij}^v - u_{ij})^2 + 2\lambda Tr(F^T L_u F) \qquad (10)$$
$$s.t. u_{ij} \geq 0, \mathbf{1}^T u_i = 1$$

where $u_i \in \mathbb{R}^{n \times 1}$ is a column vector, u_{ij} is the j-th element of u_i.

Noting that $Tr(F^T L_u F) = 1/2 \sum_{i=1}^{n} \sum_{j=1}^{n} \|f_{i,:} - f_{j,:}\|_2^2 u_{ij}$, we denote h_i be a vector with the j-th element $h_{ij} = \|f_{i,:} - f_{j,:}\|_2^2$. Through simple mathematical derivation, problem (10) can be rewritten as follows:

$$\min_{u_i} \sum_{v=1}^{V} \left\| u_i - s_i^v + \frac{\lambda}{2V} h_i \right\|_2^2 \qquad (11)$$
$$s.t. u_{ij} \geq 0, \mathbf{1}^T u_i = 1$$

We define $q^v = s_i^v - \frac{\lambda}{2V} h_i$, we can obtain:

$$\min_{u_i} \sum_{v=1}^{V} \|u_i - q^v\|_2^2 \qquad (12)$$
$$s.t. u_{ij} \geq 0, \mathbf{1}^T u_i = 1$$

Model (12) is effectively optimized by an iterative algorithm referring to the work [50].

4.3. Update F

When U and S^v are fixed, the objective function about F is represented as:

$$\min_{F} Tr(F^T L_u F) \qquad (13)$$
$$s.t. F^T F = \mathbf{I}_c$$

The optimal solution F yielded by Model (13) is formed by the c eigenvectors of L_u corresponding to the c smallest eigenvalues. Finally, the procedure for optimizing Model (5) is described in Algorithm 1.

4.4. Convergence Analysis

In this paper, we adopt an alternate updating algorithm (Algorithm 1) to solve the objective function in Model (5). Since λ is changed during the iteration to accelerate the procedure in the experiment, the objective function of Model (5) is varied during each iteration. Hence, it is difficult to guarantee convergence theoretically. However, in the experiments, the results show that Algorithm 1 for optimizing Model (5) has good convergence.

4.5. Computational Complexity Analysis

According to the optimization process described in Algorithm 1, the computational complexity of LGOMSC consists of updating S^v, U, and F. First, the update of S^v takes $O(n^3 + Vn^2)$. Second, the update of U needs $O(n^2)$. Third, the update of F costs $O(n^3)$ for compute eigenvectors of the Laplacian matrix. Overall, the complexity of Algorithm 1 is $O((2n^3 + (V+1)n^2))T)$, where T is the total number of iterations.

Algorithm 1: Optimization Algorithm for Model (5)

Input: given V view data $X^1, \ldots, X^V$ with $X^v \in \mathbb{R}^{d_v \times n}$, the number of clusters c, parameters λ_1 and λ.
Output: U with exact c connected components.
Initialize S^v by the optimization problem [51]:
$$\min_{S^v} \sum_{i=1}^n \sum_{j=1}^n \left\| x_i^v - x_j^v \right\|_2^2 s_{ij}^v + \lambda_1 \|S^v\|_F^2$$
$$s.t. s_{ii}^v = 0, 0 \leq s_{ij}^v \leq 1, \mathbf{1}^\mathrm{T} s_i^v = 1$$
Initialize U and F based on $s^1, s^2 \ldots s^V$.
Repeat
Update S^v by model (9).
Update U by model (12).
Update F by model (13).
Until $\sum_i^c \sigma_i(L_U) < 1.0e^{-13}$ and $\sum_i^{c+1} \sigma_i(L_U) > 1.0e^{-13}$

5. Experiments

In this section, we use nine popular multi-view data sets to assess the clustering performance of LGOMSC.

5.1. Data Set Descriptions

In the experiments, the nine public multi-view benchmark data sets were 3Sources, 100leaves, BBC, Caltech101, COIL-20, NottingHill, Webkb, Cornell, and Wikipedia Articles. All the data sets are summarized in Table 2.

Table 2. Summary of nine multi-view benchmark data sets (d_v denotes the dimensionality of the v-th view).

Data Set	Point	Class	View	d1	d2	d3	d4	d5	d6
3sources	169	6	3	3560	3631	3068			
100leaves	1600	100	3	64	64	64			
BBC	685	5	4	4659	4633	4665	4684		
Caltech101	1474	7	6	48	40	254	1984	512	928
COIL-20	1440	20	3	1024	3304	6750			
NottingHill	4660	5	3	6750	3304	2000			
Webkb	1051	2	2	1840	3000				
Cornell	195	5	2	195	1703				
Wikipedia	693	10	2	128	10				

5.2. Experimental Setting

In this paper, LGOMSC is compared with twelve relevant methods including

- **FeatConcate:** Concatenate the features of different views into a vector and utilize k-means to acquire the clustering result. It is regarded as the baseline method.
- **Co-reg_c** and **Co-reg_p**: Centroid-based co-regularization [52] and pairwise co-regularization [52].
- **LMSC:** Latent multi-view subspace clustering [35].
- **FMR:** Flexible multi-view representation learning for subspace clustering [36].
- **MLRSSC:** Multi-view low-rank sparse subspace clustering [17].
- **RMKMC:** Robust multi-view k-means clustering [53].
- **mPAC:** Multiple partitions aligned clustering [25].
- **LMVSC:** Large-scale multi-view subspace clustering in linear time [26].
- **PMSC:** Partition level multi-view subspace clustering [22].
- **COMVSC:** Consensus one-step multi-view subspace clustering [24].
- **GMC:** Graph-based multi-view clustering [43].
- **MVSCLG:** Multi-view subspace clustering with local and global information [45].

We conduct these comparison methods from corresponding open-source codes and follow their papers to set the optimal parameters. LGOMSC contains two parameters λ

and λ_1. For λ, it is first set with a proper value in a heuristic way, then in each iteration, λ is divided by two if the number of zero eigenvalues of L_U is greater than c and multiplied by two if it is smaller than c. For λ_1, we adopt the grid search method to empirically choose it in the range of $\{1, 10, 20, \ldots, 100\}$. In this paper, we use L2 norm for data normalization, i.e., $(x_i^v)' = \frac{x_i^v}{\|x_i^v\|_2}$.

5.3. Experiment Results and Analysis

In the experiments, four popular metrics including accuracy (ACC), normalized mutual information (NMI), F-score and adjusted Rand index (ARI) are utilized to assess the clustering result. These evaluation metrics reflect different natures of the clustering results, thus providing a comprehensive analysis from multiple perspectives. For all four of the evaluation metrics, higher values indicate better results. The comparison results are shown in Tables 3–6. The best result is highlighted in red font, and the second best result is reported in blue font.

Table 3. The clustering performance comparison in terms of ACC on nine multi-view data sets.

ACC (%)	3sources	100leaves	BBC	Caltech101	COIL-20	NottingHill	Webkb	Cornell	Wikipedia
FeatConcate	65.09	71.00	61.46	54.27	67.50	91.93	94.77	43.08	57.72
Co-reg_c	69.17	78.53	34.74	42.00	70.36	74.77	80.42	38.41	38.59
Co-reg_p	66.18	75.60	35.99	42.14	72.42	72.14	83.24	36.26	20.70
LMSC	71.60	77.00	86.28	53.80	75.35	83.78	95.34	43.59	56.85
FMR	70.41	69.25	85.11	47.69	72.01	82.85	93.24	43.08	56.85
MLRSSC	34.88	1.44	33.14	54.21	5.07	30.11	78.02	43.08	15.22
RMKMC	54.44	1.00	60.44	54.14	61.60	75.43	94.01	43.59	61.04
mPAC	76.92	47.06	58.10	59.36	73.40	90.28	78.12	45.64	56.71
LMVSC	63.31	71.06	84.38	56.72	74.17	89.25	95.62	55.90	59.16
PMSC	63.85	22.46	34.45	44.45	49.09	70.21	78.02	45.44	19.70
COMVSC	65.09	70.88	69.49	77.54	77.64	81.70	82.87	55.38	60.46
GMC	65.09	86.38	69.05	65.74	87.57	31.24	77.64	38.97	31.89
Ours	83.43	94.31	88.47	79.85	92.08	100.00	98.38	64.62	62.63

Table 4. The clustering performance comparison in terms of NMI on nine multi-view data sets.

NMIn (%)	3sources	100leaves	BBC	Caltech101	COIL-20	NottingHill	Webkb	Cornell	Wikipedia
FeatConcate	56.53	87.64	60.63	56.18	79.15	86.66	66.18	19.02	54.04
Co-reg_c	55.03	92.04	13.38	43.68	81.69	69.79	8.79	12.60	26.22
Co-reg_p	50.85	90.04	6.59	43.59	82.17	67.73	18.91	11.57	7.54
LMSC	69.18	89.22	65.70	51.85	84.54	78.57	70.36	18.89	52.60
FMR	57.34	85.44	66.04	46.77	78.53	66.38	59.59	21.44	51.81
MLRSSC	5.75	13.33	1.03	2.11	2.66	0.23	0.08	4.86	2.31
RMKMC	40.23	0.00	54.38	63.16	79.06	75.28	63.81	27.99	55.09
mPAC	64.13	73.96	47.41	50.56	85.86	83.14	16.79	15.48	47.53
LMVSC	59.35	87.49	69.45	55.70	82.24	82.33	69.41	27.84	52.81
PMSC	48.75	64.02	4.19	21.38	70.31	65.73	7.95	10.42	6.47
COMVSC	50.69	87.19	57.67	52.70	87.50	75.16	23.73	24.51	53.58
GMC	53.73	95.37	55.62	53.77	96.31	9.23	0.17	15.90	29.98
Ours	77.74	97.43	75.20	54.23	97.44	100.00	84.84	39.40	56.06

Table 5. The clustering performance comparison in terms of F-score on nine multi-view data sets.

F-score (%)	3sources	100leaves	BBC	Caltech101	COIL-20	NottingHill	Webkb	Cornell	Wikipedia
FeatConcate	66.61	63.54	60.32	56.91	61.05	88.77	92.78	33.27	50.34
Co-reg_c	69.04	74.46	37.00	42.83	68.20	72.38	79.66	32.06	26.96
Co-reg_p	66.01	69.67	37.74	42.34	69.67	69.08	81.71	31.12	13.00
LMSC	65.58	68.72	76.81	53.58	71.33	83.04	93.02	42.59	48.96
FMR	63.89	59.07	76.55	47.24	66.58	72.19	90.08	34.50	48.22
MLRSSC	37.49	1.97	37.88	55.93	9.42	36.03	79.27	42.88	19.57
RMKMC	47.89	1.86	57.46	55.64	58.84	72.58	91.87	35.94	51.85
mPAC	76.93	36.61	59.10	61.19	68.73	88.17	79.18	43.34	45.81
LMVSC	55.85	61.67	78.44	51.90	69.29	85.36	93.83	49.57	50.13
PMSC	57.05	18.77	38.48	43.46	49.56	66.84	79.27	43.55	19.67
COMVSC	56.74	62.95	62.72	71.83	71.79	80.11	81.47	46.18	51.02
GMC	52.88	66.27	63.05	61.54	85.31	36.94	78.67	37.06	23.00
Ours	78.34	88.91	82.83	75.72	91.86	100.00	97.58	57.32	52.48

Table 6. The clustering performance comparison in terms of ARI on nine multi-view data sets.

ARI (%)	3sources	100leaves	BBC	Caltech101	COIL-20	NottingHill	Webkb	Cornell	Wikipedia
FeatConcate	56.90	63.17	49.20	41.94	58.89	85.67	76.99	12.33	44.28
Co-reg_c	58.38	74.20	6.05	27.03	66.46	64.70	16.20	6.57	18.39
Co-reg_p	54.81	69.36	1.52	26.46	68.03	60.71	27.19	2.47	2.80
LMSC	56.72	68.41	69.70	37.80	69.74	78.20	80.79	10.18	42.89
FMR	53.70	58.66	69.13	31.86	64.83	64.68	72.84	12.01	42.12
MLRSSC	0.33	0.12	−0.04	0.93	0.03	0.04	−0.14	1.33	−0.09
RMKMC	34.27	0.00	45.61	41.21	56.45	64.69	73.61	15.73	45.97
mPAC	69.79	35.87	40.27	45.19	67.08	84.79	28.17	8.97	38.86
LMVSC	43.16	61.36	71.75	34.48	67.67	81.15	80.81	32.54	44.24
PMSC	36.70	17.47	1.17	17.58	46.33	57.26	0.18	10.44	1.96
COMVSC	36.94	62.55	47.36	50.17	70.28	74.50	25.52	17.68	44.69
GMC	32.87	65.86	47.46	39.64	84.45	2.21	1.02	3.47	6.07
Ours	70.63	88.79	77.06	57.54	91.42	100.00	92.92	35.08	46.13

Experiment Analysis. Through observing the clustering results from Tables 3–6, one can see that our proposed method can obtain the best results on all multi-view data sets except Caltech101. For the Caltech101 data set, LGOMSC is lower than FeatConcate and RMKMC in terms of NMI. However, in terms of ACC, our proposed method exceeds the second best results on the data sets including 3sources, 100leaves, BBC, Caltech101, COIL-20, NottingHill, Webkb, Cornell, and Wikipedia by 6.51%, 7.93%, 2.19%, 2.31%, 4.51%, 8.07%, 2.76%, 0.72% and 1.59%, respectively. For NMI, our proposed method is 8.93% lower than the best method, RMKMC, on the Caltech101 data set. For F-score, our proposed method exceeds the second best method by 1.41%, 14.45%, 4.39%, 3.89%, 6.55%, 11.23%, 3.75%, 7.75% and 0.63% for the corresponding data sets, respectively. In terms of ARI, our proposed method exceeds the second best method by 0.84%, 14.59%, 5.31%, 7.37%, 6.97%, 14.33%, 12.11%, 2.54% and 0.16% for the corresponding data set, respectively. The above results demonstrate the effectiveness and superiority of the proposed method. Hence, our LGOMSC method is a valuable multi-view subspace clustering method.

From the results in these tables, one can see that the baseline method (i.e., FeatConcate) sometimes exhibits comparable performance to the multi-view subspace method and even exceeds some multi-view clustering methods. However, in most cases, this baseline method still has a big gap in comparison with multi-view clustering methods. It confirms that multi-view clustering methods that consider the consistency or complementary information can obtain a good multi-view clustering performance. However, in some data sets, e.g., 3sources and 100leaves, the multi-view k-means method RMKMC, produces even worse results than FeatConcate. This phenomenon has been observed by some previous researchers [18,54].

Multi-view subspace clustering based on intact space learning methods like LMSC and FMR performs clustering on the latent representation space. However, there is a large gap between these methods and our approach, probably because they separate the representation learning and clustering processes, leading to suboptimal clustering results.

Compared with similar multi-view information fusion methods like GMC and LMVSC, our approach achieves a more impressive performance. This is mainly because we fuse graph learning into the self-expression model to jointly explore local and global structural information in the data. More information is used to serve the clustering task and therefore better performance is obtained.

Compared with the partition-based models to construct multi-view subspace clustering for joint optimization methods like COMVSC, mPAC, and PMSC, our approach achieves more impressive performance. This is mainly because we make the clustering structure of the multi-view data revealed while generating the common subspace representation under the rank constraint of Laplacian matrix. Thus, the end-to-end clustering approach facilitates a better clustering performance.

Compared with the MVSCLG method, the overall results from the nine data sets show that our method achieves the best results on all evaluation criteria, except for the Caltech101, Cornell and Wikipedia data sets. It states that using rank constraint can obtain an ideal graph matrix fitting for direct clustering. Our method also has superiorities in terms of running time and memory usage, which will be discussed in the next subsection. Thus, our method is more effective than MVSCLG.

Compared with these state-of-the-art clustering methods, the proposed method can achieve a more impressive clustering performance. The main reason is that it considers the local and global information from the original multi-view data to learn the subspace representation. Moreover, the proposed LGOMSC is an end-to-end model, which fuses the construction of subspace representation, multi-view information fusion, and clustering into a seamless whole. The purpose for this is to dig into their potential correlations.

Statistical Analysis. To demonstrate the statistical properties of our proposed method, we conducted the Friedman test and Nemenyi post-hoc test.

The Friedman test assumes that all the k compared methods hold the same performance on H data sets. Specifically, this model performance evaluation consists of the following two main steps. In the first step, first, sort all methods on each data set from high to low according to the clustering performance index and assign corresponding ordinal values (e.g., 1, 2, ...), and then calculate each method on all data sets average rank. In particular, the ordinal values are averaged if the performance of the two methods is the same. Finally, Γ_{χ^2} and Γ_F are calculated, and their mathematical expressions are as follows:

$$\Gamma_F = \frac{(H-1)\Gamma_{\chi^2}}{H(k-1) - \Gamma_{\chi^2}} \tag{14}$$

where $\Gamma_{\chi^2} = \frac{12H}{k(k+1)}(\sum_{i=1}^{k} r_i^2 - \frac{k(k+1)^2}{4})$, r_i represents the average rank of the i-th method over all data sets. Besides, Γ_F obeys the F-distribution with the degree of freedom $k-1$ and $(k-1)(H-1)$. The correctness of the hypothesis is eliminated by comparing the Γ_F with its corresponding threshold (the thresholds of the Friedman test can be calculated by $qf(1-\alpha, k-1, (k-1)(H-1))$ in R programming language). If the hypothesis is rejected, this indicates a significant difference in the performance of the compared methods. A further Nemenyi post-hoc test is then required to further distinguish between the methods.

In the second step, the Nemenyi post-hoc test calculates the critical distance by Equation (15) to reflect the difference between the average ordinal results of various methods.

$$CD = q_\alpha \sqrt{\frac{k(k+1)}{6H}} \tag{15}$$

where q_α can be calculate by $qtukey(1-\alpha, k, Inf)/Sqrt(2)$ in R programming language.

In our case, the number of compared methods, k, equals 13 and H equals 9. We sort the ACC, NMI, F-score, and ARI of the compared methods from high to low and obtain the average ranking of each method in terms of all data sets.

When $\alpha = 0.05$, the threshold for the Friedman test was 1.8544. According to Equation (14), the Γ_F values can be calculated for different clustering evaluation metrics (ACC, NMI, F-score, and ARI), which are 8.4348, 13.0826, 5.9892 and 11.3655, respectively. These Γ_F values are all greater than the threshold of the Friedman test, which rejects the hypothesis that all the methods being compared hold the same performance. Then, we perform the Nemenyi post-hoc test to further distinguish multiple methods. After obtaining the critical distance, CD = 6.2982, according to Equation (15), we can draw the Friedman test chart as Figure 2. For each method, the blue dot marks its average rank. The horizontal lines with the dot at the center indicate the critical distance, CD. If the lines do not have overlapping areas this indicates a significant difference in the comparison methods.

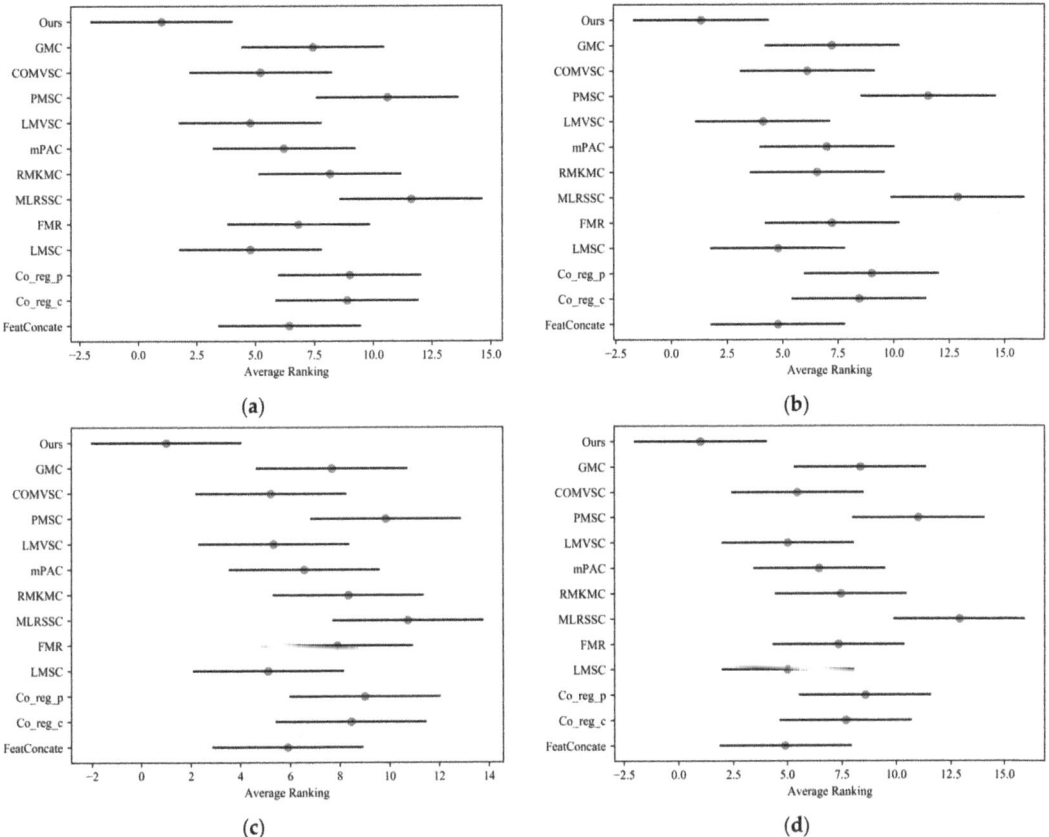

Figure 2. Friedman test charts. (**a**) Friedman test on ACC, (**b**) Friedman test on NMI, (**c**) Friedman test on F-score, (**d**) Friedman test on ARI.

From the figure, we can see that there are significant differences between the method in this paper and Co_reg_c, Co_reg_p, MLRSSC, and PMSC, and that the other methods do not differ from one another significantly. Compared with other methods, our method has the best average ranking regardless of the clustering evaluation index. In summary, our proposed method holds statistical advantages.

5.4. Running Time Analysis

We used MATLAB 2018b to run each clustering method independently and recorded the running time of each clustering method under one hyperparameter combination on each data set in Figure 3. From these results, one can see that FeatConcate is the fastest among most of multi-view data sets. To explore the consistency or complementary information of the multi-view data, these multi-view clustering methods generally need a relatively long running time to produce the final clustering. Our LGOMSC method is faster than Co_reg_c, Co_reg_p, LMSC, FMR, MLRSSC, RMKMC, mPAC, PMSC, and COMVSC on most data sets. On the Caltech101 data set, our LGOMSC method is 4.35 s slower than MLRSSC, and on the Wikipedia data set, our LGOMSC method is 1.89 s slower than MLRSSC and 0.21 s slower than RMKMC. On these data sets, LMVSC and GMC are the fastest, and they are more suitable for solving large-scale clustering problems. They are more concerned with efficiency rather than effectiveness. Therefore, the clustering performance is relatively poor. Although our proposed method is slower than the methods designed specifically for large-scale scenarios, our method is still comparable to LMVSC and GMC for the NottingHill data set. In Table 7, MVSCLG costs more running times than our method. In the NottingHill data set, LGOMSC costs 4744.49 s under one hyperparameter combination. However, since MVSCLG contains three hyperparameters, we need to conduct the grid search strategy to select the optimal one from 420 hyperparameter combinations, which may need 553.52 h. Hence, we ignore it.

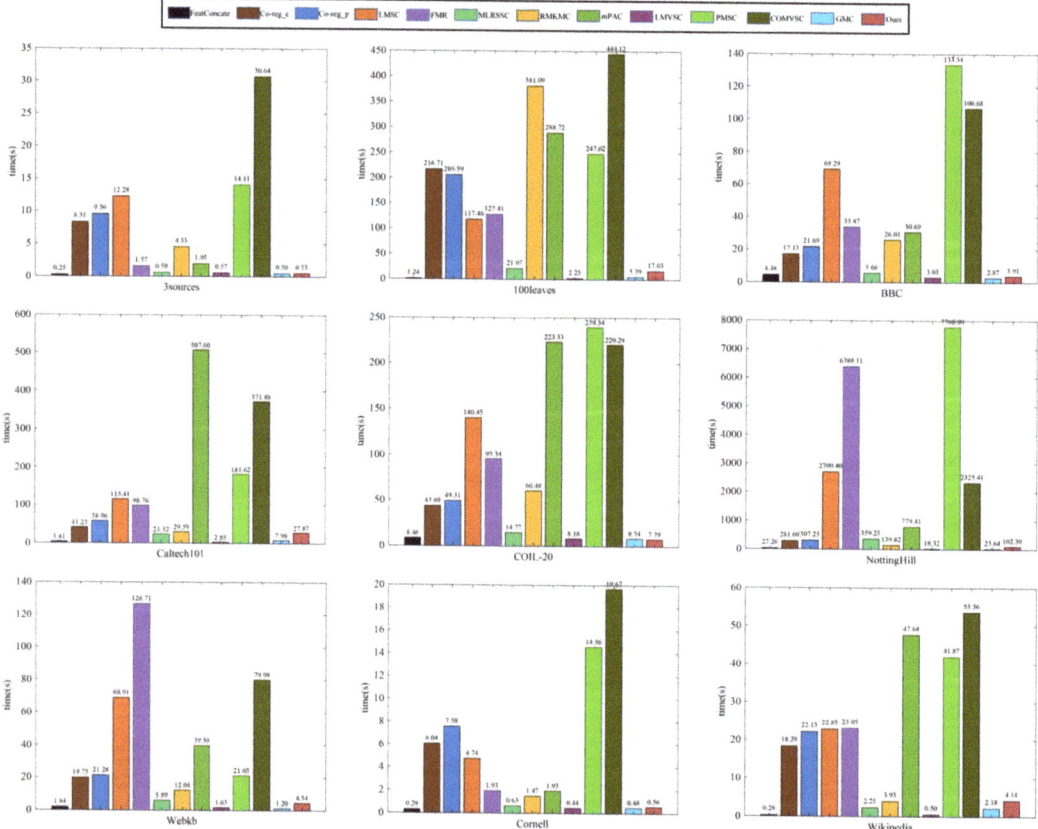

Figure 3. Running time of different methods on nine multi-view data sets.

Table 7. Comparing LGOMSC with MVSCLG on nine multi-view data sets.

Methods		3sources	100leaves	BBC	Caltech101	COIL-20	NottingHill	Webkb	Cornell	Wikipedia
MVSCLG	ACC	78.70	73.38	80.88	82.56	78.61	-	91.82	70.77	63.49
	NMI	65.13	88.54	58.14	58.06	90.28	-	56.49	49.88	59.09
	F-score	76.13	66.10	69.95	79.07	72.94	-	88.04	65.40	56.16
	ARI	68.10	65.74	61.18	61.76	71.34	-	67.89	51.35	49.69
	Time (s)	59.77	947.71	209.86	764.76	443.21	4744.49	163.05	40.27	107.10
Ours	ACC	83.43	94.31	88.47	79.85	92.08	100.00	98.38	64.62	62.63
	NMI	77.74	97.43	75.20	54.23	97.44	100.00	84.84	39.40	56.06
	F-score	78.34	88.91	82.83	75.72	91.86	100.00	97.58	57.32	52.48
	ARI	70.63	88.79	77.06	57.54	91.42	100.00	92.92	35.08	46.13
	Time (s)	0.53	17.03	3.91	27.87	7.79	102.3	4.54	0.56	4.15

"-" indicates that MVSCLG requires more than 553.52 h on the NottingHill data set.

5.5. Memory Usage Analysis

We use MATLAB 2018b to run each clustering method independently and recorded the memory usage of each clustering method. Specifically, we recorded the current matlab memory usage once before we ran the program. After the program has finished, we recorded the current matlab memory usage again. Subtracting the first reading from the second reading gives us the memory usage of the method. For example, on the 3sources data set, as can be seen in Figure 4, the memory usage of our method is small compared with the LMVSC, PMSC, COMVSC and MVSCLG methods. The memory usage of our method is slightly larger than the other remaining methods, but there is no significant difference between our method and the other remaining methods in terms of memory usage by an order of magnitude. Therefore, our method has an appropriate space complexity.

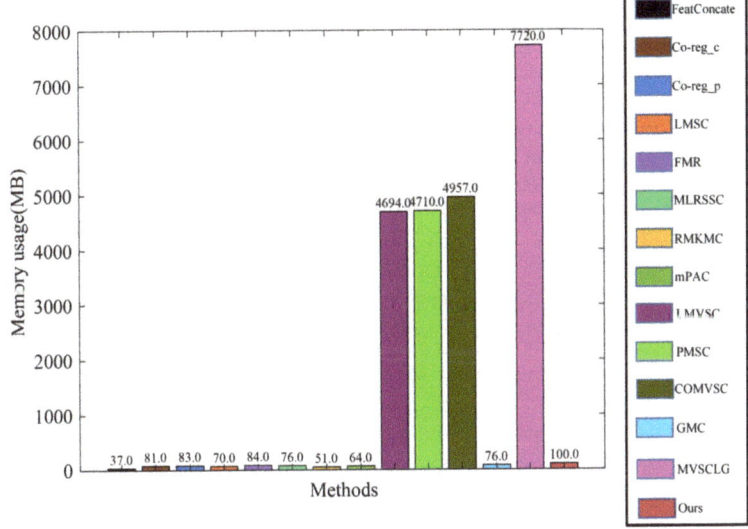

Figure 4. The memory usage representation of compared methods on 3sources.

5.6. Convergence Study

The objective function of LGOMSC has multiple variables and constraints. We have developed an effective iterative optimization algorithm to solve the proposed method. We conduct the convergence experiment in Figure 5, which provides the convergence curves of LGOMSC on nine multi-view data sets. The x-axis displays the number of iterations, and the y-axis displays the corresponding objective function value. From these results, one can see that the proposed method is well convergent and converges quickly. Within 10 iterations, the objective function value can converge to a stable value.

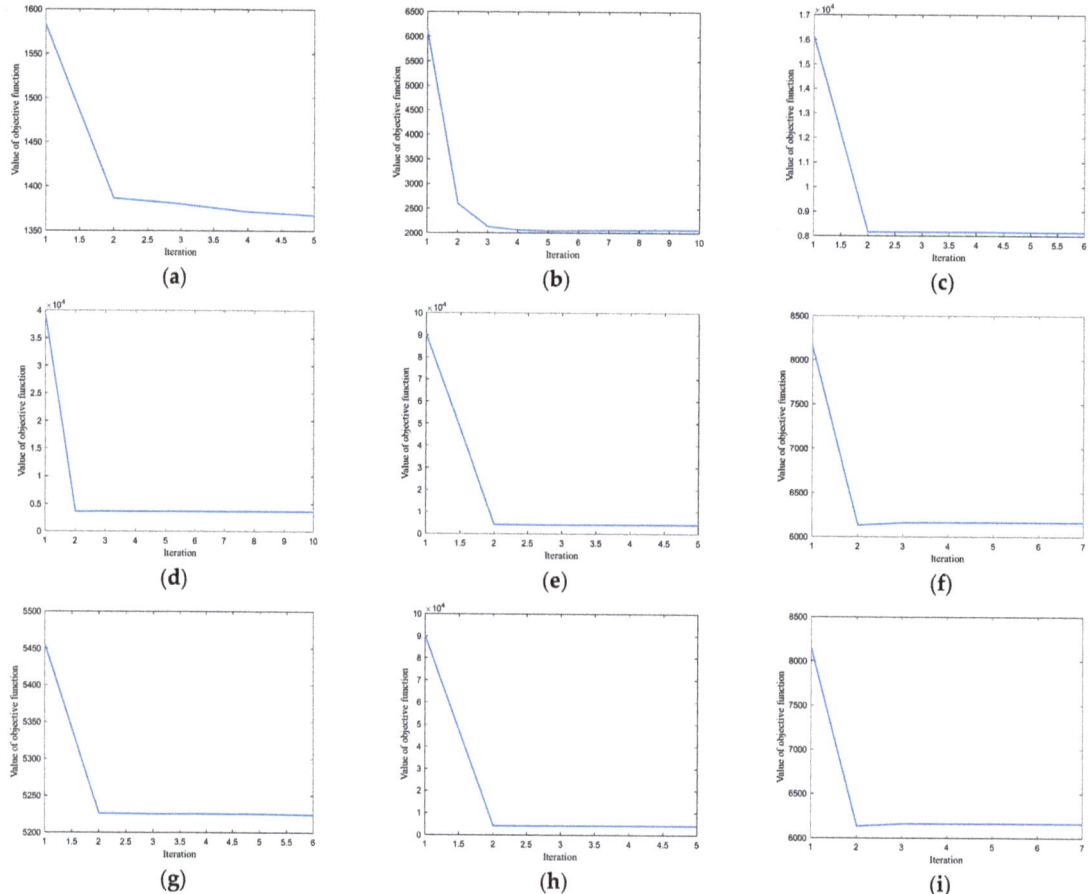

Figure 5. Convergence performance on nine multi-view data sets. (**a**) 3sources, (**b**) 100leaves, (**c**) BBC, (**d**) Caltech101, (**e**) COIL-20, (**f**) NottingHill, (**g**) Webkb, (**h**) Cornell, (**i**) Wikipedia.

5.7. Parameter Tuning

We conducted an experiment to analyze the hyperparameter λ_1 in this paper. In the experiment, we tuned λ_1 from a candidate set of $\{1, 10, 20, \ldots, 100\}$. As shown in Figure 6, we give the parameter tuning of LGOMSC on nine multi-view data sets and show the clustering performance under different values. One can see that the proposed method keeps a relatively robust clustering performance under a large range of λ_1. This helps us to easily select a proper parameter.

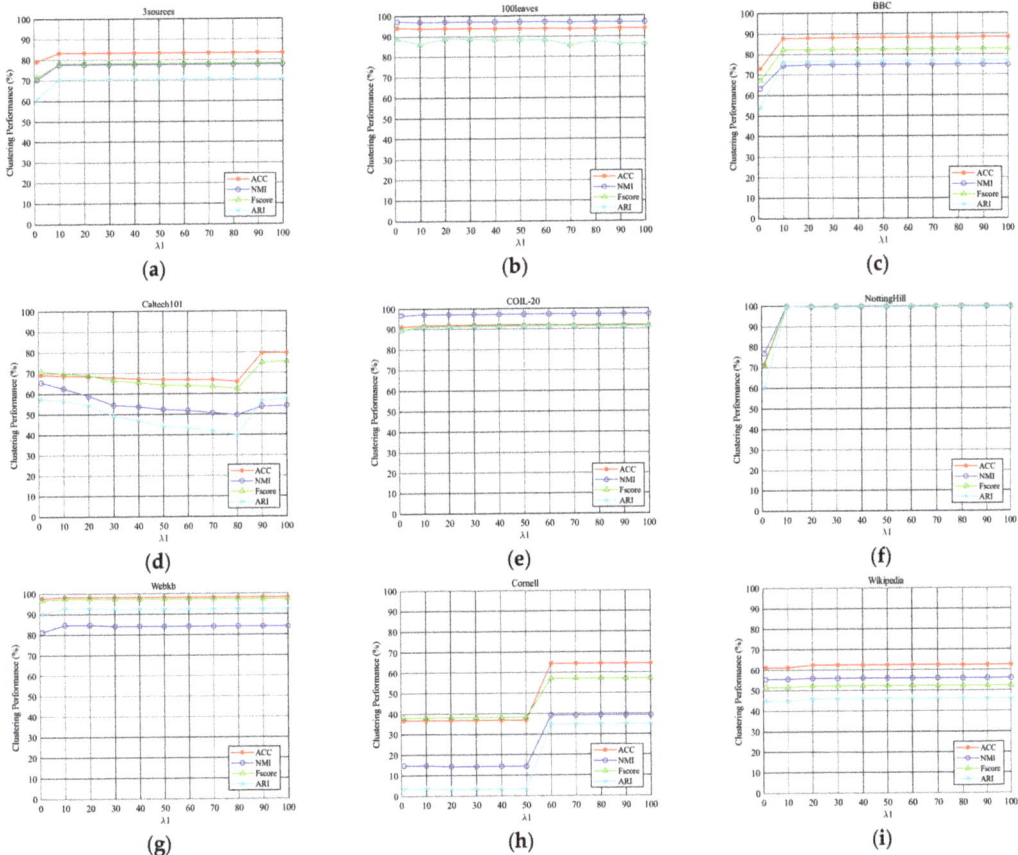

Figure 6. The parameter sensitivity of λ_1 on nine multi-view data sets. (**a**) 3sources, (**b**) 100leaves, (**c**) BBC, (**d**) Caltech101, (**e**) COIL-20, (**f**) NottingHill, (**g**) Webkb, (**h**) Cornell, (**i**) Wikipedia.

6. Conclusions

In this paper, we propose a novel one-step multi-view subspace clustering method, which integrates the self-expression model and graph learning to simultaneously exploit the local and global information of subspace representations from multi-view data. Moreover, to further exploit the hidden relationships between different steps to achieve an end-to-end clustering, our method integrates the subspace representation learning, multi-view information fusion, and clustering tasks into a joint framework. Experimental results on nine popular multi-view data sets confirm the effectiveness of our method by comparing with many baseline methods.

In the future, we have two directions to improve our method. First, since our method is a linear model, we will consider expanding our model to non-linear cases to deal with complex multi-view data. Second, for better fitting for large-scale multi-view subspace clustering, we will adopt anchor-based ideas to improve our method.

Author Contributions: Conceptualization, Y.D. and H.Y.; methodology, H.Y. and C.S.L.; software, Y.D.; experiment, validation and analysis, Y.D. and H.Y.; investigation, Y.D. and H.Y.; resources, H.Y.; data curation, Y.D.; writing—original draft preparation, Y.D. and H.Y.; writing—review and editing, Y.D., H.Y., C.S.L. and L.L.L. All authors have read and agreed to the published version of the manuscript.

Funding: This research was funded by National Natural Science Foundation of China under Grant 61903091; Guangdong Basic and Applied Basic Research Foundation (No. 2020A1515010801).

Institutional Review Board Statement: Not applicable.

Informed Consent Statement: Not applicable.

Data Availability Statement: The original contributions presented in the study are included in the article, further inquiries can be directed to the corresponding author.

Conflicts of Interest: The authors declare no conflict of interest.

References

1. Goldberg, D.; Holland, J. Genetic algorithms and machine learning. *Mach. Learn.* **1988**, *3*, 95–99. [CrossRef]
2. Liu, S.; Liang, X.; Liu, L.; Shen, X.; Yang, J.; Xu, C.; Lin, L.; Cao, X.; Yan, S. Matching-cnn meets knn: Quasi-parametric human parsing. In Proceedings of the IEEE Conference on Computer Vision and Pattern Recognition (CVPR), Boston, MA, USA, 7–12 June 2015; pp. 1419–1427.
3. Witten, I.; Frank, E. *Data Mining: Practical Machine Learning Tools and Techniques*; Morgan Kaufmann: San Mateo, CA, USA, 2005; pp. 35–40.
4. Berkhin, P. A survey of clustering data mining techniques. In *Grouping Multidimensional Data*; Kogan, J., Nicholas, C., Teboulle, M., Eds.; Springer: Berlin/Heidelberg, Germany, 2006; pp. 25–71.
5. Astolfi, D.; Pandit, R. Multivariate wind turbine power curve model based on data clustering and polynomial LASSO regression. *Appl. Sci.* **2021**, *12*, 72. [CrossRef]
6. Dinh, D.; Fujinami, T.; Huynh, V. Estimating the optimal number of clusters in categorical data clustering by silhouette coef-ficient. In Proceedings of the Twentieth International Symposium on Knowledge and Systems Sciences (ISKSS), Da Nang, China, 29 November–1 December 2019; pp. 1–17.
7. Liu, G.; Lin, Z.; Yan, S.; Sun, J.; Yu, Y.; Ma, Y. Robust recovery of subspace structures by low-rank representation. *IEEE Trans. Pattern Anal. Mach. Intell.* **2013**, *35*, 171–184. [CrossRef] [PubMed]
8. Chao, G.; Sun, S.; Bi, J. A survey on multiview clustering. *IEEE Trans. Artif. Intell.* **2021**, *2*, 146–168. [CrossRef]
9. Yang, Y.; Wang, H. Multi-view clustering: A survey. *Big Data Min. Anal.* **2018**, *1*, 83–107.
10. Fu, L.; Lin, P.; Vasilakos, A.; Wang, S. An overview of recent multi-view clustering. *Neurocomputing* **2020**, *402*, 148–161. [CrossRef]
11. Yan, S.; Xu, D.; Zhang, B.; Zhang, H.; Yang, Q.; Lin, S. Graph embedding and extensions: A general framework for dimensionality reduction. *IEEE Trans. Pattern Anal. Mach. Intell.* **2006**, *29*, 40–51. [CrossRef]
12. Gao, H.; Nie, F.; Li, X.; Huang, H. Multi-view subspace clustering. In Proceedings of the 2015 IEEE International Conference on Computer Vision (ICCV), Santiago, Chile, 7–13 December 2015; pp. 4238–4246.
13. Cao, X.; Zhang, C.; Fu, H.; Liu, S.; Zhang, H. Diversity-induced multi-view subspace clustering. In Proceedings of the IEEE Conference on Computer Vision and Pattern Recognition (CVPR), Boston, MA, USA, 7–12 June 2015; pp. 586–594.
14. Luo, S.; Zhang, C.; Zhang, W.; Cao, X. Consistent and specific multi-view subspace clustering. In Proceedings of the Thirty-Second AAAI Conference on Artificial Intelligence (AAAI-18), New Orleans, LA, USA, 2–7 February 2018; pp. 3730–3737.
15. Wang, X.; Guo, X.; Lei, Z.; Zhang, C.; Li, S. Exclusivity-consistency regularized multi-view subspace clustering. In Proceedings of the IEEE Conference on Computer Vision and Pattern Recognition (CVPR), Long Beach, CA, USA, 15–20 June 2019; pp. 923–931.
16. Guo, J.; Sun, Y.; Gao, J.; Hu, Y.; Yin, B. Rank Consistency induced multiview subspace clustering via low-rank matrix factorization. *IEEE Trans. Neural Netw. Learn. Syst.* **2021**, 1–14. [CrossRef]
17. Brbić, M.; Kopriva, I. Multi-view low-rank sparse subspace clustering. *Pattern Recognit.* **2018**, *73*, 247–258. [CrossRef]
18. Zhang, C.; Fu, H.; Liu, S.; Liu, G.; Cao, X. Low-rank tensor constrained multiview subspace clustering. In Proceedings of the 2015 IEEE International Conference on Computer Vision (ICCV), Santiago, Chile, 7–13 December 2015; pp. 1582–1590.
19. Xie, Y.; Tao, D.; Zhang, W.; Liu, Y.; Zhang, L.; Qu, Y. On unifying multi-view self-representations for clustering by tensor multi-rank minimization. *Int. J. Comput. Vis.* **2018**, *126*, 1157–1179. [CrossRef]
20. Zhang, C.; Fu, H.; Wang, J.; Li, W.; Cao, X.; Hu, Q. Tensorized multi-view subspace representation learning. *Int. J. Comput. Vis.* **2020**, *128*, 2344–2361. [CrossRef]
21. Yin, M.; Gao, J.; Xie, S.; Guo, Y. Multiview subspace clustering via tensorial t-product representation. *IEEE Trans. Neural Netw. Learn. Syst.* **2018**, *30*, 851–864. [CrossRef]
22. Kang, Z.; Zhao, X.; Peng, C.; Zhu, H.; Zhou, J.; Peng, X.; Chen, W.; Xu, Z. Partition level multiview subspace clustering. *Neural Netw.* **2020**, *122*, 279–288. [CrossRef]
23. Lv, J.; Kang, Z.; Wang, B.; Ji, L.; Xu, Z. Multi-view subspace clustering via partition fusion. *Inf. Sci.* **2021**, *560*, 410–423. [CrossRef]
24. Zhang, P.; Liu, X.; Xiong, J.; Zhou, S.; Zhao, W.; Zhu, E.; Cai, Z. Consensus one-step multi-view subspace clustering. *IEEE Trans. Knowl. Data Eng.* **2020**, 1–14. [CrossRef]
25. Kang, Z.; Guo, Z.; Huang, S.; Wang, S.; Chen, W.; Su, Y.; Xu, Z. Multiple partitions aligned clustering. In Proceedings of the Twenty-Eighth International Joint Conference on Artificial Intelligence (IJCAI), Macao, China, 10–16 August 2019; pp. 2701–2707.
26. Kang, Z.; Zhou, W.; Zhao, Z.; Shao, J.; Han, M.; Xu, Z. Large-scale multi-view subspace clustering in linear time. In Proceedings of the Thirty-Fourth AAAI Conference on Artificial Intelligence (AAAI-20), New York, NY, USA, 7–12 February 2020; pp. 4412–4419.

27. Wang, S.; Liu, X.; Zhu, X.; Zhang, P.; Zhang, Y.; Gao, F.; Zhu, E. Fast parameter-free multi-view subspace clustering with consensus anchor guidance. *IEEE Trans. Image Process.* **2022**, *31*, 556–568. [CrossRef]
28. Sun, M.; Zhang, P.; Wang, S.; Zhou, S.; Tu, W.; Liu, X.; Zhu, E.; Wang, C. Scalable multi-view subspace clustering with unified anchors. In Proceedings of the 29th ACM International Conference on Multimedia, Chengdu, China, 20–24 October 2021; pp. 3528–3536.
29. Liu, J.; Wang, C.; Gao, J.; Han, J. Multi-view clustering via joint nonnegative matrix factorization. In Proceedings of the 2013 SIAM International Conference on Data Mining (SDM), Austin, TX, USA, 2–4 May 2013; pp. 252–260.
30. Guo, Y. Convex subspace representation learning from multi-view data. In Proceedings of the Twenty-Seventh AAAI Conference on Artificial Intelligence Bellevue, Washington, DC, USA, 14–18 July 2013; pp. 387–393.
31. Wang, J.; Tian, F.; Yu, H.; Liu, C.; Zhan, K.; Wang, X. Diverse non-negative matrix factorization for multiview data representation. *IEEE Trans. Cybern.* **2018**, *48*, 2620–2632.
32. Nie, F.; Shi, S.; Li, X. Auto-weighted multi-view co-clustering via fast matrix factorization. *Pattern Recognit.* **2020**, *102*, 107207. [CrossRef]
33. Liu, X.; Song, P.; Sheng, C.; Zhang, W. Robust multi-view non-negative matrix factorization for clustering. *Digit. Signal Process.* **2022**, *123*, 103447. [CrossRef]
34. Xu, C.; Tao, D.; Xu, C. Multi-view intact space learning. *IEEE Trans. Pattern Anal. Mach. Intell.* **2015**, *37*, 2531–2544. [CrossRef]
35. Zhang, C.; Hu, Q.; Fu, H.; Zhu, P.; Cao, X. Latent multi-view subspace clustering. In Proceedings of the IEEE Conference on Computer Vision and Pattern Recognition (CVPR), Honolulu, HI, USA, 21–26 July 2017; pp. 4279–4287.
36. Li, R.; Zhang, C.; Hu, Q.; Zhu, P.; Wang, Z. Flexible multi-view representation learning for subspace clustering. In Proceedings of the Twenty-Eighth International Joint Conference on Artificial Intelligence (IJCAI), Macao, China, 10–16 August 2019; pp. 2916–2922.
37. Xie, D.; Zhang, X.; Gao, Q.; Han, J.; Xiao, S.; Gao, X. Multiview clustering by joint latent representation and similarity learning. *IEEE Trans. Cybern.* **2019**, *50*, 4848–4854.
38. Tang, W.; Lu, Z.; Dhillon, I.S. Clustering with multiple graphs. In Proceedings of the 2009 Ninth IEEE International Conference on Data Mining, Miami, FL, USA, 6–9 December 2009; pp. 1016–1021.
39. Nie, F.; Cai, G.; Li, X. Multi-view clustering and semi-supervised classification with adaptive neighbours. In Proceedings of the Thirty-First AAAI Conference (AAAI-17), San Francisco, CA, USA, 4–9 February 2017; pp. 2408–2414.
40. Nie, F.; Li, J.; Li, X. Self-weighted multiview clustering with multiple graphs. In Proceedings of the Twenty-Sixth International Joint Conference on Artificial Intelligence, Melbourne, Australia, 19–25 August 2017; pp. 2564–2570.
41. Zhan, K.; Zhang, C.; Guan, J.; Wang, J. Graph learning for multiview clustering. *IEEE Trans. Cybern.* **2017**, *48*, 2887–2895. [CrossRef]
42. Zhan, K.; Nie, F.; Wang, J.; Yang, Y. Multiview consensus graph clustering. *IEEE Trans. Image Process.* **2018**, *28*, 1261–1270. [CrossRef]
43. Wang, H.; Yang, Y.; Liu, B. GMC: Graph-based multi-view clustering. *IEEE Trans. Knowl. Data Eng.* **2019**, *32*, 1116–1129. [CrossRef]
44. Zhao, Y.; Yun, Y.; Zhang, X.; Li, Q.; Gao, Q. Multi-view spectral clustering with adaptive graph learning and tensor schatten p-norm. *Neurocomputing* **2022**, *468*, 257–264. [CrossRef]
45. Duan, Y.; Yuan, H.; Lai, L.; He, B. Multi-view subspace clustering with local and global information. In Proceedings of the International Conference on Wavelet Analysis and Pattern Recognition (ICWAPR), Adelaide, Australia, 3–5 December 2021; pp. 1–6.
46. Nie, F.; Wang, X.; Jordan, M.; Huang, H. The constrained Laplacian rank algorithm for graph-based clustering. In Proceedings of the Thirtieth AAAI Conference on Artificial Intelligence (AAAI-16), Phoenix, AZ, USA, 12–17 February 2016; pp. 1969–1976.
47. Mohar, B.; Alavi, Y.; Chartrand, G.; Oellermann, O. The Laplacian spectrum of graphs. *Graph Theory Comb. Appl.* **1991**, *2*, 871–898.
48. Fan, K. On a theorem of Weyl concerning eigenvalues of linear transformations I. *Proc. Natl. Acad. Sci. USA* **1949**, *35*, 652–655. [CrossRef]
49. Huang, J.; Nie, F.; Huang, H. A new simplex sparse learning model to measure data similarity for clustering. In Proceedings of the the Twenty-fourth International Joint Conference on Artificial Intelligence, Buenos Aires, Argentina, 25–31 July, 2015; pp. 3569–3575.
50. Wang, H.; Yang, Y.; Liu, B.; Fujita, H. A study of graph-based system for multi-view clustering. *Knowl.-Based Syst.* **2019**, *163*, 1009–1019. [CrossRef]
51. Nie, F.; Wang, X.; Huang, H. Clustering and projected clustering with adaptive neighbors. In Proceedings of the 20th ACM SIGKDD International Conference on Knowledge Discovery and Data Mining, New York, NY, USA, 24–27 August 2014; pp. 977–986.
52. Kumar, A.; Rai, P.; Daumé, H. Co-regularized multi-view spectral clustering. In Proceedings of the 25th Annual Conference on Neural Information Processing Systems, Granada, Spain, 12–15 December 2011; pp. 1413–1421.
53. Cai, X.; Nie, F.; Huang, H. Multi-view k-means clustering on big data. In Proceedings of the Twenty-Third International Joint Conference on Artificial Intelligence, Beijing, China, 3–9 August 2013; pp. 724–730.
54. Yang, Y.; Song, J.; Huang, Z.; Ma, Z.; Sebe, N.; Hauptmann, A.G. Multi-feature fusion via hierarchical regression for multimedia analysis. *IEEE Trans. Multimed.* **2013**, *15*, 572–581. [CrossRef]

Article

A Multi-Leak Identification Scheme Using Multi-Classification for Water Distribution Infrastructure

Yang Wei, Kim Fung Tsang *, Chung Kit Wu, Hao Wang and Yucheng Liu

Department of Electrical Engineering, City University of Hong Kong, Hong Kong 999077, China;
ywei22-c@my.cityu.edu.hk (Y.W.); chungkwu4-c@my.cityu.edu.hk (C.K.W.); hwang272-c@my.cityu.edu.hk (H.W.);
yucliu4-c@my.cityu.edu.hk (Y.L.)
* Correspondence: ee330015@cityu.edu.hk

Abstract: Water distribution infrastructure (WDI) is well-established and significantly improves living quality. Nonetheless, aging WDI has posed an awkward worldwide problem, wasting natural resources and leading to direct and indirect economic losses. The total losses due to leaks are valued at USD 7 billion per year. In this paper, a multi-classification multi-leak identification (MC-MLI) scheme is developed to combat the captioned problem. In the MC-MLI, a novel adaptive kernel (AK) scheme is developed to adapt to different WDI scenarios. The AK improves the overall identification capability by customizing a weighting vector into the extracted feature vector. Afterwards, a multi-classification (MC) scheme is designed to facilitate efficient adaptation to potentially hostile inhomogeneous WDI scenarios. The MC comprises multiple classifiers for customizing to different pipelines. Each classifier is characterized by the feature vector and corresponding weighting vector and weighting vector pertinent to system requirements, thus rendering the developed scheme strongly adaptive to ever-changing operating environments. Hence, the MC scheme facilitates low-cost, efficient, and accurate water leak detection and provides high practical value to the commercial market. Additionally, graph theory is utilized to model the realistic WDIs, and the experimental results verify that the developed MC-MLI achieves 96% accuracy, 96% sensitivity, and 95% specificity. The average detection time is about 5 s.

Keywords: adaptive kernel; multiple classifiers; graph theory; hydraulic model; multi-criteria decision-making

Citation: Wei, Y.; Tsang, K.F.; Wu, C.K.; Wang, H.; Liu, Y. A Multi-Leak Identification Scheme Using Multi-Classification for Water Distribution Infrastructure. *Appl. Sci.* **2022**, *12*, 2128. https://doi.org/10.3390/app12042128

Academic Editor: Bart Van der Bruggen

Received: 31 December 2021
Accepted: 14 February 2022
Published: 18 February 2022

Publisher's Note: MDPI stays neutral with regard to jurisdictional claims in published maps and institutional affiliations.

Copyright: © 2022 by the authors. Licensee MDPI, Basel, Switzerland. This article is an open access article distributed under the terms and conditions of the Creative Commons Attribution (CC BY) license (https://creativecommons.org/licenses/by/4.0/).

1. Introduction

Freshwater is one of the most important components to maintain human survival. The reliability and sustainability of water distribution infrastructure (WDI) are always the fundamental issues determining the livability of cities. The WDIs established in the past are now facing an aging problem. The number of leakages and bursts has been ever-escalating year-by-year. This results in not only economic losses but also the dissipation of natural resources. Half of the freshwater is wasted worldwide due to leaking pipelines [1]. In the United States, over 200,000 water bursts are recorded, yielding USD 2 billion of economic loss every year [2]. In Europe, WDIs in England, France, and Italy are leaking around 25% of freshwater annually [2]. It is also highlighted that 15% of the freshwater is wasted due to leaking pipelines in Hong Kong [3]. To combat this problem, current research on water leak detection technologies with high detection accuracy shows great effectiveness in protecting water resources and mitigating economic loss.

A variety of water leak detection approaches for WDIs were developed by researchers. One of the most common leak detection methods is manual listening through portable detection devices [4]. These devices are usually placed on the surface of the pipelines, and the leaking point is determined by listening for changes in sound or vibration produced by leaky pipes. Paper [5] proposed an effective detection scheme based on acoustic emission and pattern recognition. Paper [6] proved the leak detection efficiency via laboratory validation experiments with ground penetrating radar. Other detection approaches based on portable devices have been developed [7–9].

However, low penetration capability and ambient noise are two critical challenges for external portable detection. Moreover, lack of autonomy is another disadvantage for external portable detection. As a result, internal automatic leak detection methods are being explored. An automatic leak detection system integrates internal-sensor water network and automatic leak detection algorithms, which enable capacities in real-time motoring and early detection. In an internal-sensor water network, wireless sensors (e.g., flow meters, pressure sensors) are embedded into pipelines for remote data measurement and collection. Based on the data from internal sensors in the water distribution network, automatic leak detection algorithms have been developed to analyze variations in hydraulic behavior, indicate pipe status, and produce early alarms. Given their high efficiency, internal automatic leak detection systems have been reported extensively in the literature [10–13]. The paper [11] proposed a small-town water system combining water balance and minimum night flow approaches to detect water leakage efficiently. However, only the influence on water flow was considered. The detection accuracy can be improved by integrating other types of sensors (e.g., pressure sensors). In paper [13], an accelerometer-based automatic leak detection system was developed for early detection of single-event leaks in water pipelines with high accuracy; however, it shows relative low efficiency in the detection of multiple leak. To reduce the computational complexity, in [14] a model reduced by integrating multiple pipe branches into a single node was proposed to convert complicated WDI into a network. In terms of the practicability of internal automatic leak detection, various automatic leak detection algorithms have been developed, including state estimation algorithms [15,16], signal analysis algorithms [2,17], machine learning algorithms [12,18,19], etc. The paper [15] proposed a burst detection approach that utilizes an adaptive Kalman filter to perform hydraulic measurement of flow and pressure in a district metered area (DMA). In research [17], a burst detection scheme using principal component analysis was proposed, which enables a sensitive and quick analysis of water flow in DMAs with low computational complexity. By integrating CNN-SVM and graph-based localization algorithms, the proposed leakage detection scheme in paper [18] achieved more than 90% detection accuracy and positioning accuracy within less than 3 m.

Most of the aforementioned leakage detection schemes and algorithms showed relatively high detection accuracy in experimental settings or a certain part of WDI. Nevertheless, these methodologies would reveal certain limitations in flexibility and adaptiveness whenever water distribution infrastructure is modified or expanded. In general, WDIs inevitably undergo modification or expansion due to urban reconstruction or pipeline system optimization. When a new pipe branch is added to the existing system, most of the above detection algorithms need to be reconfigured to apply to the new pipeline architecture, which increases the complexity of operation. In addition, different hydraulic behaviors in various pipeline conditions are not considered in these algorithms. These hydraulic behaviors greatly impact the accuracy of detection. For instance, water pressure decreases gradually as the distance from the water pump increases. The corresponding water pressure due to leakage would also drop. Hence, it is necessary to assign different levels of importance to different hydraulic behaviors to ensure leak detection accuracy for each pipeline. Further, whether these detection methods are applicable to multiple leaks remains to be investigated.

This paper aims at developing a multi-leak identification (MLI) system for WDI. At this point in time, the real challenges in leak detection and the related research consist of the following: (i) lack of systematic design for automated leak detection systems; (ii) low practicability limited by inhomogeneous operating environments, i.e., the MLI system might perform differently in different parts of WDI; and (iii) lack of adaptiveness to modifications of WDI (referred to as the addition/removal/repair/replacement of pipelines).

In view of the need for an MLI system, a multi-classification MLI system (MC-MLI) for WDI is developed in this paper. An adaptive kernel (AK), which is the core of classification, is designed to incorporate the weighting vector into the extracted feature vector. The weighting vector revealing the features' importance levels will improve the overall detection performance. The problem of inhomogeneous working environments is always critical, and now it will be solved by the proposed multi-classification (MC) scheme. The MC scheme is composed of multiple unique classifiers assigned to the specific pipe sections. The classifiers are configured by different classifier scenarios, which are subject to feature combinations, feature weighting vector, and performance weighting vector. Every scenario has different detection performance, and will be chosen to meet certain requirements (e.g., sensitivity > detection time > cost). Thus, the proposed MC scheme can adapt to inhomogeneous working environments and system requirements. High system flexibility to the modification of WDI is another crucial advantage attributed to the MC scheme. When WDI is modified, only several affected classifiers will be re-trained, or new classifiers will be trained for new pipe sections. This mechanism significantly improves system feasibility.

The novelty of this study is as follows. A new multi-leak detection scheme, namely multi-classification-based multi-leak identification (MC-MLI), is developed for WDI. In this scheme, an adaptive kernel (AK) and a multi-classification (MC) algorithm are developed to adapt to inhomogeneous working environments and maintain high detection accuracy.

2. Methodology of the Multi-Classification Multi-Leak Identification (MC-MLI) Scheme

In this paper, an MC-MLI for WDI is proposed, and its development flow is illustrated in Figure 1. First, pressure sensors and flow meters are deployed and initialized in WDI. Data collection from practical WDI is then performed. The measured detection parameters, such as mean value, peak-to-peak value, and the variance in flow rates and pressures, are collected. Furthermore, the water behavior of the WDI determined by the WDI structure, the pipe properties and sensor locations, etc., are modeled as a hydraulic model. The estimated detection parameters can then be obtained based on the hydraulic model. Afterward, data analysis and feature extraction are performed to design the adaptive kernel, which is the core of classification. It is worthwhile to point out that the proposed multi-classification is a parallel structure in which the status of each pipe section is identified by a unique classifier. Feature combination will influence the design of the adaptive kernel and the detection performance. The proposed scheme is adaptive in that various classifiers might have different feature combinations and kernels to accomplish the desired detection performance. For instance, sensitivity and detection time take priority over accuracy and specificity for an important pipe section. Multiple classifier scenarios with varying feature combinations are created and evaluated by multi-criteria decision-making (MCDM). Afterward, the overall scoring of each classifier scenario is obtained and becomes the reference to select the best classifier under certain system requirements. Finally, MLI is obtained by integrating all selected classifiers in a parallel structure.

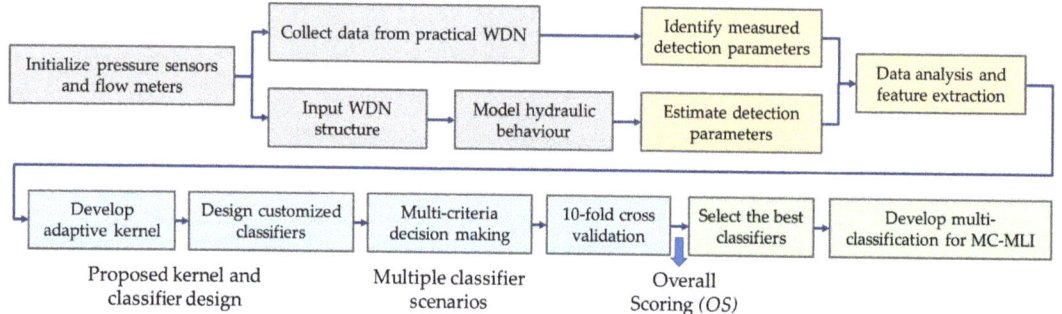

Figure 1. The development flow of the proposed multi-classification multi-leak identification (MC-MLI) scheme.

2.1. Hydraulic Model Analysis

A hydraulic model is a mathematical model to represent the water behavior in a given hydraulic system. The model facilitates the analysis of general WDIs. In this paper, the model revealing hydraulic behavior in the pipeline system is used to estimate the identification parameters for feature vector construction.

Obviously, there is no ideal (frictionless) fluid in a practical situation. Energy is required to push fluid moving along a pipe against the friction due to the fluid viscosity. The energy, also known as the loss in pressure energy, is defined as the hydraulic head loss in a pipe. The factors affecting hydraulic head loss include flow rate, the friction of the inner wall related to pipe material, and pipe diameter.

The Hazen–William equation is an empirical formulation to describe the water flow inside the pipeline with practical considerations such as pipe material, pipe diameter, pressure drop due to friction, etc. It has lower complexity compared to other hydraulic models, and therefore, it is suitable for low-cost and real-time applications. According to the Hazen–Williams equation, the water flow rate Q (m^3/s) in a pipe is expressed as follows [20]:

$$Q = 0.278 \times C \times D^{2.63} \times S^{0.54} \tag{1}$$

where C is the Hazen–Williams friction coefficient dependent on the pipe material [21]. D is the pipe diameter in meters, and S represents the energy slope, which is also known as the head loss per pipe length.

The proposed scheme considers not only the water flow rate but also pressure drop. Similarly, pressure drop PD (kPa/m) as defined in the Hazen–Williams equation is expressed as follows [20]:

$$PD = 1.192 \times 10^{21} \times \left(\frac{Q}{C}\right)^{1.85} \times \frac{S_G}{D^{4.87}} \tag{2}$$

where Q, C, and D are consistent with previous definitions. S_G is the specific gravity of the liquid, and the S_G of water equals 1.

Note that the Hazen–Williams equation is generally accurate under the conditions of >50 mm pipe diameter, 5–25 °C water temperature, and <1.2 MPa of inner pressure. Otherwise, a large error will occur, and a more complicated hydraulic model, namely the D'Arcy–Weisbach formula, should be considered.

2.2. Feature Extraction for the MC-MLI

The features are defined as the parameters that can be commonly found in all situations but demonstrate distinctive characteristics. Based on the data analysis, ten (10) features are extracted for pattern recognition and classification. They are listed in Table 1.

Table 1. Ten features for the proposed MC-MLI scheme.

Feature	Symbol	Detail
Feature 1	ΔQ_M	The difference between mean measured flow rate ($Q_{M,m}$) and mean estimated flow rate ($Q_{M,e}$)
Feature 2	ΔP_M	The difference between mean measured pressure ($P_{M,m}$) and mean estimated pressure ($P_{M,e}$)
Feature 3	ΔQ_{p-p}	The difference in peak-to-peak values between measured flow rate ($Q_{p-p,m}$) and estimated flow rate ($Q_{p-p,e}$)
Feature 4	ΔP_{p-p}	The difference in peak-to-peak values between measured pressure ($P_{p-p,m}$) and estimated pressure ($P_{p-p,e}$).
Feature 5	σ_Q	The variance of measured flow rate σ_{Qm}.
Feature 6	σ_P	The variance of measured pressure σ_{Pm}.
Feature 7	$xcorr_Q$	The correlation of measured flow rate and estimated flow rate.
Feature 8	$xcorr_P$	The correlation of measured pressure and estimated pressure.
Feature 9	$xcorr_{Q+P}$	The correlation of measured flow rate and measured pressure.
Feature 10	ΔQT	The difference in total flow volume between measured and estimated values.

The feature vector X with features entry x_i for $i = 1, 2, \ldots, M$, is formulated as follows:

$$X = [x_1\ x_2 \ldots x_M] \quad (3)$$

where M denotes the number of features and is equal to 10 in this case.

Note that the extracted features will be further analyzed and weighted at the stage of designing an adaptive kernel. Further, the proposed MC-MLI is said to be adaptive because it combines multiple unique classifiers that are customized to the assigned pipe sections. Eventually, the overall detection performance will be enhanced. The variation of data quality among different pipes will influence detection performance. For example, the pressure in a pipe at a faraway location might remain at a low level. False alarms might frequently occur if all classifiers utilize the same feature vector. In this paper, the customization of the feature vector for each pipe section will overcome the captioned challenge and thus improve overall detection performance.

2.3. The Adaptive Kernel Design for MC-MLI

The proposed MC-MLI scheme involves a classification problem to identify multiple leak points. As such, a support vector machine (SVM), recognized for its low computational cost and good classification performance, is suitable for MLI in WDI. Therefore, SVM-based multi-classification is designed and customized with a newly designed adaptive kernel.

In most practical problems, data are not linearly separable. Non-linear hyperplanes separating the data of different classes will significantly increase computational cost, which is not practical. Kernel trick is a solution to model linear hyperplanes through mapping the input feature vector into a high-dimension feature space.

Conventionally, all features share the same weighting in the kernel trick method. However, the feature importance should not be equal in reality. This means that various features should contribute to the classification to different degrees. The kernel function is not able to discover the feature importance levels. Therefore, an adaptive kernel is designed via incorporating a weighting vector and the feature vector. Correlation is performed on the features to determine their importance levels and weighting values. A higher correlation coefficient value for a feature indicates that it has a higher impact on the classification output. Therefore, a larger weighting value is assigned to that feature.

The designed adaptive kernel $K_a(x_i, x_j)$ is formulated as:

$$K_a(x_i, x_j) = \sum_{n=1}^{M} w_n k_n(x_i^n, x_j^n) \quad (4)$$

where x_i and x_j denote the feature vectors of i-th and j-th data samples, respectively. $k_n(.)$ is the radial basis function (RBF) kernel with respect to the n-th feature entry, for $n = 1, 2, \ldots, M$ of the feature vector. w_n represents the weighting of the n-th feature entry of the feature vector, and it is calculated by:

$$w_n = \frac{\sum_{q=1}^{N} \left(x_q^n - \overline{x^n}\right)\left(y_q - \overline{y}\right)}{\sqrt{\sum_{q=1}^{N} \left(x_q^n - \overline{x^n}\right)^2 \left(y_q - \overline{y}\right)^2}} \quad (5)$$

where ($^-$) denotes taking the mean value and y is the class label. N is the total number of training data samples. Note that the number of training samples of normal cases and leak cases are equal to avoid the classifier being biased.

The data for different classes can be linearly separated by hyperplanes after the feature vectors are transformed into high-dimensional feature space using a kernel trick. The feature vectors lying on the hyperplanes are defined as support vectors. The gap between the support vectors is defined as a margin. A larger separation distance of the margin will increase classification accuracy because the data for other classes are less likely to cross the large-distance margin. The high leak identification accuracy will be achieved by solving the maximizing margin problem. The customized optimization problem is formulated as follows:

$$L(\alpha, w_n) = \operatorname{argmax} \left\{ \sum_{i=1}^{L} \alpha_i - \frac{1}{2} \sum_{i=1}^{L} \sum_{j=1}^{L} \alpha_i \alpha_j y_i y_j K_a(x_i, x_j) \right\}$$
$$s.t. \begin{cases} \alpha_i \geq 0 \\ \sum_{i=1}^{L} \alpha_i y_i = 1 \quad \forall i = 1, 2, \ldots, L \\ \sum_{n=1}^{M} w_n = 1 \end{cases} \quad (6)$$

where α denotes the Lagrange multiplier and $K_a(x_i, x_j)$ is the designed adaptive kernel.

2.4. The Development of Multi-Classification for MLI

Inspired by biometric authentication, each pipe section is assigned with a unique classifier to identify the pipe status. Figure 2 shows the parallel detection structure of the proposed MC-MLI. The input of MC-MLI is the feature vector composed of the measured parameters from practical WDI and the estimated parameters from the hydraulic model. For a classifier C_h of the h-th pipe section, the output "1" means the pipe section operates normally, where the output "0" means leakage exists in the pipe section. Each pipe status is identified by a unique classifier. It is worthwhile to point out that the parallel structure facilitates the expansion of WDI. For example, when a new pipe section (e.g., (G + E)th pipe section) is connected to the WDI, a new classifier $C_{(G+E)}$ will be trained for the newly added pipe section. Other existing classifiers do not need to be re-trained unless the new pipe section influences the hydraulic behaviors of other pipe sections. The advantages of the proposed multi-classification structure are summarized as:

$$S_{a,b} = \frac{V_{a,b}}{\sum_{l=1}^{NS} V_{a,l}} \text{ with } \sum_{l=1}^{NS} S_{a,l} = 1 \quad (7)$$

where V is the value of Acc, Se, Sp, and T. NS is the number of classifier scenarios.

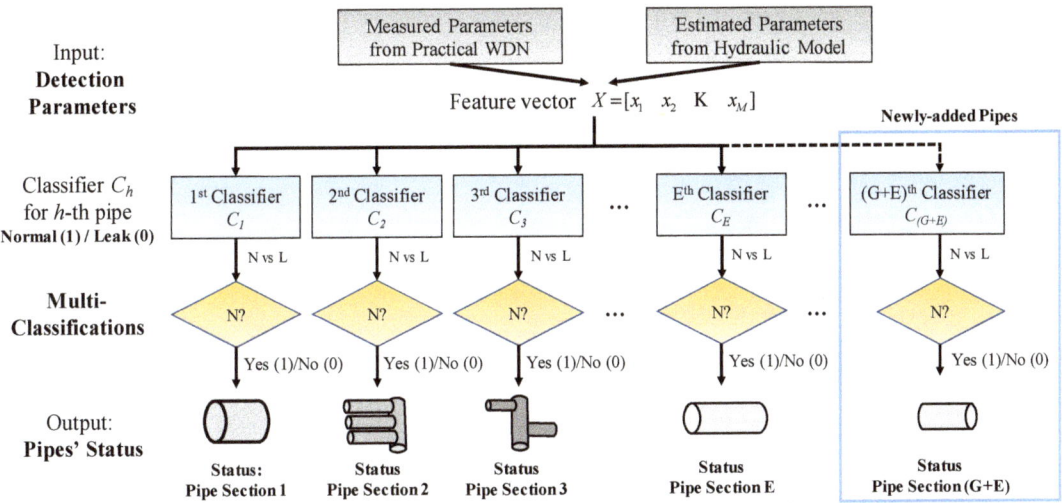

Figure 2. The parallel detection structure of MC-MLI.

The overall scoring OS_b of classifier C_h in the b-th scenario is computed as:

$$OS_b = \sum_{a=1}^{4} S_{a,b} w_a \qquad (8)$$

where w_a is the weighting of the a-th criterion obtained from AHP (with pairwise comparison [22]).

The overall scoring obtained from MCDM is a good indicator to evaluate classifier performance. The values of the four criteria are obtained from 10-fold cross-validation. Ten-fold cross-validation is a commonly adopted method for training and validating classifiers.

The proposed MC-MLI was implemented experimentally and evaluated practically. Around 500,000 pieces of data were collected and analyzed. The huge amount of data was sufficient to develop and validate the proposed MC-MLI.

3. Performance Evaluation of Different Classifier Scenarios

3.1. Experimental Setup for Evaluating MC-MLI

In real WDI, the underground pipeline network consists of multiple branches for water distribution. In this paper, a multi-leak scenario is considered, and thus a multi-branch pipe network is established. Recently, plastic pipe has been more common in practical WDIs due to its low-cost, light-weight, chemically and electrically neutral, anti-rusting and anti-corrosion properties. Therefore, the pipes utilized in the experimental setup are made of PVC. The proposed scheme is applicable to other WDIs using pipes made of other materials by adjusting several parameters in the hydraulic model. The experimental setup of the multi-branch pipe network (i.e., 3 branches: B1, B2, B3) is shown in Figure 3. Each sensor group contains a flow meter (Q) and a pressure sensor (P). All sensors are connected to Arduino broad for data transmission. In the middle of each branch, one adjustable valve is installed to act as a leak point (L). This facilitates the evaluation of different leak sizes/rates by controlling the degree of valve opening.

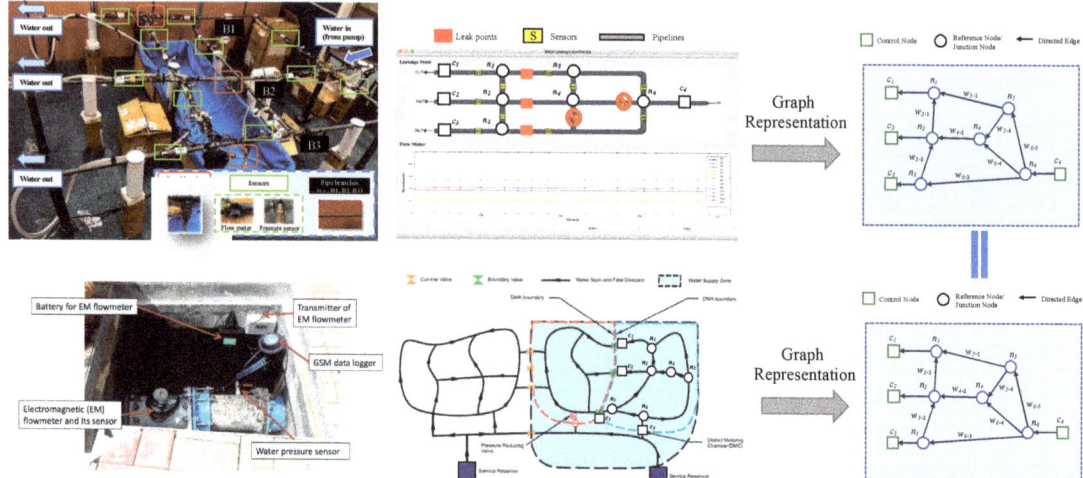

Figure 3. The experimental setup for multi-leakage detection.

Graph theory is the best candidate to represent pairwise linkages between entities. In this paper, the WDI is transformed into a directed graph using graph theory, thus identifying the behavior of WDI and exploring the correlation between a real-world model and small-scale model. An example of transforming a realistic WDI model and the developed small-scale WDI model is shown in Figure 3. Water supply zones can be clustered by deploying cut-line valves and boundary valves. The valves control the amount of water flowing into the zones, and thus they can be transformed into control nodes c_h in the directed graph. Denoted n_i and w_{i-j} are the i-th node and the edge between the i-th and j-th nodes in the directed graph, respectively. The node n represents the reference and/or conjunction point in a realistic WDI, whereas the edge represents the pipe connecting two points. The pairwise relations between nodes can be formulated as an adjacent matrix, i.e.,

$$A = \begin{bmatrix} a_{11} & \cdots & a_{1M} \\ \vdots & \ddots & \vdots \\ a_{M1} & \cdots & a_{MM} \end{bmatrix} \qquad (9)$$

where the entry a_{ij} represents the directed connection from the i-th node to the j-th node. If there is a connection between the i-th node and the j-th node, a_{ij} equals 1. Otherwise, there is no connection between them, and a_{ij} equals 0.

The essential metrics in graph theory include average node degree K, average node betweenness NB, edge betweenness EB, and the distribution of edge weighting. In order to compare these metrics between realistic a WDI and the small-scale WDI, the developed small-scale model is transformed into a graph representation, as shown in Figure 3. The transformed small-scale model is configured the same as one of the water supply zones. The correlation between the two transformed models is evaluated as follows.

$$R_m = \frac{n(\sum r_m \times p_m) - (\sum r_m)(\sum p_m)}{\sqrt{[n \sum r_m^2 - (\sum r_m)^2][n \sum p_m^2 - (\sum p_m)^2]}} \qquad (10)$$

where n denotes the amount of data, r_m denotes the m-th metric of the transformed realistic WDI model, and p_m denotes the m-th metric of the transformed small-scale model.

The results reveal that the transformed realistic WDI model is highly correlated to the transformed small-scale WDI model, reflected on the resultant coefficient of correlation (close to 1).

3.2. Performance Evaluation of the MC-MLI

The performance of the proposed MC-MLI is evaluated in terms of average accuracy, sensitivity, specificity, and detection time. In this experiment, all classification schemes are evaluated under the same scenario, i.e., the same training data, same testing data, and the same leak scenario. Three (3) schemes are compared, namely prime multi-class classification (prime MCC), multi-class classification with the adaptive kernel (MCC + AK), prime multi-classification (prime MC), and multi-classification with the adaptive kernel (MC + AK) (the proposed MC-MLI). The multi-class classification refers to one multi-class classifier. The multi-classification refers to multiple binary classifiers. The prime classification means the classifier does not comprise any adaptive kernel or MCDM.

As shown in Figure 4, the proposed MC-MLI achieves the best performances: 96.1% accuracy, 96.9% sensitivity, 95.3% specificity, and 5.3 s detection time. The comparison of prime MCC and MCC + AK reveals that the adaptive kernel significantly improves detection accuracy, sensitivity and specificity by >15%. The main drawback of multi-class classification is low accuracy, and thus the performance of MCC + AK is limited. The performance of multi-class classification degrades with the increasing number of classes. This renders serious challenges for multi-leak detection. For instance, if there are 10 pipe sections and 10 potential leak points, the number of leak combinations will be 1023, and thus 1023 classes will be needed for the multi-class classifier. Furthermore, multi-class approaches are inflexible in that the entire detection system need to be re-trained when a new pipe section is connected.

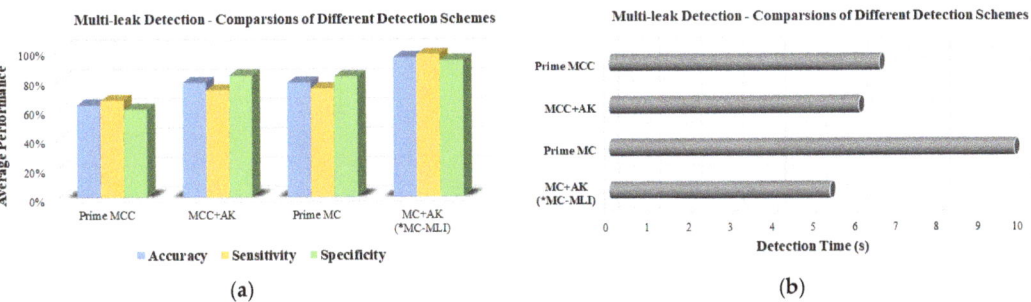

Figure 4. Performance evaluation of the proposed MC-MLI. (a) Comparisons of average accuracy, sensitivity, and specificity; (b) Comparisons of average detection time. * Note: MC + AK is actually the proposed MC-MLI scheme.

Multi-classification is proposed to overcome the captioned challenges. The multi-classification can break down an entire pipeline system into numerous pipe sections. A unique classifier is assigned to each pipe section. This facilitates classifier customization for every pipe. Furthermore, it is not necessary to re-train the entire detection system when the new pipe section is connected. A new classifier will be assigned and customized to the new pipe section. This facilitates the management of WDI. The result demonstrates that prime MC achieves higher accuracy, sensitivity and specificity than that of prime MCC. The performance of the prime MC is close to MCC + AK. This proves that multi-classification accomplishes better identification performances than multi-class classification. However, the detection time of prime MC is much longer than that of other schemes due to the complex feature vector. The proposed MC-MLI solves the problem by determining the best feature combination for every classifier by MCDM. Owing to weighting assignment for feature importance levels, MC-MLI using adaptive kernel improves accuracy, sensitivity and specificity by >15% compared to the prime MC.

3.3. Performance Evaluation of Different Classifier Scenarios

The proposed MC-MLI customizes every classifier for the assigned pipe sections. The hydraulic behavior of each pipe section can be different. It depends on various factors, such as pipe locations, user behaviors, pipe properties, network structure, etc. The classifier customization will improve not only detection performance but also operating efficiency by selecting a proper feature combination. In addition, the classifier customization facilitates meeting certain performance requirements. For example, the classifier scenario with short detection time and high sensitivity will be chosen for the important pipe sections. The classifier scenario with high specificity will be chosen for the pipes with a lower importance level. This reduces the probability of false alarms as well as unnecessary inspection costs.

The performance of different classifier scenarios of MC-MLI is shown in Figure 5. The y-axis denotes average accuracy, average sensitivity, average specificity, and average detection, respectively. The x-axis denotes the classifier scenarios. Each classifier scenario refers to a feature combination. The number of feature combinations is calculated as:

$$NC = \sum_{m=1}^{M} C_m^M \qquad (11)$$

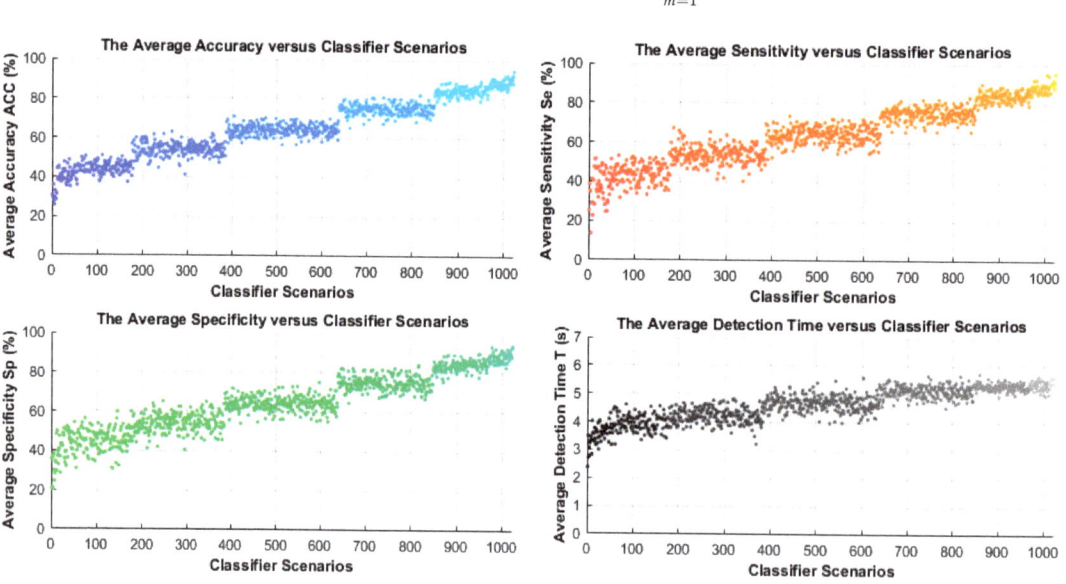

Figure 5. Performance evaluation of various classifier scenarios of MC-MLI.

Note that there are 10 features for classifier development, and thus M equals 10. The total number of classifier scenarios is 1023. The 1st to 10th classifier scenarios refer to the 1st to 10th feature combinations comprising one feature. The 11th to 55th classifier scenarios refer to the 11th to 55th feature combinations comprising two features, and so on. The result reveals that the accuracy, sensitivity, and specificity are generally improved with the increasing number of features. Meanwhile, the detection time becomes longer. The result is consistent with the previous discussion that increasing feature dimensions (i.e., the number of feature elements) usually improves the detection performance, but it leads to longer detection time. In brief, the classifier scenarios with the feature vectors made of >7 features (i.e., >967th classifier scenarios) achieve over 85% accuracy, over 80% sensitivity and over 80% specificity. The detection time is about 5 s. The classifier performances can be converted to overall scoring using the weighted sum of the performance indicators. To summarize, the customization in MC-MLI provides high adaptiveness to meet various system requirements. The best classifier for each pipe section can be evaluated using overall scoring.

3.4. The Performance of MC-MLI under Different Leak Rates

During the experiment, various leak scenarios were considered. The leak scenarios refer to different leak rates and leak locations. There are three adjustable valves to act as leak points. The leak rate is adjusted by controlling the valve openness. The leak rates vary from 0 mL/s to 40 mL/s with an interval of 5 mL/s. The number of leaks in the scenarios can be either a single leak or multiple leaks.

The result, as shown in Figure 6, is averaged from all leak scenarios (i.e., both single leak and multiple leaks are involved). The detection accuracies of three pipe branches B1, B2, and B3 are represented by three curves, respectively. Note that the curves of sensitivity and specificity are similar to the accuracy curve. The result demonstrates that the identification accuracy of MC-MLI is directly proportional to the leak rate. This is because a higher leak rate will show more significant changes in feature values, and thus a higher probability of correct classification. In the normal cases (0 mL/s leak rate), the average accuracies of all three classifiers are higher than 80%. Furthermore, the average accuracies of all classifiers are >80% under >15 mL/s leak rate, and >90% under >30 mL/s leak rate. Several factors lead to unsatisfactory accuracies at a low leak rate, including sensor sensitivity and network structure. For instance, if the noise level of a sensor is sufficiently high to interrupt measured data, a large error will occur.

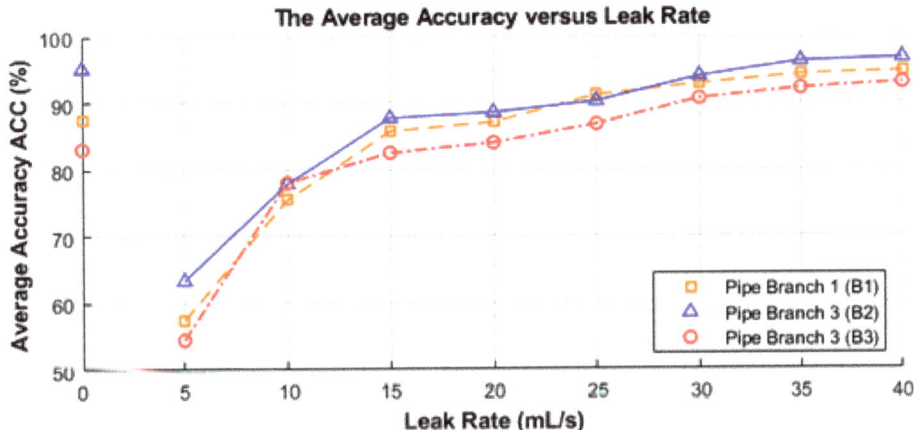

Figure 6. Performance evaluation of MC-MLI under different leak rates.

4. Conclusions

Aged water distribution infrastructure (WDI) is a global issue for water wastage due to leakage. It is necessary to develop automated leak detection to avoid economic loss and save freshwater resources. In this paper, a new multi-classification-based multi-leak identification (MC-MLI) scheme was proposed. After the data analysis, ten (10) features were extracted to develop classifiers. A new adaptive kernel was developed to transform the input features into high dimensional feature space. The features were weighted in the adaptive kernel with respect to their importance levels to improve identification performances. Moreover, a multi-classification with a parallel structure was designed to integrate multiple binary classifiers. Each pipe section was monitored by a unique binary classifier. Owing to the parallel structure of multi-classification, single/multiple detection(s) can be performed without activating all classifiers. Additionally, it facilitates the integration of a newly added pipe section. An extra classifier is trained for the newly added pipe section without reconfiguration of the whole identification system. Multiple classifier scenarios were designed to adapt to various performance requirements (e.g., accuracy, detection time, etc.). All scenarios had different performances in terms of accuracies, sensitivities, specificities and detection times. The performance requirements can be varied according to the pipes' importance levels. Therefore, the proposed MC-MLI has high detection efficiency, high system flexibility, and high system adaptiveness. An experiment involving multiple leak scenarios was performed to evaluate the proposed MC-MLI. The results demonstrate that MC-MLI achieves 96% accuracy, 96% sensitivity, and 95% specificity. The average detection time was about 5 s. The improvement in identification performance was in the range of 15% to 30% compared to prime classification schemes.

In the experiment, the same types of flow sensors and same types of pressure sensors were adopted. They had the same connection interface (i.e., serial port) and showed similar measurement precision. In reality, different types of sensors from different manufacturers may be utilized, which may pose difficulty when replacing sensors in the detection system. Considering this situation, the interoperability of different sensors based on the IEEE P2668 standard will be studied in future work to make sensors "plug-and-play". Besides that, the accuracy of sensor data needs to be guaranteed. By evaluating the accuracy of sensor data based on the IEEE P2668 standard, the reliability of the detection system will be further improved.

Author Contributions: Conceptualization, writing—original draft preparation, data collection, software, Y.W.; supervision and project administration, K.F.T.; methodology and formal analysis, C.K.W.; literature collection, H.W.; validation and investigation, Y.L. All authors have read and agreed to the published version of the manuscript.

Funding: This research was funded by Electrical and Mechanical Services Department of Hong Kong, grant number 9211292.

Institutional Review Board Statement: Not applicable.

Informed Consent Statement: Not applicable.

Data Availability Statement: Not applicable.

Conflicts of Interest: The authors declare no conflict of interest.

References

1. Goulet, J.-A.; Coutu, S.; Smith, I.F. Model falsification diagnosis and sensor placement for leak detection in pressurized pipe networks. *Adv. Eng. Inform.* **2013**, *27*, 261–269. [CrossRef]
2. Srirangarajan, S.; Allen, M.; Preis, A.; Iqbal, M.; Lim, H.B.; Whittle, A.J. Wavelet-based burst event detection and localization in water distribution systems. *J. Signal Process. Syst.* **2013**, *72*, 1–16. [CrossRef]
3. Loss of Water Due to Water Mains Leakage. Available online: https://www.info.gov.hk/gia/general/201802/07/P2018020700639.htm (accessed on 12 November 2021).
4. Yang, J.; Wen, Y.; Li, P. Leak Acoustic Detection in Water Distribution Pipelines. In Proceedings of the 2008 7th World Congress on Intelligent Control and Automation, Chongqing, China, 8 August 2008.
5. Li, S.; Song, Y.; Zhou, G. Leak detection of water distribution pipeline subject to failure of socket joint based on acoustic emission and pattern recognition. *Measurement* **2018**, *115*, 39–44. [CrossRef]
6. Huang, Z.; Lin, J.; Meng, X.; Liu, H. Characteristics analysis of ground penetrating radar signals for groundwater pipe leakage detection. In Proceedings of the IET International Radar Conference, Online Conference, 4–6 November 2020.
7. Almeida, F.C.; Brennan, M.J.; Joseph, P.F.; Gao, Y.; Paschoalini, A.T. The effects of resonances on time delay estimation for water leak detection in plastic pipes. *J. Sound Vib.* **2018**, *420*, 315–329. [CrossRef]
8. Cataldo, A.; De Benedetto, E.; Cannazza, G.; Leucci, G.; De Giorgi, L.; Demitri, C. Enhancement of leak detection in pipelines through time-domain reflectometry/ground penetrating radar measurements. *IET Sci. Meas. Technol.* **2017**, *11*, 696–702. [CrossRef]
9. De Coster, A.; Medina, J.P.; Nottebaere, M.; Alkhalifeh, K.; Neyt, X.; Vanderdonckt, J.; Lambot, S. Towards an improvement of GPR-based detection of pipes and leaks in water distribution networks. *J. Appl. Geophys.* **2019**, *162*, 138–151. [CrossRef]
10. Loureiro, D.; Amado, C.; Martins, A.; Vitorino, D.; Mamade, A.; Coelho, S.T. Water distribution systems flow monitoring and anomalous event detection: A practical approach. *Urban Water J.* **2016**, *13*, 242–252. [CrossRef]

11. Farah, E.; Shahrour, I. Leakage detection using smart water system: Combination of water balance and automated minimum night flow. *Water Resour. Manag.* **2017**, *31*, 4821–4833. [CrossRef]
12. Romano, M.; Kapelan, Z.; Savić, D.A. Automated detection of pipe bursts and other events in water distribution systems. *J. Water Resour. Plan. Manag.* **2014**, *140*, 457–467. [CrossRef]
13. Zahab, S.E.; Mosleh, F.; Zayed, T. An accelerometer-based real-time monitoring and leak detection system for pressurized water pipelines. In Proceedings of the Pipelines 2016 Conference, Kansas City, MO, USA, 17–20 July 2016.
14. Moser, G.; Paal, S.G.; Smith, I.F. Performance comparison of reduced models for leak detection in water distribution networks. *Adv. Eng. Inform.* **2015**, *29*, 714–726. [CrossRef]
15. Choi, D.Y.; Kim, S.-W.; Choi, M.-A.; Geem, Z.W. Adaptive Kalman filter based on adjustable sampling interval in burst detection for water distribution system. *Water* **2016**, *8*, 142. [CrossRef]
16. He, Y.; Li, S.; Zheng, Y. Distributed state estimation for leak detection in water supply networks. *IEEE/CAA J. Autom. Sin.* **2017**, *7*, 1–9. [CrossRef]
17. Palau, C.; Arregui, F.; Carlos, M. Burst detection in water networks using principal component analysis. *J. Water Resour. Plan. Manag.* **2012**, *138*, 47–54. [CrossRef]
18. Kang, J.; Park, Y.-J.; Lee, J.; Wang, S.-H.; Eom, D.-S. Novel leakage detection by ensemble CNN-SVM and graph-based localization in water distribution systems. *IEEE Trans. Ind. Electron.* **2017**, *65*, 4279–4289. [CrossRef]
19. Mounce, S.; Mounce, R.; Jackson, T.; Austin, J.; Boxall, J. Pattern matching and associative artificial neural networks for water distribution system time series data analysis. *J. Hydroinform.* **2014**, *16*, 617–632. [CrossRef]
20. Lin, C.-C. A hybrid heuristic optimization approach for leak detection in pipe networks using ordinal optimization approach and the symbiotic organism search. *Water* **2017**, *9*, 812. [CrossRef]
21. Mays, L.W. Water supply security: An introduction. In *Water Supply Systems Security*; Larry, W.M., Ed.; McGraw-Hill: New York, NY, USA, 2004; pp. 1.1–1.12.
22. Duleba, S.; Moslem, S. Examining Pareto optimality in analytic hierarchy process on real Data: An application in public transport service development. *Expert Syst. Appl.* **2019**, *116*, 21–30. [CrossRef]

Article

Solar Irradiance Forecasting Using a Data-Driven Algorithm and Contextual Optimisation

Paula Bendiek [1,2], Ahmad Taha [3], Qammer H. Abbasi [3] and Basel Barakat [1,4,*]

1. School of Engineering and Built Environment, Edinburgh Napier University, Edinburgh EH14 1DJ, UK; paula.bendiek.21@ucl.ac.uk
2. Bartlett School of Environment, Energy and Resources, University College London, Central House, 14 Upper Woburn Pl, London WC1H 0NN, UK
3. James Watt School of Engineering, University of Glasgow, Glasgow G12 8QQ, UK; Ahmad.Taha@Glasgow.ac.uk (A.T.); Qammer.Abbasi@glasgow.ac.uk (Q.H.A.)
4. School of Computer Science, University of Sunderland, Sir Tom Cowie Campus, St Peters Way, Sunderland SR6 0DD, UK
* Correspondence: basel.barakat@sunderland.ac.uk

Citation: Bendiek, P.; Taha, A.; Abbasi, Q.H.; Barakat, B. Solar Irradiance Forecasting Using a Data-Driven Algorithm and Contextual Optimisation. *Appl. Sci.* 2022, 12, 134. https://doi.org/10.3390/app12010134

Academic Editor: Chun Sing Lai

Received: 21 October 2021
Accepted: 7 December 2021
Published: 23 December 2021

Publisher's Note: MDPI stays neutral with regard to jurisdictional claims in published maps and institutional affiliations.

Copyright: © 2021 by the authors. Licensee MDPI, Basel, Switzerland. This article is an open access article distributed under the terms and conditions of the Creative Commons Attribution (CC BY) license (https://creativecommons.org/licenses/by/4.0/).

Abstract: Solar forecasting plays a key part in the renewable energy transition. Major challenges, related to load balancing and grid stability, emerge when a high percentage of energy is provided by renewables. These can be tackled by new energy management strategies guided by power forecasts. This paper presents a data-driven and contextual optimisation forecasting (DCF) algorithm for solar irradiance that was comprehensively validated using short- and long-term predictions, in three US cities: Denver, Boston, and Seattle. Moreover, step-by-step implementation guidelines to follow and reproduce the results were proposed. Initially, a comparative study of two machine learning (ML) algorithms, the support vector machine (SVM) and Facebook Prophet (FBP) for solar prediction was conducted. The short-term SVM outperformed the FBP model for the 1- and 2- hour prediction, achieving a coefficient of determination (R^2) of 91.2% in Boston. However, FBP displayed sustained performance for increasing the forecast horizon and yielded better results for 3-hour and long-term forecasts. The algorithms were optimised by further contextual model adjustments which resulted in substantially improved performance. Thus, DCF utilised SVM for short-term and FBP for long-term predictions and optimised their performance using contextual information. DCF achieved consistent performance for the three cities and for long- and short-term predictions, with an average R^2 of 85%.

Keywords: solar irradiance forecasting; short-term and long-term predictions; machine learning; support vector machine; Facebook Prophet; contextual optimisation

1. Introduction

Greenhouse gases are major drivers of climate change [1] and are primarily produced by energy generation from fossil fuels [2]. Substantial research and political attention have been devoted to renewable energies in order to reduce the consumption of fossil fuels [3]. According to Huybrechts [4], renewable solar energy generation has continuously increased in the context of attempts to transition to a net-zero carbon economy, as shown in Figure 1. However, major challenges arise when a higher percentage of renewable energy is connected to the grid, due to its volatile nature [5]. If supply and demand are not of a similar magnitude, energy grids become unstable, potentially leading to blackouts [6]. Load balancing, ensuring that equal amounts of energy are generated and consumed, is one of the most important and difficult of these challenges [7]. This has conventionally been achieved by adjusting energy generation to demand patterns and scaling up power generation whenever necessary. Currently, the backup capacity for load balancing is mostly provided by fossil fuels, generation of which can be ramped up on demand [8].

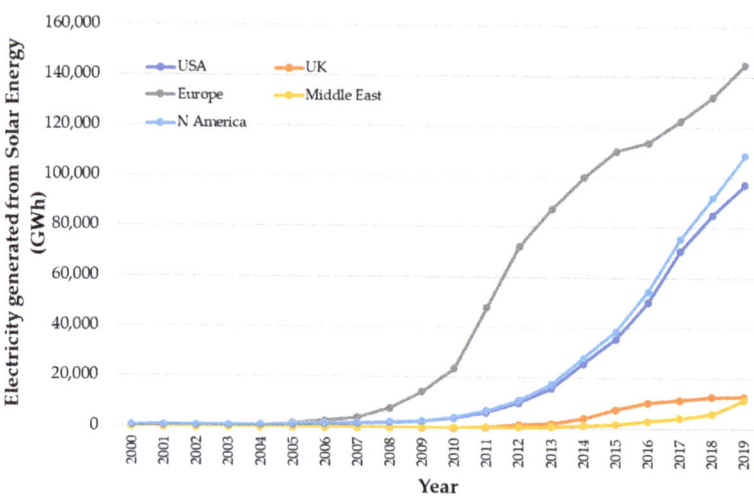

Figure 1. Increase in solar power generation worldwide [9].

Renewable energy depends on environmental factors [10,11], and is, therefore, harder to match to demand patterns. This stipulates the need for appropriate energy management, including the organisation of generation, storage, and consumption. Understanding energy generation patterns plays a key part in developing effective management strategies. Therefore, the prediction of renewable power output is necessary to integrate more renewable energy into the grid and thus reduce the emission of greenhouse gases [12].

In order to forecast the power output of any solar technology, the amount of available potential energy must be known. If prediction models are specific to one type of device, it is harder to adapt them to other use cases. The potential energy generated by many technologies, e.g., PV panels, depends on the amount of solar global horizontal irradiance received at a certain location. Global horizontal irradiance is the sum of direct and diffuse radiation on a horizontal plane and is also used to calculate the radiation on an inclined plane, such as a solar panel [13]. The prediction of solar radiation allows us to infer the power output of devices, such as photovoltaic cells or solar water heaters. Throughout this paper, global horizontal irradiance will also be referred to as simply irradiance or radiation.

In recent years, solar prediction in particular has become more sophisticated. Much of this advancement is attributed to the development of machine learning (ML) algorithms [14]. There has been a tremendous increase in the use of ML for solar predictions in the last decade. It has been successfully employed and is extensively discussed in review papers by Sobri et al. [12] and Wang et al. [14]. This paper builds on these insights and proposes a forecasting algorithm that predicts solar irradiance using ML algorithms and contextual optimisation.

Motivations and Impact

The need for ML-driven energy management solutions is increasing with the net-zero carbon by 2050 target set by the UK government [15]. Several contributing parameters to managing energy in our society include demand, energy usage behaviour, environmental factors, etc. In this paper, we addressed the question of how to accurately forecast solar irradiance. This plays a crucial role in choosing the most optimal energy system management strategy, and optimising the integration of solar cells [16]. Moreover, we aim to present a methodological foundation of algorithm and feature selection, and evaluation metrics for other studies to follow.

The main contributions of this paper are as follows:

- Data-driven and contextual optimisation forecasting (DCF) algorithm, which accurately predicts solar irradiance in the short- and long term. DCF is a hybrid algorithm that utilises state-of-the-art ML algorithms and optimises their accuracy using contextual information;
- A comparative study of two ML algorithms (Support Vector Machine and Facebook Prophet), in which an investigation was made into the effect of adding extraterrestrial radiation as a feature;
- Comprehensive validation of the forecasting accuracy for short- and long-term predictions in three cities to ensure that the model is not specific to one location. This was evaluated by computing the coefficient of determination (R^2), mean absolute error (MAE), and root-mean-squared error (RMSE).

The rest of the paper is organised as follows: Section 2 reviews previously proposed algorithms for solar forecasting. Section 3 presents the dataset used for training the ML algorithms and the evaluation methods, respectively. The DCF algorithm is introduced in Section 4, while Section 5 discusses the forecasting results. Finally, Section 6 concludes this paper, and Section 7 suggests potential future research.

2. Literature Review

There is a range of ML algorithms that have been used in solar irradiance prediction, such as regression, Markov chain [17], autoregressive integrated moving average (ARIMA) [18], and neural networks [19]. One of the most commonly used ML algorithms is the support vector machine (SVM) [12,20–22]. The SVM model is a conventional algorithm that has been used for more than a decade to predict solar irradiance [21]. There are several advantages to using an SVM; for example, it is able to model complex nonlinear models with considerably high accuracy and robustness, and it is usually immune to overfitting. Furthermore, there are novel algorithms, which are not yet established in solar prediction but have the potential to increase forecasting accuracy, such as the Facebook Prophet (FBP) algorithm. FBP was proposed for forecasting time series where nonlinear trends fit with yearly, weekly, and daily seasonality. It achieves high accuracy with time series that have strong seasonal effects and several seasons of historical data. Additionally, it is robust in handling missing data and shifts in the trend and typically reduces the effect of outliers as shown in Section 2.2.

2.1. Support Vector Machines

SVM is a statistical learning algorithm originally designed for classifying data [23]. It can also be used for regression tasks such as predicting solar radiation [24]. A kernel function transforms a nonlinear input space into a higher-dimensional space [25]. It allows efficient computation of the scalar products of multiple vectors in this higher-dimensional space. Common kernel functions include the polynomial, radial basis (RBF), and sigmoid functions [21]. In the higher-dimensional space, the optimal hyperplane, which separates the margins of errors in regression and classes in classification, can be identified.

The use of SVMs in renewable forecasting has increased drastically in recent years [21]. The SVM is an established method, used across the renewable energy sector, especially for solar forecasting, because of its accurate prediction ability for nonlinear data. Further advantages include its fast computational speed, as no iterative tuning is required, and its capability to produce accurate predictions with a small volume of data [26]. SVMs solve a convex programming problem resulting in the global optimum, avoiding being trapped in local optima (local optimum is either the highest or lowest point, compared with nearby data points. The global optimum is the highest or lowest point in the whole function or dataset. Further reading on convex optimisation problems can be found in [27]).

Zeng and Qiao proposed a least-square SVM to forecast global horizontal irradiance for 1-, 2- and 3-hour ahead [28]. Their model significantly outperformed an autoregressive (AR) model, as well as a radial basis function neural network. However, their evaluation was performed for a short period (10 days) without cross-validating the model performance.

VanDeventer et al. developed an SVM model in hybrid with a genetic algorithm to forecast the power output of residential PV systems [29]. The model demonstrated good adaptability to different locations, weather patterns, and climatic conditions. As, PV power output depends on the system parameters and technologies, prediction of the power source (irradiance) is more useful in the long term. An SVM with radial basis function to global solar irradiance in a single location (Tehran) was used by Ramedani et al. [25]. The radial basis function was chosen because it outperformed the polynomial as a kernel function. Furthermore, it outperforms an ANN in terms of root-mean-squared error (RMSE) while being computationally more efficient.

2.2. Facebook Prophet

Facebook Prophet (FBP) is a decomposable time series model, based on additive modelling [30]. Recently, it has gained significant attention due to its capability to accurately forecast time series data. For instance, Lim et al. compared FBP to autoregressive integrated moving average (SARIMA) and concluded that FBP outperformed SARIMA for the prediction of electricity and natural gas demand [31]. Additionally, Shawon et al. predicted PV short circuit current for the next day, deeming it to be a reliable forecasting method [32].

FBP delivers its peak performance when dealing with a time series with strong seasonal effects [33]. This applies to solar irradiance and is one of the main reasons to believe that this algorithm is suitable for solar irradiance forecasting. However, in the literature, FBP has not yet been utilised for solar irradiance prediction.

FBP models the time series data as follows:

$$y(t) = g(t) + s(t) + h(t) + \epsilon_t \tag{1}$$

where the trend is $g(t)$, the seasonality is $s(t)$, and the holidays are $h(t)$. It is worth mentioning that holidays and weekly trends were not accounted for, as these have no influence on solar irradiance, ϵ_t indicates the changes not represented by the model and is assumed to be normally distributed. It has intuitively adaptable parameters, designed to be used by analysts that have domain knowledge rather than statistical expertise. Therefore, it is important to know the characteristics of the subject that is being predicted, in this case, the behaviour of solar radiation.

3. Dataset and Evaluation

3.1. Dataset

The data for this paper were acquired from the National Solar Radiation Database (NSRDB) [34] for solar irradiance values in Denver, Seattle, and Boston, as shown in Table 1. These were selected due to their different geographical and meteorological conditions. Thus, the forecasting algorithm would not be specific to one location.

Table 1. Datasets are from the National Solar Radiation Database [34].

City	Station Name	ID	Latitude	Longitude
Denver	Denver/Centennial	724666	39.742°	−105.179°
Boston	Boston Logan	725090	42.367°	−71.017°
Seattle	Seattle Seattle-Tacoma	727930	47.46°	122.317°

The datasets contained hourly data for 8 years (1998–2005), including global horizontal irradiance and extraterrestrial radiation on a horizontal surface. Extraterrestrial radiation on a horizontal surface is the amount of solar radiation received at the top of the atmosphere on a horizontal surface. This will be referred to as extraterrestrial radiation throughout this paper (this is not to be confused with the solar constant. Further reading on solar radiation can be found in Kalogirou's book *Solar Energy Engineering* [35]. These datasets were used to predict hourly values for the global horizontal irradiance.

By averaging every hour of the day over the given 8 years, 1D and 2D plots were created and are shown in Figure 2, respectively. While the 1D plot only captures the seasonal trend, the 2D representation also displays the daily seasonality which depends on the latitude of the location.

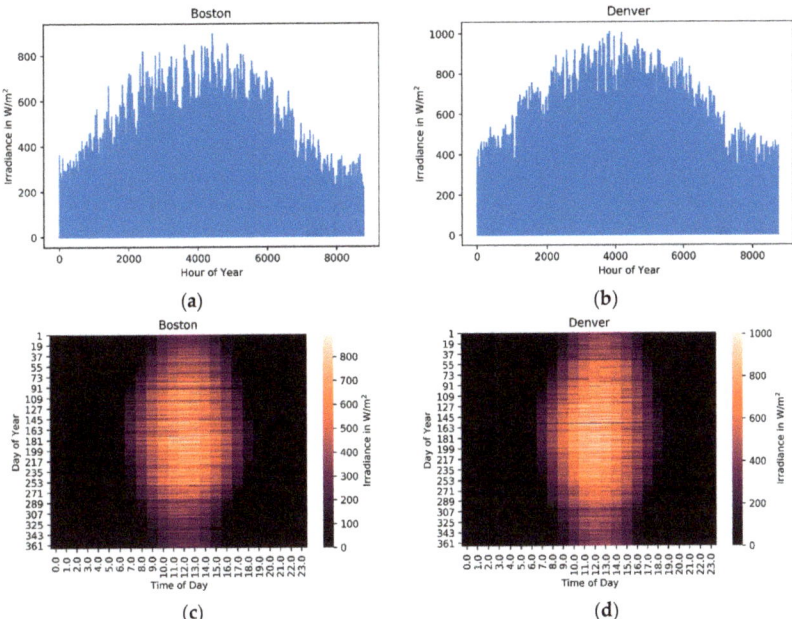

Figure 2. The 1D and 2D representations of average irradiance in Boston and Denver [34]. (**a**) 1D representation of average irradiance in Boston (**b**) 1D representation of average irradiance in Denver (**c**) 2D representation of average irradiance in Boston (**d**) 2D representation of average irradiance in Denver.

3.2. Evaluation

The DCF algorithm was assessed for short- and long-term forecasting. The short-term forecasts for 1-, 2- and 3-hour ahead were generated, as is common in the literature [12,29,36,37]. Forecasts for a few hours ahead help to manage and schedule the start-up of power plants (load scheduling) [37]. Furthermore, short-term forecasts of 30 min to 6 h are important for load dispatch and scheduling [24]. Load dispatch means that electricity can be dispatched on demand, and load scheduling is the management of this electricity and its usage.

The long-term prediction capabilities were investigated by forecasting irradiance data for 1 year (24 × 365 h) ahead. Long-term forecasting of several months up to a year is useful for scheduling maintenance and has value when bidding on the energy market [38]. There are few studies on long-term predictions in the literature using statistical methods [12]. It might relate to the fact that physical models based on meteorological expertise are generally more accurate at predicting long-term solar radiation [39]. The long-term prediction of this ML model does not detect any change in weather and only gives an approximate idea of the radiation values. However, this model is useful, as its implementation is easier and quicker than the implementation of a physical model and still gives a good indication of the amount of radiation that will be received

All models were tested on hourly data for a whole year (2005). These results were affirmed using fivefold cross validation for the SVM model. Cross validation for FBP cannot be performed like common k-fold validation, as the time series should not be randomly separated. Therefore, the 1-, 2-, and 3-hour predictions were made for FBP using every hour of the year as the starting point, thus generating 8760 × 3 forecasts. Based on these predictions and target values, several evaluation metrics were calculated. As for k-fold cross validation, the more starting points there are (the higher the k), the more generalised the result will be.

The forecasting was evaluated and compared using the coefficient of determination (R^2), mean absolute error (MAE), and root-mean-squared error (RMSE).

The R^2 value is obtained as follows [40]:

$$R^2 = \frac{\Sigma_i(y_i - \hat{y}_i)}{\Sigma_i(y_i - \overline{y}_i)} \quad (2)$$

where y_i are the actual values, $\overline{y}_i$ is the mean of the actual values, and $\hat{y}_i$ are the predicted values.

MAE has the same units as the predicted value and thus represents the expected absolute error, which is calculated by [41].

$$MAE = \frac{1}{N}\sum_{i=1}^{N}|y_i - \hat{y}_i| \quad (3)$$

where N is the total number of samples.

The RMSE value squares the difference between actual and predicted values, emphasising larger errors. This is appropriate for solar prediction as larger errors lead to disproportionally higher costs [42]. RMSE can be calculated as follows [43]:

$$RMSE = \sqrt{\sum_{i=1}^{N}\frac{(y_i - \hat{y}_i)^2}{N}} \quad (4)$$

To evaluate the prediction accuracy, the data were trained on radiation data from 1998 to 2004 and tested on data from 2005. Cross validation was performed, showing that the models generalise well. Furthermore, grid search was applied to tune the hyperparameters. After training and making predictions, these were adjusted using contextual optimisation.

4. Data-Driven and Contextual Optimisation Forecasting Algorithm

The DCF algorithm consists of two parts, i.e., data-driven and optimisation using contextual information, as shown in Figure 3. The data-driven part purely depends on the algorithm and the input data, e.g., the selection of the input features. The optimisation part uses contextual information to enhance the forecasting of the data-driven models, such as the elimination of negative predictions. Using this approach, we can harvest the strengths of both machine learning and the contextual understanding of the data.

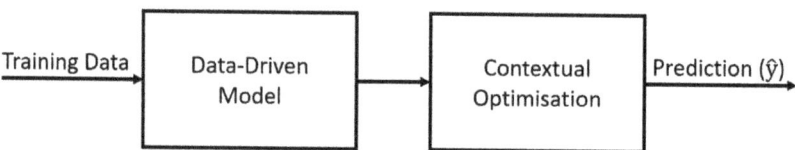

Figure 3. Block diagram of DCF, showing its two main parts: data-driven model and contextual optimisation.

4.1. Data-Driven Model

In the data-driven part, two promising ML algorithms (SVM and FBP) were utilised to generate the predictions. It was implemented in Python [44] using Scikit-learn [45] and Prophet Libraries [30]. Initially, a comparative study of the SVM and FBP algorithms was conducted to assess their accuracy. Subsequently, the effects of adding extraterrestrial radiation as an input feature to the model were investigated.

For the SVM short-term prediction, three variables were used as initial features, all past values of the global horizontal irradiance. These are the radiation of the same day 1 h ago, the same hour 1 day ago, and the same hour 2 days ago, as shown in Table 2. Zeng and Qiao found that the same hour of previous days has a stronger correlation with the target variable than radiation data from 1h ago [28]. For the long-term prediction of the SVM model, radiation values of the same hour and same day one year ago were used as its initial feature (see Table 2), as these have a strong correlation [38].

Table 2. Initial and additional features for SVM short- and long-term forecast.

Variable Name	Description
	Short-term Forecast
	Initial Input Features
1H Radiation	Radiation values for the same day 1 h ago
1D Radiation	Radiation values for the same hour 1 day ago
2D Radiation	Radiation values for the same hour 2 days ago
	Additional Input Features
1H Extraterr	Extraterrestrial values for the same day 1 h ago
1D Extraterr	Extraterrestrial values for the same hour 1 day ago
2D Extraterr	Extraterrestrial values for the same hour 2 days ago
	Long-term Forecast
	Initial Input Features
1Y Radiation	Radiation from the same hour and day a year ago
	Additional Input Features
2Y Radiation	Radiation at the same hour and day two years ago
1Y Extraterr	Extraterrestrial radiation of same hour and day a year ago
2Y Extraterr	Extraterrestrial radiation of same hour and day two years ago

For the Facebook Prophet short-term prediction, the same variable as for SVM was used, the global horizontal irradiance. However, as FBP has a different algorithm structure, the feature is the time series of solar radiation up to the values that are predicted. There is no differentiation of global horizontal radiation (1H-, 1D-, 2D radiation) as for the SVM model. For example, all values from 00:00 on 1 January 1998 up to 08:00 on 24 June 2005 were used to predict 09:00 + 10:00 + 11:00 on 24 June 2005. Similarly, for the long-term prediction, the entire past time series up to the predicted year was used. The past time series should contain at least one year of data so that seasonalities can be captured. Both the long- and short-term prediction features are shown in Table 3. These will only differ in their predicted output values (3 h or 1 year).

Table 3. Initial and additional features for FBP short- and long-term forecast.

Variable Name	Description
	Short- and Long-term Forecast
	Initial Input Features
$X_{t=0} \dots X_{t=N}$	Time series of radiation values from 1 January 1998 to 31 December 2004
	Additional Input Features
$E_{t=0} \dots E_{t=N}$	Time series of extraterrestrial radiation values from 1 January 1998 to 31 December 2004

After choosing the initial features for the data-driven model, further features were added and their effectiveness evaluated. Adding features to a model can improve its performance [28]. However, there is no inherent benefit to increasing the model complexity. Additional features can also lead to worse results or have no impact on performance [46].

Therefore, additional features must be carefully evaluated and only added if shown to have a positive impact.

For the SVM short-term forecast, three inputs were added to the initial features, as shown in Table 2: extraterrestrial radiation for the previous hour of the same day, for the same hour 1 day ago, and for the same hour 2 days prior. For the long-term forecast, the irradiance of the same hour and the same day two years ago, as well as the extraterrestrial radiation were added. The long-term forecast further included the global horizontal irradiance of the same hour and the same day two years ago, as well as the extraterrestrial radiation, as shown in Table 2.

It is only possible to add features to FBP if the future values for these are known. This is not the case for most additional features, such as extraterrestrial radiation. However, extraterrestrial radiation is approximately the same for every time of the year at a given location, so it can be predicted precisely. Thus, a time series of predicted extraterrestrial radiation was added for FBP as additional regressors, for both short- and long-term predictions, as shown in Table 3.

Hyperparameters are different from "normal" parameters, e.g., the weights (ω) and biases (b). They are the parameters that cannot be learned by the SVM model but must be chosen. The hyperparameter were tuned after evaluating the results of the basic algorithm operations for the default values in Scikit-learn. Hyperparameters should be selected to give the best results and can be tuned using several different methods. These include grid search [47], random search [48], and bio-inspired techniques, e.g., swarm optimisation [49].

The hyperparameters for this SVM model were tuned by the grid search cross validation (grid-search cross-validation searches for the best combination of the given parameters using cross validation to evaluate each combination of hyperparameters). For this, a grid of possible hyperparameters was provided. Firstly, the radial basis function (RBF), shown in Equation (5), was chosen, as it produces the best results in the literature [50]. This was verified for these solar models. When using an SVM for regression with an RBF kernel, three parameters must be found: C, the regularisation parameter; ε, the term defining the size of the error tube; γ, the width of the RBF kernel.

$$RBF = exp\left(-\gamma \|x - x'\|^2\right) \quad (5)$$

One drawback of grid search cross validation is its computational cost. Other optimisation techniques should be investigated, as discussed in Section 7. To avoid excessive computations, a log-scale was initially used for all hyperparameters, e.g., 0.1, 1, 10, and 100 for C. Depending on the outcome, the range was adjusted (e.g., 5, 10, and 50). It was found that C had the greatest influence on the results of this model.

4.2. Contextual Optimisation

The second part of the DCF algorithm optimised the accuracy of the data-driven predictions using the contextual information of solar irradiance. This information was derived from comparing the forecasted values to the measured values, thus not relying on a specific location/time. As shown in Figure 4, optimisation had three steps. It was observed that the data-driven approaches forecasted negative values, so these negative values were eliminated. Then, the forecasted values were amended based on the time of sunrise and sunset, (a similar approach were taken in [19] daytime forecasting). Here, we used two approaches: one static, in which night hours were defined from 8 p.m. to 6 a.m., and one dynamic which determined the hours of sunset and sunrise. The static approach was implanted by Zeng and Quiao, producing good results [28]. The dynamic approach is a more accurate representation of reality and thus can be more flexibly implemented in any location. However, it requires additional computational power. The last step was the seasonal adaptation in which we amended the forecasted values in the long-term model according to the month of the year.

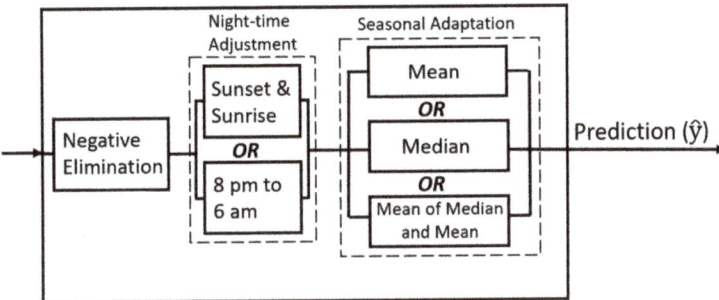

Figure 4. Contextual optimisation block diagram showing the three main steps.

FBP generated large negative values for both long- and short-term predictions. For all negative predictions (which only occurred in winter), the target value was zero. This shows that FBP only forecasted negative values during the night hours, as shown in Figure 5. In summer, all night hour predictions were positive. As there could not be negative irradiance and most negative predictions occurred at night, all negative values were eliminated and set to zero. The SVM model also predicted some negative values (around 5% for short-term and 50% for long-term). For most predictions with negative values, the target value was zero. For the non-zero target values, the radiation was very low (maximum of 15 W/m^2). Therefore, here too, all negative values were set to zero.

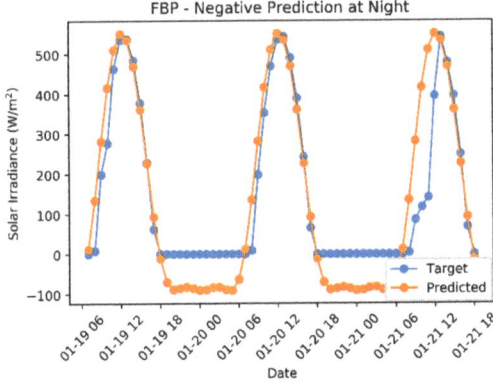

Figure 5. Three days of long-term FBP prediction displaying negative values at night.

After eliminating the negative values, all values between 8 pm and 6 am were set to zero, as they were considered night hours [28]. However, this static approach does not represent that sunrise and sunset hours vary over the year. Therefore, the sunset and sunrise for every day of the year were determined and subsequently used to set all values between sunset and sunrise to zero. Both static and dynamic methods were implemented to compare their impact on the model accuracy.

A seasonal adaptation was created for the long-term models, as a general trend was detected. For instance, the long-term FBP model would overpredict in summer and underpredict in winter, especially for the model without extraterrestrial radiation. Further, there was over- and underprediction trends in both seasonal and daily forecasts. For example, in some months, morning and evening hours were underpredicted, while the noon hours were overpredicted, as shown in Figure 6. The seasonal adaptation aimed to prevent these general trends of over- and underpredicting. The model with extraterrestrial radiation displayed less of a yearly seasonal trend; however, the daily trend still existed.

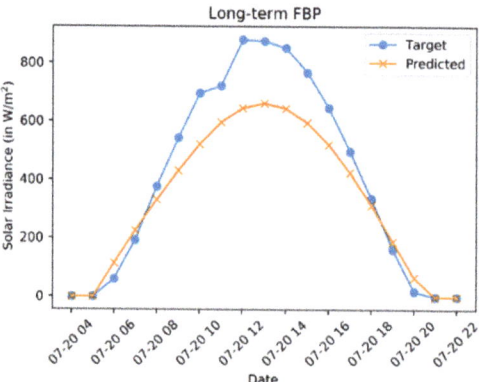

Figure 6. FBP, displaying overprediction in morning and evening and underprediction at noon.

For the seasonal adaptation, for every hour of the day within each month (e.g., the 6th hour of every day in January), all values from previous years were collected. The average of these target values for the particular hour was taken for each month, as shown in Figure 7. The same was carried out for the predicted values. Three different versions of average were used: the mean (V1), the median (V2), and the mean of median and mean (V3).

January - y					
Hour \ Day	1	2	3	...	31
1	0	0	0	0	0
...	...	...	...	...	...
10	338	365	210	...	325
11	418	463	413	...	438
12	463	488	519	...	461
...	...	...	...	...	...
24	0	0	0	0	0

January - y		
Mean	Median	Mean of Median and Mean
0	0	0
...	...	...
308	325	316.5
418	438	428
490	488	489
...	...	...
0	0	0

Figure 7. Example of working principle: grouping the average into one value per hour per month.

The seasonal adaptation adjusted the values according to the month of the year by increasing/decreasing every predicted value that was on average lower/higher than the target values of the same hour of the day of that month in past years. The seasonal adaptation (SA) is calculated as follows:

$$\hat{y}_{SA} = \hat{y} \times \left(1 + \frac{\bar{y} - \bar{\hat{y}}}{\bar{\hat{y}}}\right) \tag{6}$$

where $\hat{y}$ refers to the predicted value, y is the target value, and the $\bar{\hat{y}}$ is the average predicted value. The average here refers to either the mean, median, or mean of median and mean, depending on the version.

In the final DCF, SVM was used for first- and second-hour predictions. Beyond this, FBP would be used as the core algorithm. Furthermore, the best outcome of every comparative step was used. In the data-driven part, extraterrestrial radiation was added as an input feature to the DCF algorithm. The most influential hyperparameter was the regularisation parameter C, which was chosen to be 120 for the short-term model and 0.5 for the long-term DCF. In the contextual optimisation, the negative values were eliminated and dynamic sunset- and sunrise adjustments were performed. For the long-term prediction, seasonal adaptation was applied. From the seasonal adaptation variations, V3 (mean of median and mean) was chosen for the SVM model, while V1 (mean) was selected for the FBP. This was verified by the results, presented in Section 5.

5. Results and Discussion

This section consists of three main parts. First, the data-driven part of the model is evaluated, followed by a discussion of the improvements brought about by contextual optimisation. Subsequently, the final DCF model is presented and validated by the short- and long-term models in all three cities.

5.1. Data-Driven Model Results

The initial model was based on historical solar radiation data and the respective algorithm. SVM outperformed FBP in the 1-hour ahead prediction in terms of R^2 and RMSE (Table 4). It also had the lowest MAE for all three horizons. For 2-hour prediction, the FBP yielded similar results in R^2 and MAE to SVM, while beyond this horizon, it outperformed the SVM model. This is because SVM displayed a stark decline in accuracy with the increase in prediction horizon. For the long-term forecast, FBP resulted in a better R^2 and RMSE, while SVM yielded a better MAE (Table 5). Adding extraterrestrial radiation to the model enhanced the performance of SVM and FBP for both the short- and long-term predictions (Tables 4 and 5). For the short-term prediction, R^2 increased by ca. 7% for FBP, and between 5% (for 1 hour ahead) and 10.5%, (for 3 hours ahead) for the SVM model. MAE decreased noticeably for FBP, by ca. 34 W/m^2, and also, but less drastically, for the SVM model. RMSE also decreased for both algorithms. The SVM model, which included global and extraterrestrial radiation of the same hour and day, 1 and 2 years ago, yielded the best results. The R^2 value in the long-term model increased by 7% for FBP and 17% for SVM. Furthermore, MAE and RMSE were reduced substantially. Overall, the addition of extraterrestrial radiation resulted in considerable improvements of all models. Extraterrestrial radiation on a horizontal surface is a good indicator of potential global horizontal irradiance, stating how much solar radiation is received at the top of the atmosphere for a certain location [51].

Table 4. Short-term results using data-driven and contextual optimisation.

Algorithm	Forecast Horizon	Data-Driven			Contextual	
		Initial Features	Additional Features	Tuned	Negative Elimination and Night Hours	Overall Improvement
				R^2		
SVM	1 h	83.27%	87.45%	87.64%	87.64%	5.25%
	2 h	76.02%	82.74%	83.07%	83.07%	9.27%
	3 h	72.05%	79.56%	80.02%	80.02%	11.07%
FBP	1 h	77.84%	83.46%	83.46%	83.55%	7.34%
	2 h	77.82%	83.37%	83.37%	83.46%	7.24%
	3 h	77.82%	83.33%	83.33%	83.41%	7.18%
				MAE		
SVM	1 h	73.42	55.2	46.88	46.70	36.40%
	2 h	84.71	67.5	58.86	58.86	30.51%
	3 h	89.98	74.65	65.99	65.96	26.69%
FBP	1 h	99.49	65.39	65.39	60.69	39.00%
	2 h	99.53	65.62	65.62	60.97	38.74%
	3 h	99.54	65.79	65.79	61.23	38.49%
				RMSE		
SVM	1 h	124.93	108.2	107.37	107.37	14.06%
	2 h	149.55	126.90	125.68	125.68	15.96%
	3 h	161.48	138.07	136.51	136.51	15.47%
FBP	1 h	135.03	116.47	116.47	116.30	13.87%
	2 h	135.07	116.78	116.78	116.63	13.65%
	3 h	135.09	116.94	116.94	116.81	13.53%

Table 5. Long-term results using data-driven and contextual optimisation.

Algorithm	Data-Driven				Contextual			
	Initial Features	Additional Features	Tuned	Negative Elimination	Negative Elimination and Night Hours	Seasonal Adaptation	Overall Improvement
				R^2				
SVM	68.72%	80.32%	80.78%	80.78%	80.78%	82.56%	20.13%
FBP	77.83%	83.11%	83.11%	83.19%	83.22%	83.97%	7.89%
				MAE				
SVM	72.67	54.02	55.61	55.57	55.56	56.67	22.01%
FBP	99.57	67.07	67.07	63.96	62.42	57.81	41.94%
				RMSE				
SVM	160.33	127.19	125.67	125.67	125.67	119.74	25.32%
FBP	135.02	117.85	117.85	117.59	117.46	114.83	14.95%

The hyperparameters were tuned for the SVM model, using grid search cross validation. The tunable parameters were the regularisation parameter C, the size of the error tube ε, and the width of the RBF kernel γ. The influence of ε and γ were minimal, leading to improvements of less than 0.0004% in R^2. Therefore, it was focused on tuning the regulation parameter C. SVMs are generally strongly dependent on their hyperparameters [10]. However, tuning the hyperparameters for these models did not lead to significant improvements. For the short-term prediction, $C = 120$ led to the best results. This, however, only improved R^2 by 0.5%, MAE by 8.6 W/m^2, and RMSE by 1.6 W/m^2. These improvements were low, compared with the addition of features. For the long-term prediction, the best C was 0.5. The improvements for this were even smaller.

The results of the data-driven model can be seen in Figure 8, displaying the same trend as described for the initial model (untuned, without added features).

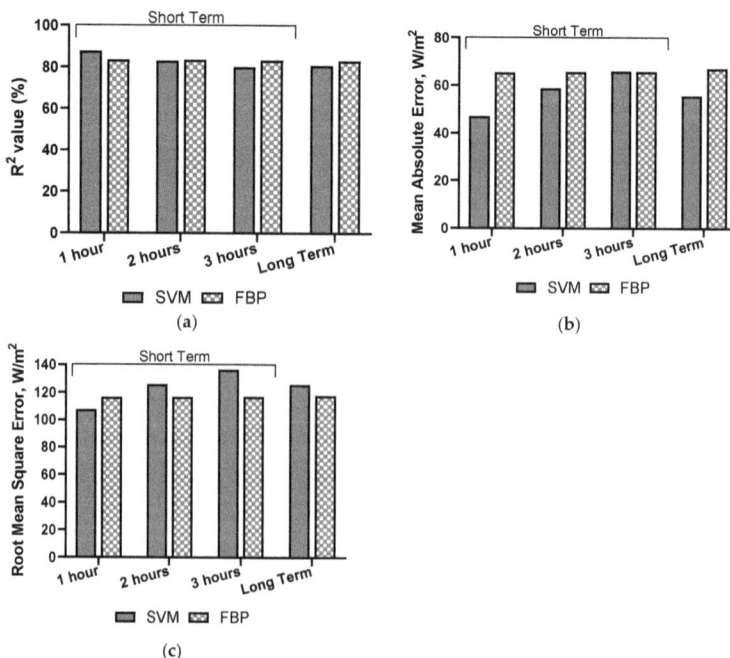

Figure 8. Error metrics comparing SVM and FBP for data-driven short- and long-term forecasts in Denver. (**a**) Coefficient of determination (R^2) (**b**) Mean absolute error (**c**) Root-mean-square error.

5.2. Contextual Optimisation Results

The results of further contextual optimisation are presented in this section. Setting all negative values to zero slightly improved the SVM model. It further enhanced the model, as it does not confuse the user with the prediction of impossible (negative) values. As the FBP short-term model had larger negative predictions, eliminating these led to greater improvements. The R^2 increased by 3% and MAE and RMSE decreased by 26 W/m^2 and 8 W/m^2, respectively. The long-term model improvements were less significant. As neither of the models predicted negative solar radiation during the day, setting all values to zero was appropriate. A model that predicted zero values at night, instead of negative values, was a closer reflection of reality.

There were some positive predictions at night. As this was not possible, sunrise and sunset adjustments were applied. Setting all values from sunset to sunrise to zero gave slightly better prediction results than defining all night hours as 8 p.m.–6 a.m. This was to be expected and true for short- and long-term predictions, in both SVM and FBP models. Including the flexible sunrise and sunset in the model allowed it to be easily applied to a location with different geographical conditions. This is particularly important in locations that are far from the equator, as sunset and sunrise vary more over the year in those places. However, it must be noted that including this adjustment into the model requires extra computational power. In locations where there is no significant variation in sunset and sunrise times during the year, this step may not be worth the marginally improved performance.

Seasonal adaptation only applied to the long-term forecast. There were three versions of this amendment, using the mean (V1), the median (V2), and the mean of the mean and median (V3). For SVM, the seasonal adaptation had a greater impact on the model with additional features. Version 1 performed best for the R^2 value, reducing the error by 11% and decreasing RMSE by 7 W/m^2, as shown in Figure 9. However, MAE increased by 6 W/m^2, which should be avoided. Version 2 performed better for MAE, decreasing it. However, the R^2 value decreased by 0.2% and RMSE increased slightly, which is also not desirable. Version 3 combines aspects of both preceding versions, offering more continuity and stable results. The R^2 and RMSE values for this version were better in comparison with the previous amendment (sunrise and sunset), while MAE was very similar. Therefore, version 3 of the seasonal adaptation, using the mean of the median and mean, was chosen as the last amendment for the long-term SVM model. The improvement of applying the seasonal adaptation can clearly be observed in Figure 10.

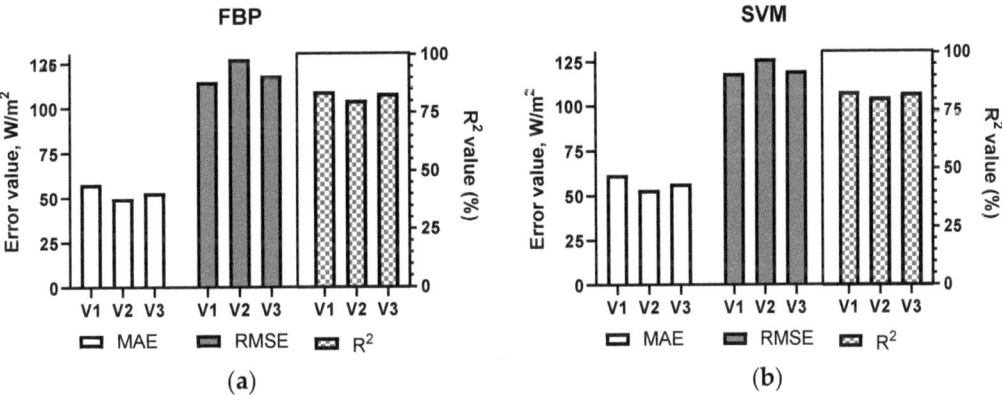

Figure 9. Comparison of (**a**) FBP and (**b**) SVM of seasonal adaptation versions.

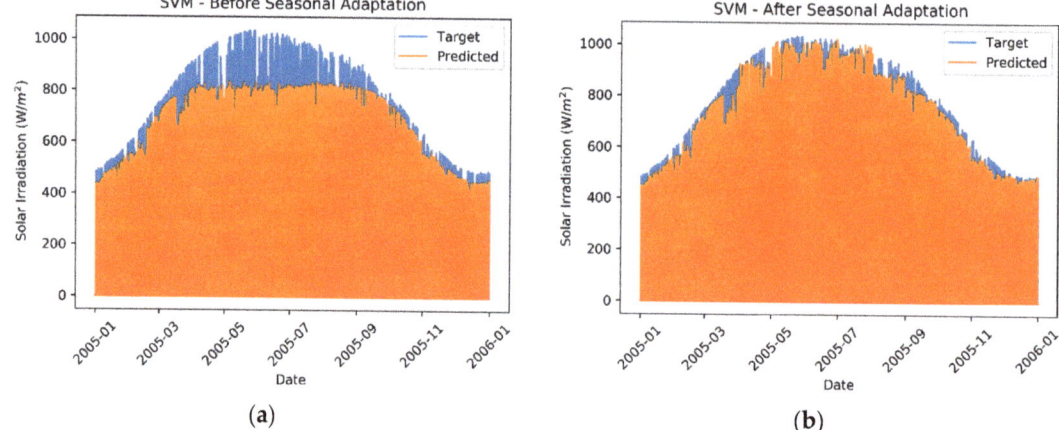

Figure 10. SVM (**a**) before and (**b**) after seasonal adaptation.

For FBP, the improvement on the model with additional features was marginal. As version 1 (using the mean as the average) led to improvements for all metrics, it was chosen for the FBP model. Interestingly, applying the seasonal adaptation to the FBP model without the extraterrestrial radiation led to results in R^2, MAE, and RMSE that were only slightly different from the model with extraterrestrial radiation. The seasonal adaptation had a greater positive impact on the model without extraterrestrial radiation, as shown in Table 6, with the addition of correcting the daily seasonality. The impact on this model was larger because the yearly and daily seasonality were both corrected, while for the model with extraterrestrial radiation mostly daily seasonality was adjusted. Thus, using a model without extraterrestrial radiation could be considered if these data are not available.

Table 6. Comparison of influence on seasonal adaptation on FBP models with different features.

	Initial Features			Initial + Additional Features		
	Sunset and Sunrise	Seasonal Adaptation	Improvement	Sunset and Sunrise	Seasonal Adaptation	Improvement
R^2	80.56%	83.35%	2.79%	83.22%	83.97%	0.74%
MAE	70.1	61.51	8.60	62.42	57.81	4.61
RMSE	126.4	117.01	9.42	117.46	114.83	2.62

Tables 4 and 5 display the results of all steps of data-driven and contextual parts for short- and long-term forecasts. It is clear that the accuracy was enhanced at each step of the algorithm, starting from the initial features training to the SA. The proposed model changes improved R^2 of the short-term model by 5% (1 h) to 11% (3 h) for SVM and 7% for FBP. The MAE for the FBP model decreased by 39 W/m^2 and by ca. 25 W/m^2 for SVM. RMSE was also decreased by 17 to 24 W/m^2 for SVM and 18 W/m^2 for FBP. The overall R^2 improvement associated with model changes for the long-term forecast is 20% for SVM and 8% for FBP, as shown in Table 5. MAE decreased by 42 W/m^2 for FBP but only by 16 W/m^2 for SVM. For SVM, however, RMSE decreased by 41 W/m^2, whereas for FBP, it decreased by 20 W/m^2.

The insights of the individual model results for different horizons were taken to determine which algorithm to use for which horizon in the final DCF. For DCF, the highest accuracy for the 1- and 2-hour predictions was achieved using SVM with extraterrestrial radiation as an additional input feature, using the dynamic night-time adjustment and version 3 of the seasonal adaptation. Figure 11 shows that the 1-hour prediction SVM displayed a compact trend line with only a few normally distributed errors. For FBP, most values were on a line that was slightly too steep, indicating an overprediction for those values. However, there were also many points below the dense line, signalling underprediction. For the 3-hour and long-term predictions, the FBP using V1 of the seasonal adaptation outperformed all the other versions and algorithms. It can be concluded that the SVM model should be used for 1- and 2-hour ahead predictions, while beyond that, the FBP model should be utilised in the final DCF.

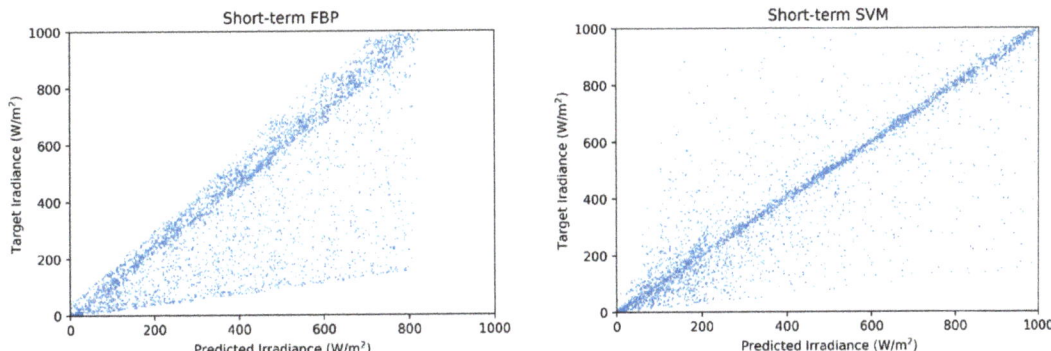

Figure 11. Short-term FBP, SVM predicted, and target values: 1 h ahead.

The performance of FBP suffered less from an increase in horizon than the SVM model. This is due to the underlying characteristics of the algorithm; FBP is specifically designed for time-series prediction [30]. An advantage is that the performance declines less over time. However, inputting the whole past time series into the model did not allow emphasising values that had a higher correlation and were more relevant to the particular prediction. For SVM, this could be differentiated.

5.3. DCF Performance

In this section, the DCF performance for short- and long-term forecasting is presented. To validate its performance and ensure that DCF is a generic model that can be utilised for different locations, forecasts were conducted for three cities, i.e., Denver, Boston, and Seattle.

The results for all three cities and both algorithms are presented in Table 7. It can be seen that the SVM model performed even better on the short-term prediction in Seattle and Boston than for Denver, while the general trend remained the same as for the Denver results. For the long-term prediction, Denver displayed the best results in terms of R^2; however, both MAE and RMSE were as low or lower for Boston and Seattle than for Denver. Again, the SVM model mostly outperformed FBP in the 1- and 2-hour forecasts, while the FBP model generally generated better results for 3-hour prediction and in the long term. This was observed similarly in the results and its trend validated the chosen DCF model.

Two days of short-term predictions by the DCF algorithm are displayed in Figure 12. It shows that the model was noticeably accurate for sunny days (first day), with smooth irradiance transitions. Furthermore, it captured trends for changes in weather, as can be observed on the second day. Despite the rapid change in irradiance, the model still generated accurate predictions.

As shown in Figure 13, DCF was applicable to different locations, conserving the general pattern of performance. This validated the DCF algorithm and provided us with confidence that this model will perform well in other not-yet-tested locations. Results of around 90% (91.2%, 90.6%, and 87.6%) for the 1-hour predictions were achieved for R^2, while MAE ranged from 36 W/m^2 for Seattle to 47 W/m^2 for Denver and RMSE from 75 W/m^2 for Seattle to 107 W/m^2 for Denver. For the 2-hour forecast, the R^2 value declined by about 5%, and MAE and RMSE increased by ca. 12 and 18 W/m^2, respectively, for all locations. The 3-hour prediction still generated R^2 of 78% (Seattle) to about 83% (Denver and Boston), while MAE ranged from 56 (Seattle) to about 61 W/m^2 (Denver and Boston) and RMSE from 103 W/m^2 (Seattle) to 116 W/m^2 (Denver and Boston). Even the long-term prediction for one year ahead still generated good results for all cities, with high R^2 values and low error values, as shown in Figure 13.

Table 7. Comparison of SVM and FBP performance in all cities.

Algorithm	Horizon		Denver	Seattle	Boston
				R^2	
SVM	Short term	1 h	87.64%	90.62%	91.19%
		2 h	83.07%	85.80%	86.77%
		3 h	80.02%	82.32%	81.94%
	Long term	8760 h	82.56%	78.41%	75.92%
FBP	Short term	1 h	83.55%	78.35%	83.53%
		2 h	83.46%	78.32%	83.44%
		3 h	83.41%	78.32%	83.39%
	Long term	8760 h	83.97%	80.29%	78.32%
				MAE	
SVM	Short term	1 h	46.70	37.40	36.15
		2 h	58.86	49.27	47.48
		3 h	65.96	58.39	57.95
	Long term	8760 h	56.67	52.90	58.67
FBP	Short term	1 h	60.69	55.94	60.96
		2 h	60.97	56.01	61.23
		3 h	61.23	56.02	61.48
	Long term	8760 h	57.81	51.04	57.86
				RMSE	
SVM	Short term	1 h	107.37	75.36	77.05
		2 h	125.68	92.72	94.42
		3 h	136.51	103.48	110.29
	Long term	8760 h	119.74	106.39	116.94
FBP	Short term	1 h	116.30	103.54	116.37
		2 h	116.63	103.60	116.69
		3 h	116.81	103.60	116.87
	Long term	8760 h	114.83	98.77	110.99

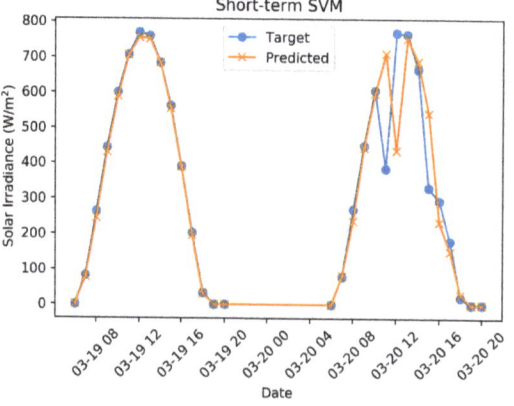

Figure 12. Two days of 1-hour ahead SVM prediction in Boston.

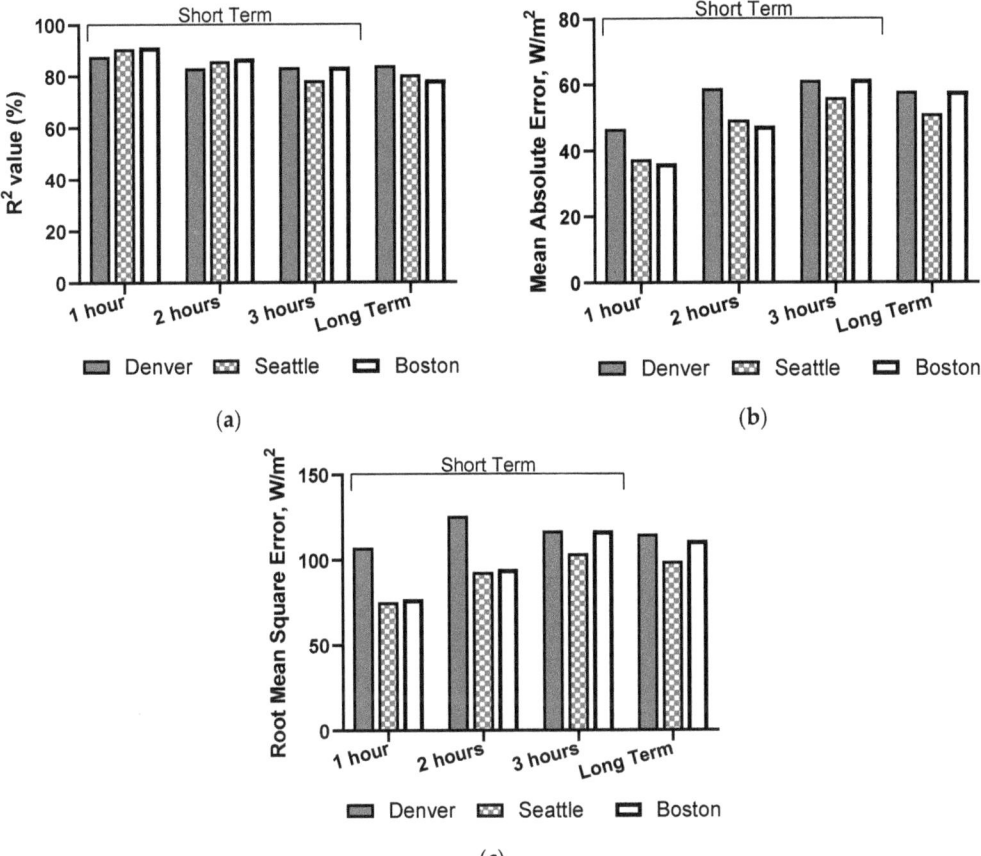

Figure 13. DCF accuracy, evaluated in three cities using all evaluation metrics. (**a**) Coefficient of determination (R^2) (**b**) Mean absolute error (**c**) Root-mean-squared error.

6. Conclusions

This paper presented the DCF algorithm, a forecasting algorithm that accurately predicts solar irradiance. Unlike other state-of-the-art models, the forecast accuracy was validated for short- and long-term predictions in three cities. The DCF algorithm had two main parts. Initially, it utilised the most accurate data-driven (ML) algorithms and then optimised their performance using contextual information. SVM and FBP were used as the data-driven models. SVM has been used for solar forecasting for over a decade. FBP, in contrast, is a novel algorithm that has rarely been used in the field of solar prediction. Nevertheless, its design characteristics seemed inherently promising for solar prediction.

Firstly, a basic model was constructed for both algorithms with only hourly solar irradiance as input. The data were taken from the National Solar Radiation Database (NSRDB). Adding extraterrestrial radiation led to the largest improvement in R^2, MAE, and RMSE, for both SVM and FBP models. For the SVM model, the regularisation parameter C was tuned using grid search cross validation. This did not have a significant impact on the performance of the model. After training the model with the additional input features and the tuned hyperparameters, solar irradiance was predicted. The prediction was subject to several adjustments. All negative values and all values between sunset and sunrise were set to zero. This had a greater impact on FBP than on SVM, as FBP would generate larger non-zero predictions at night. Furthermore, a seasonal adaptation was applied. This increased or decreased every hour of the day for each month if it was above or below the average of the last years. It led to a significant improvement, as shown in Table 6.

For the 1-hour short-term prediction, the final SVM model outperformed FBP and, thus, was utilised for the DCF algorithm. As shown in Table 7, it achieved an R^2 value of 87.6% for Denver, 90.6% for Seattle, and 91.2% for Boston. An MAE value of 36 W/m^2 was attained for Boston and similar values for Seattle and Denver. RMSE varied from 75 W/m^2 (Seattle) and 77 W/m^2 (Boston) to 107 W/m^2 (Denver). For the 2-hour prediction, SVM mostly outperformed FBP. On occasions in which this was not the case, the results were very similar. However, the SVM model displayed a strong decrease in forecasting accuracy with the increase in the forecast horizon. Therefore, for the 3-hour prediction, the FBP model yielded better results and thus was used beyond the 3-hour forecast in the DCF algorithm. The FBP performance only decreased very slightly over time, compared to the SVM. The reason for its sustained performance is its specific design for time-series predictions. The FBP model performed better for the long-term forecast than the SVM model. This was true for all cities and thus validated the use of the suggested model.

7. Future Research

Improvements may arise from analysing and adding further meteorological input features. This could, for example, be a measure of cloud cover or temperature. Care must be taken that no features are included that either worsen the prediction or have no positive impact while making the model more complicated. Adding features could be advantageous for the SVM model, as for SVM, any features can be added, while for FBP, only features that are known in the future can be added.

The SVM model might be improved by further analysing the correlation of the irradiance with past values. This could reveal correlations with hours that have not yet been used as input features. Adding these would be a promising path to further enhance the model. This also suggests another set of experiments that could be executed to examine the mid-term horizon for both SVM and FBP models. FBP might be better at mid-term forecasts, e.g., 3 months. However, this has not been experimentally investigated. A correlation analysis would be of great use for a mid-term SVM model and would therefore lend itself to being carried out in parallel with a comparative analysis of mid-term SVM and FBP models.

The long-term FBP model showed that applying the seasonal adaptation to Denver nearly made the extraterrestrial radiation redundant. Both models, with and without extraterrestrial radiation, displayed similar results. This could be useful for datasets that do not possess measurements of extraterrestrial radiation. Therefore, the benefits of only seasonal adaptation instead of adding extraterrestrial radiation to the model should be explored further.

Author Contributions: Conceptualization, P.B. and B.B.; Formal analysis, P.B. and B.B.; Funding acquisition, A.T. and Q.H.A.; Investigation, P.B. and B.B.; Software, P.B. and B.B.; Visualization, P.B. and B.B.; Writing—original draft, P.B. and B.B.; Writing—review & editing, P.B., A.T. and B.B. All authors have read and agreed to the published version of the manuscript.

Funding: This study was supported in part by the Engineering and Physical Sciences Research Council (EPSRC) Grants, EP/T517896/1.

Informed Consent Statement: Not applicable.

Data Availability Statement: Datasets related to this article can be found at https://nsrdb.nrel.gov/data-sets/archives.html, hosted by the National Solar Radiation Database (NSRDB) [35], accessed on 20 October 2021.

Acknowledgments: We would like to thank Aiste Steponenaite, from the University of Kent, UK, for her help in plotting the graphs.

Conflicts of Interest: The authors declare no conflict of interests.

References

1. Thompson, L.G. Climate change: The evidence and our options. *Behav. Anal.* **2010**, *33*, 153–170. [CrossRef]
2. EPA-United States Environmental Protection Agency. Sources of Greenhouse Gas Emissions. Available online: https://www.epa.gov/ghgemissions/sources-greenhouse-gas-emissions (accessed on 20 October 2021).
3. Newell, P.; Simms, A. How Did We Do That? Histories and Political Economies of Rapid and Just Transitions. *New Political Econ.* **2020**, *26*, 907–922. [CrossRef]
4. Huybrechts, B. Social Enterprise, Social Innovation and Alternative Economies: Insights from Fair Trade and Renewable Energy. In *Alternative Economies and Spaces: New Perspectives for a Sustainable Economy*; Transcript Verlag: Bielefeld, Germany, 2013; pp. 113–130.
5. Jia, Y.; Lyu, X.; Lai, C.S.; Xu, Z.; Chen, M. A retroactive approach to microgrid real-time scheduling in quest of perfect dispatch solution. *J. Mod. Power Syst. Clean Energy* **2019**, *7*, 1608–1618. [CrossRef]
6. Perera, K.S.; Aung, Z.; Woon, W.L. Machine Learning Techniques for Supporting Renewable Energy Generation and Integration: A Survey. In *International Workshop on Data Analytics for Renewable Energy Integration*; Springer: Cham, Switzerland, 2014; pp. 81–96. [CrossRef]
7. Fouilloy, A.; Voyant, C.; Notton, G.; Motte, F.; Paoli, C. Solar irradiation prediction with machine learning: Forecasting. *Energy* **2018**, *165*, 620–629. [CrossRef]
8. IEA. *Fossil Fuel Energy Consumption*; International Energy Agency: Paris, France, 2020.
9. International Renewable Energy Agency. IRENA—Download Data. 2020. Available online: https://www.irena.org/Statistics/Download-Data (accessed on 12 January 2021).
10. Van der Wiel, K.; Bloomfield, H.C.; Lee, R.W.; Stoop, L.P.; Blackport, R.; Screen, J.A.; Selten, F.M. The influence of weather regimes on European renewable energy production and demand. *Environ. Res. Lett.* **2019**, *14*, 94010. [CrossRef]
11. Staffell, I.; Pfenninger, S. The increasing impact of weather on electricity supply and demand. *Energy* **2017**, *145*, 65–78. [CrossRef]
12. Sobri, S.; Koohi-Kamali, S.; Rahim, N.A. Solar photovoltaic generation forecasting methods: A review. *Energy Convers. Manag.* **2018**, *156*, 459–497. [CrossRef]
13. Arraez-Cancelliere, O.A.; Muñoz-Galeano, N.; López-Lezama, J.M. Computing the Global Irradiation over the Plane of Photovoltaic Arrays: A Step-by-Step Methodology. In *Renewable Energy—Technologies and Applications*; Taner, T., Tiwari, A., Ustun, T.S., Eds.; IntechOpen: London, UK, 2020.
14. Wang, H.; Liu, Y.; Zhou, B.; Li, C.; Cao, G.; Voropai, N.; Barakhtenko, E. Taxonomy research of artificial intelligence for deterministic solar power forecasting. *Energy Convers. Manag.* **2020**, *214*, 112909. [CrossRef]
15. Net Zero Strategy: Build Back Greener October 2021. Available online: https://assets.publishing.service.gov.uk/government/uploads/system/uploads/attachment_data/file/1033990/net-zero-strategy-beis.pdf (accessed on 20 October 2021).
16. Kahwash, F.; Barakat, B.; Taha, A.; Abbasi, Q.H.; Imran, M.A. Optimising Electrical Power Supply Sustainability Using a Grid-Connected Hybrid Renewable Energy System—An NHS Hospital Case Study. *Energies* **2021**, *14*, 7084. [CrossRef]
17. Sanjari, M.J.; Gooi, H.B. Probabilistic Forecast of PV Power Generation Based on Higher Order Markov Chain. *IEEE Trans. Power Syst.* **2016**, *32*, 2942–2952. [CrossRef]
18. Kavasseri, R.G.; Seetharaman, K. Day-ahead wind speed forecasting using f-ARIMA models. *Renew. Energy* **2009**, *34*, 1388–1393. [CrossRef]
19. Guariso, G.; Nunnari, G.; Sangiorgio, M. Multi-Step Solar Irradiance Forecasting and Domain Adaptation of Deep Neural Networks. *Energies* **2020**, *13*, 3987. [CrossRef]
20. Jiang, H.; Dong, Y. Forecast of hourly global horizontal irradiance based on structured kernel support vector machine: A case study of Tibet area in China. *Energy Convers. Manag.* **2017**, *142*, 307–321. [CrossRef]
21. Zendehboudi, A.; Baseer, M.; Saidur, R. Application of support vector machine models for forecasting solar. *J. Clean. Prod.* **2018**, *199*, 272–285. [CrossRef]
22. Bae, K.Y.; Jang, H.S.; Sung, D.K. Hourly solar irradiance prediction based on support vector machine and its error analysis. *IEEE Trans. Power Syst.* **2016**, *32*, 935–945. [CrossRef]
23. Mueller, K.R.; Smola, A.J.; Raetsch, G.; Schoelkopf, B.; Kohlmorgen, J. *Using Support Vector Machines for Time Series Prediction*; GMD FIRST: Berlin, Germany, 2000.
24. Fentis, A.; Bahatti, L.; Mestari, M.; Chouri, B. Short-term solar power forecasting using Support Vector Regression and feed-forward NN. In Proceedings of the 2017 15th IEEE International New Circuits and Systems Conference (NEWCAS), Strasbourg, France, 25–28 June 2017; pp. 405–408. [CrossRef]
25. Ramedani, Z.; Omid, M.; Keyhani, A.; Shamshirband, S.; Khoshnevisan, B. Potential of radial basis function based support vector regression for global solar radiation prediction. *Renew. Sustain. Energy Rev.* **2014**, *39*, 1005–1011. [CrossRef]
26. Meenal, R.; Selvakumar, A.I. Assessment of SVM, empirical and ANN based solar radiation prediction models with most influencing input parameters. *Renew. Energy* **2018**, *121*, 324–343. [CrossRef]
27. Boyd, S. *Convex Optimization*; Cambridge University Press: Cambridge, UK, 2004.
28. Zeng, J.; Qiao, W. Short-term solar power prediction using a support vector machine. *Renew. Energy* **2013**, *52*, 118–127. [CrossRef]
29. VanDeventer, W.; Jamei, E.; Thirunavukkarasu, G.S.; Seyedmahmoudian, M.; Soon, T.K.; Horan, B.; Mekhilef, S.; Stojcevski, A. Short-term PV power forecasting using hybrid GASVM technique. *Renew. Energy* **2019**, *140*, 367–379. [CrossRef]
30. Taylor, S.J.; Letham, B. Forecasting at Scale. *Am. Stat.* **2017**, *72*, 37–45. [CrossRef]

31. Lim, J.Y.; Safder, U.; How, B.S.; Ifaei, P.; Yoo, C.K. Nationwide sustainable renewable energy and Power-to-X deployment planning in South Korea assisted with forecasting model. *Appl. Energy* **2020**, *283*, 116302. [CrossRef]
32. Shawon, M.H.; Akter, S.; Islam, K.; Ahmed, S.; Rahman, M. Forecasting PV panel output using prophet time. In Proceedings of the 2020 IEEE REGION 10 CONFERENCE (TENCON), Osaka, Japan, 16–19 November 2020.
33. Žunić, E.; Korjenić, K.; Hodžić, K.; Dženana, Ð. Application of Facebook's Prophet Algorithm for Successful Sales Forecasting Based on Real-world Data. *Int. J. Comput. Sci. Inf. Technol.* **2020**, *12*, 23–36. [CrossRef]
34. National Renewable Energy Laboratory. *National Solar Radiation Database 1991–2005 Update: User's Manual*; National Renewable Energy Laboratory: Golden, CA, USA, 2007.
35. Kalogirou, S. *Solar Energy Engineering: Processes and Systems*; Academic Press: Cambridge, MA, USA, 2013.
36. Malvoni, M.; De Giorgi, M.G.; Congedo, P.M. Forecasting of PV Power Generation using weather input data-preprocessing techniques. *Energy Procedia* **2017**, *126*, 651–658. [CrossRef]
37. Voyant, C.; Notton, G.; Kalogirou, S.; Nivet, M.-L.; Paoli, C.; Motte, F.; Fouilloy, A. Machine learning methods for solar radiation forecasting: A review. *Renew. Energy* **2017**, *105*, 569–582. [CrossRef]
38. Sreekumar, S.; Bhakar, R. Solar Power Prediction Models: Classification Based on Time Horizon, Input, Output and Ap-plication. In Proceedings of the 2018 International Conference on Inventive Research in Computing Applications (ICIRCA), Coimbatore, India, 3 January 2019.
39. Martín-Pomares, L.; Martínez, D.; Polo, J.; Perez-Astudillo, D.; Bachoura, D. Analysis of the long-term solar potential for electricity generation in Qatar. *Renew. Sustain. Energy Rev.* **2017**, *73*, 1231–1246. [CrossRef]
40. Olatomiwa, L.; Mekhilef, S.; Shamshirband, S.; Mohammadi, K.; Petković, D.; Sudheer, C. A support vector machine–firefly algorithm-based model for global solar radiation prediction. *Sol. Energy* **2015**, *115*, 632–644. [CrossRef]
41. Quej, V.H.; Almorox, J.; Arnaldo, J.A.; Saito, L. ANFIS, SVM and ANN soft-computing techniques to estimate daily global solar radiation in a warm sub-humid environment. *J. Atmos. Sol.-Terr. Phys.* **2017**, *155*, 62–70. [CrossRef]
42. Wolff, B. Statistical Learning for Short-Term Photovoltaic Power Predictions. In *Computational Sustainability*; Springer: Berlin/Heidelberg, Germany, 2016.
43. Long, H.; Zhang, Z.; Su, Y. Analysis of daily solar power prediction with data-driven approaches. *Appl. Energy* **2014**, *126*, 29–37. [CrossRef]
44. Van Rossum, G. *Python Tutorial*; Centrum voor Wiskunde en Informatica (CWI): Amsterdam, The Netherlands, 1995.
45. Pedregosa, F.; Varoquaux, G.; Gramfort, A.; Michel, V.; Thirion, B.; Grisel, O.; Blondel, M.; Prettenhofer, P.; Weiss, R.; Dubourg, V.; et al. Scikit-learn: Machine Learning in Python. *J. Mach. Learn. Res.* **2011**, *12*, 2825–2830.
46. Guyon, I.; Elisseeff, A. An Introduction to Variable and Feature Selection. *J. Mach. Learn. Res.* **2003**, *3*, 1157–1182.
47. Abuella, M.; Chowdhury, B. Solar Power Forecasting Using Support Vector Regression. In Proceedings of the American Society for Engineering Management International Annual Conference Charlotte, NC, USA, 26–29 October 2016.
48. Mantovani, R.G.; Rossi, A.L.D.; Vanschoren, J.; Bischl, B.; de Carvalho, A.C.P.L.F. Effectiveness of Random Search in SVM hyper-parameter tuning. In Proceedings of the 2015 International Joint Conference on Neural Networks (IJCNN), Killarney, Ireland, 12–17 July 2015; pp. 1–8. [CrossRef]
49. Dong, Z.; Yang, D.; Reindl, T.; Walsh, W.M. A novel hybrid approach based on self-organizing maps, support vector regression and particle swarm optimization to forecast solar irradiance. *Energy* **2015**, *82*, 570–577. [CrossRef]
50. Piri, J.; Shamshirband, S.; Petković, D.; Tong, C.W.; Rehman, M.H.U. Prediction of the solar radiation on the Earth using support vector regression technique. *Infrared Phys. Technol.* **2015**, *68*, 179–185. [CrossRef]
51. Maleki, S.A.M.; Hizam, H.; Gomes, C. Estimation of Hourly, Daily and Monthly GlobalSolar Radiation on Inclined Surfaces: Models Re-Visited. *Energies* **2017**, *10*, 134. [CrossRef]

Article

Transmission Line Fault-Cause Identification Based on Hierarchical Multiview Feature Selection

Shengchao Jian [1], Xiangang Peng [1,*], Haoliang Yuan [1,*], Chun Sing Lai [1,2,*] and Loi Lei Lai [1,*]

[1] Department of Electrical Engineering, School of Automation, Guangdong University of Technology, Guangzhou 510006, China; 2111904005@mail2.gdut.edu.cn
[2] Brunel Interdisciplinary Power Systems Research Centre, Department of Electronic and Electrical Engineering, Brunel University London, London UB8 3PH, UK
* Correspondence: epxg@gdut.edu.cn (X.P.); haoliangyuan@gdut.edu.cn (H.Y.); chunsing.lai@brunel.ac.uk (C.S.L.); l.l.lai@gdut.edu.cn (L.L.L.)

Abstract: Fault-cause identification plays a significant role in transmission line maintenance and fault disposal. With the increasing types of monitoring data, i.e., micrometeorology and geographic information, multiview learning can be used to realize the information fusion for better fault-cause identification. To reduce the redundant information of different types of monitoring data, in this paper, a hierarchical multiview feature selection (HMVFS) method is proposed to address the challenge of combining waveform and contextual fault features. To enhance the discriminant ability of the model, an ε-dragging technique is introduced to enlarge the boundary between different classes. To effectively select the useful feature subset, two regularization terms, namely $l_{2,1}$-norm and Frobenius norm penalty, are adopted to conduct the hierarchical feature selection for multiview data. Subsequently, an iterative optimization algorithm is developed to solve our proposed method, and its convergence is theoretically proven. Waveform and contextual features are extracted from yield data and used to evaluate the proposed HMVFS. The experimental results demonstrate the effectiveness of the combined used of fault features and reveal the superior performance and application potential of HMVFS.

Keywords: fault-cause identification; transmission line; sparse learning; multiview learning; feature selection

1. Introduction

Transmission lines cover a wide area and work in diverse outdoor environments to achieve long-distance, high-capacity power transmission. In order to maintain stable power supply, high-speed fault diagnosis is indispensable for line maintenance and fault disposal.

Traditional fault diagnosis technologies concerning fault detecting, fault locating, and phase selection are well developed [1,2], while diagnosis on external causes is still underdeveloped. Operation crews attach great importance to fault location for line patrol and manual inspection. However, on-site inspection is labor-intensive and depends on subjective judgment. Moreover, cause identification after inspection is too late for dispatchers to give better instructions according to the external cause, such as forced energization. Fault-cause identification is expected to help dispatch and maintenance personnel make a proper and speedy fault response.

Transmission line faults are more often triggered by external factors due to environmental change or surrounding activities. Though the cause categories are slightly different between regions or institutions, the common causes can be listed as lighting, tree, animal contact, fire, icing, pollution and external damage [3]. Considering complexity and variability of open-air work, it is hard to model fault scenarios for diverse root causes [4,5]. Thus, these existing studies on line fault-cause identification have been developed based on data-driven methods rather than physical modeling.

The early identification methods were rule-based, such as statistical analysis, CN2 rule induction [6] and fuzzy inference system (FIS) [7–9]. Their identification frameworks are finally presented in the form of logic flow, demanding a great degree of robustness and generality for their rules or thresholds. In recent years, various machine learning (ML) techniques that attach great importance to hand-crafted features have been applied to diagnose external causes [10–14], such as logistic regression (LR), artificial neural network (ANN), k-nearest neighbor (KNN) and support vector machine (SVM). Deep learning (DL) provides a more efficient way in the field of fault identification. In [15], deep belief network (DBN) is used as the classification algorithm after extracting time–frequency characteristics from traveling wave data. Even when using DL methods, feature engineering is still an inevitable part to achieve high accuracy.

Feature signature study provides knowledge about fault information and plays a critical role in fault-cause identification. On the one hand, when fault events happen, power quality monitors (PQMs) enable us to have easy access to electrical signals and time stamps [16]. Time-domain features extracted from fault waveform and time stamp were used to construct logic flow to classify lightning-, animal- and tree-induced faults [6]. To exploit transient characteristics in the frequency domain, signal processing techniques such as wavelet transform (WT) and empirical mode decomposition (EMD) are used for further waveform characteristic analysis [17–20]. In [21], a fault waveform was characterized based on the time and frequency domain to develop an identification logic. However, a fault waveform is easily affected by the system operation state, and there is no direct connection between these characteristics and external causes. On the other hand, weather condition is directly relevant to many fault-cause categories such as lightning, icing and wind. With the development of monitoring equipment and communication technology, dispatchers now can make judgments with more and more outdoor information [22]. These nonwaveform characteristics such as time stamps, environment attributes and other textual data are called contextual characteristics in this paper. Table 1 lists and compares the characterization and classification methods in existing works.

Table 1. A summarized list of characterization and classification methods used for fault-cause identification.

Article	Waveform Characteristics				Time Characteristics	External Characteristics	Classification Methods
	Signal Amplitude	Sequence Component	Spectrum Analysis	Phase or Phase Angle			
* Núñez, Meléndez [6]	✓	✓		✓	✓		CN2
Liang, Li [7]	✓	✓					FIS
* Xu, Chow [8–10]			✓		✓	✓	FIS/LR/ANN
* Cai, Chow [11]			✓		✓	✓	LR
Chang, Hong [12]	✓		✓				SVM
* Jiang, Liu [14]	✓		✓	✓	✓	✓	KNN
Liang, Liu [15]	✓		✓		✓		DBN
Asman, Aziz [20]	✓		✓				decision tree
* Qin, Wang [21]	✓	✓	✓	✓	✓		logic flow
* Dehbozorgi, Rastegar [22]					✓	✓	decision tree
Minnaar, Nicolls [23]	✓	✓	✓		✓	✓	KNN

Articles with * concern faults on distribution network but their work is still inspiring for transmission network.

Studies have shown that waveform and contextual features can achieve high accuracy without each other, but there are high data requirements. For economic and operational reasons, data condition will not change significantly in the short term. It is necessary to study performance improvement for fault-cause identification based on current data conditions. One of the challenges is determining how to combine waveform features and multisource contextual features. This is an information fusion problem, and the simplest approach is feature concatenation. The authors of [23] tried to combine contextual features and waveform features as a mixed vector, but concatenated features reduce performance. Moreover, in contrast to focusing on either side, a few studies use both waveform and contextual characteristics for higher classification performance.

To tackle the fusion challenge, multiview learning (MVL) is introduced in this paper because waveform and contextual features describe the same fault event in different views. MVL aims to integrate multiple-view data properly and overcome biases between multiple views to obtain satisfactory performance. One of typical MVL methods is canonical correlation analysis (CCA), which maps multiview features into a common feature space [24]. Instead of mapping features, multiview feature selection that selects features from each view is preferred in fault-cause identification. Unlike traditional feature selection, multiview feature selection treats multiview data as inherently related and ensures that complementary information from other views is exploited [25,26]. In [27], a review on real-time power system data analytics with wavelet transform is given. The use of discrete wavelet transform was used to identify the high impedance fault and heavy load conditions [28]. The authors of [29] propose a fault diagnosis approach for the main drive chain in a wind turbine based on data fusion. To deal with the kind of multivariable fault diagnosis problem for which input variables need to be adjusted for different typical faults, the deep autoencoder model is adopted for the fault diagnosis model training for different typical fault types.

In this paper, we propose a hierarchical multiview feature selection (HMVFS) method for transmission line fault-cause identification. Two view datasets are composed of the waveform features and the contextual features. Our proposed HMVFS is applied to conduct the feature selection for the optimal feature combination. In our model, to enhance the discriminant ability of regression, an ε-dragging technology is used to enlarge the margin between classes. Next, two regularization terms, namely $l_{2,1}$-norm and Frobenius norm (F-norm) penalty, are adopted to perform the hierarchical feature selection. Here, the $l_{2,1}$-norm realizes the row sparsity to reduce the unimportant features of each view and the F-norm realizes the view-level sparsity to reduce the diversity between these two-view data. Hence, these two penalties can be viewed as low-level and high-level feature selection, respectively. At last, the fault-cause identification is carried out using ML classifiers and integrated features. The contributions of this paper are highlighted as follows:

- To the best of our knowledge, this is the first time that multiview learning is introduced for transmission line fault-cause identification in view of the nature of multiview fault data.
- We propose a novel approach, HMVFS, based on the ε-dragging and two regularization terms to select the discriminative features across views. We also develop an iterative algorithm to solve the optimization problem and prove its convergence theoretically.
- The performance of HMVFS is evaluated on field data and compared with classical feature selection methods. Experimental results prove the effectiveness of combining waveform and contextual features and demonstrate the feasibility and superiority of HMVFS.

The rest of this paper is organized as follows: Section 2 presents the proposed HMVFS algorithm and its convergence analysis. Section 3 outlines the real-life line fault dataset and extracts features in terms of waveform and nonwaveform. The empirical study is provided and discussed in Section 4. Section 5 presents concluding remarks.

2. Hierarchical Multiview Feature Selection (HMVFS)

2.1. Notation

Sparsity-based multiview feature selection can be formulated as an optimization problem and denoted by loss functions and regularization items. Before introducing our formulation, the notation is stated.

Matrices are denoted by boldface uppercase letters, and vectors are denoted by boldface lowercase letters. Given original feature matrix $\mathbf{X} = [\mathbf{x}_1, \mathbf{x}_2, \ldots, \mathbf{x}_n]^T \in \mathbb{R}^{n \times d}$, each row of which corresponds to a fault instance, n is the total number and d denotes the size of features. $\mathbf{X}^{(v)} \in \mathbb{R}^{n \times d^{(v)}}$ and $\mathbf{x}_i^{(v)} \in \mathbb{R}^{d^{(v)}}$ denote a feature matrix and a vector in the

vth view. There are two views in this paper; thus, $\mathbf{X} = [\mathbf{X}^{(1)}, \mathbf{X}^{(2)}]$. Suppose there are c categories, the label matrix will be represented as $\mathbf{Y} = [\mathbf{y}_1, \mathbf{y}_2, \ldots, \mathbf{y}_n]^T \in \{0,1\}^{n \times c}$. Weight matrix $\mathbf{W}$ can be derived as $\mathbf{W} = [\mathbf{W}^{(1)}, \mathbf{W}^{(2)}]^T = [\mathbf{w}_1, \mathbf{w}_2, \ldots, \mathbf{w}_d]^T \in \mathbb{R}^{d \times c}$.

2.2. The Objective Function

Given the notation defined and a fault dataset $(\mathbf{X}, \mathbf{Y})$, the problem of HMVFS is transformed into determining weight matrix $\mathbf{W}$ and then ranking features for selection. We formulate the optimization problem as

$$\min_{\mathbf{W},\mathbf{M}} \Psi(\mathbf{W},\mathbf{M}) + \alpha \Phi(\mathbf{W}) + \beta \Omega(\mathbf{W}) = \min_{\mathbf{W},\mathbf{M}} \|\mathbf{X}\mathbf{W} - \mathbf{Y} - \mathbf{B} \otimes \mathbf{M}\|_F^2 + \alpha \|\mathbf{W}\|_{2,1} + \beta \sum_{v=1}^m \|\mathbf{W}_v\|_F, \quad (1)$$

where m is the view number; $m = 2$ in this paper.

In this formulation, $\Psi(\mathbf{W}, \mathbf{M})$ is the loss function that measures the calculation distance to achieve minimum regression error, which is derived from the least square loss function. Furthermore, the ε-dragging is introduced to drag binary outputs in $\mathbf{Y}$ away along two opposite directions. The outputs for positive digits will become $1 + \varepsilon_i$ and the outputs for negative digits will be $-\varepsilon_i$, in which all of the εs are nonnegative. The treatment that enlarges the distance between data points from different classes helps to develop a compact optimization model for classification [30]. $\mathbf{B} \in \{-1,1\}^{n \times c}$ in the formulation is a constant matrix, and its element B_{ij} is defined as

$$B_{ij} = \begin{cases} +1, & Y_{ij} = 1 \\ -1, & Y_{ij} = 0. \end{cases} \quad (2)$$

B_{ij} denotes the dragging direction for elements in label matrix $\mathbf{Y}$. $\mathbf{M} \in \mathbb{R}^{n \times c}$ is a nonnegative matrix that records all εs. The operator $\otimes$ is the Hadamard product operator of matrices. Thus, $\mathbf{B} \otimes \mathbf{M}$ represents the dragging distance, and we have a new label matrix after the ε-dragging:

$$\mathbf{Y}' = \mathbf{Y} + \mathbf{B} \otimes \mathbf{M}. \quad (3)$$

With the least square loss function defined as

$$\Psi(\mathbf{W}) = \|\mathbf{X}\mathbf{W} - \mathbf{Y}\|_F^2, \quad (4)$$

we can attain our loss function $\Psi(\mathbf{W}, \mathbf{B}, \mathbf{M})$.

$$\Psi(\mathbf{W}, \mathbf{B}, \mathbf{M}) = \|\mathbf{X}\mathbf{W} - \mathbf{Y} - \mathbf{B} \otimes \mathbf{M}\|_F^2. \quad (5)$$

Next, regularization items used in the formulation are $l_{2,1}$-norm and F-norm, and we take row-wise feature selection and view-wise feature selection into account.

$$\Phi(\mathbf{W}) = \|\mathbf{W}\|_{2,1} = \sum_{i=1}^d \sqrt{\sum_{j=1}^c w_{ij}^2}. \quad (6)$$

$$\Omega(\mathbf{W}) = \sum_{v=1}^m \|\mathbf{W}_v\|_F = \sum_{v=1}^m \sqrt{\sum_{i=1}^d \sum_{j=1}^c w_{ij}^2}. \quad (7)$$

$l_{2,1}$-norm measures the distance of features as a whole and forces the weights of unimportant features to be assigned small values so that it can perform feature selection among all features. Similarly, F-norm measuring the distance between views forces the weights of unimportant views to be assigned small values [31]. The weight matrix $\mathbf{W}$ is regulated by these penalty terms, and hierarchical feature selection is completed with row-wise and view-wise selection. $l_{2,1}$-norm penalty corresponds to the low-level feature selection, and F-norm penalty corresponds to the high-level feature selection.

Therefore, the objective function of the HMVFS model is obtained and represented as (1). α and β are nonnegative constants that tune hierarchical feature selection. This model is also available with more than two views.

2.3. Optimization

In order to solve $l_{2,1}$-norm minimization and F-norm minimization problems, the regularization terms $\|\mathbf{W}\|_{2,1}$ and $\sum_{v=1}^{m}\|\mathbf{W}_v\|_F$ need to be respectively relaxed by $Tr(\mathbf{W}^T\mathbf{C}\mathbf{W})$ and $Tr(\mathbf{W}^T\mathbf{D}\mathbf{W})$ [32]. The objective function is rewritten as

$$\min_{\mathbf{W},\mathbf{M},\mathbf{C},\mathbf{D}} \|\mathbf{X}\mathbf{W} - \mathbf{Y} - \mathbf{B} \otimes \mathbf{M}\|_F^2 + \alpha Tr(\mathbf{W}^T\mathbf{C}\mathbf{W}) + \beta Tr(\mathbf{W}^T\mathbf{D}\mathbf{W}), \quad (8)$$
$$s.t. \ C_{ii} = \frac{1}{2\|\mathbf{w}_i\|_2}, D_{jj} = \frac{1}{2\|\mathbf{W}_v\|_F},$$

where $\mathbf{C} \in \mathbb{R}^{d \times d}$ and $\mathbf{D} \in \mathbb{R}^{d \times d}$ are diagonal matrices and derived from $\mathbf{W}$. For D_{ii}, $\mathbf{w}_i$ is the row vector of $\mathbf{W}_v$.

Though two more variables are introduced, we obtain a convex function, and we can solve the optimization problem iteratively. In each iteration, we update one variable while others are fixed, and all variables can be optimized in order. In view of $\mathbf{C}$ and $\mathbf{D}$ derived from $\mathbf{W}$, we fix $\mathbf{M}$ and update $\mathbf{W}$ at first. The derivative of (8) w.r.t. $\mathbf{W}$ is calculated as

$$2\mathbf{X}^T(\mathbf{X}\mathbf{W} - \mathbf{Y} - \mathbf{B} \otimes \mathbf{M}) + 2\alpha \mathbf{C}\mathbf{W} + 2\beta \mathbf{D}\mathbf{W}. \quad (9)$$

Let (9) equal zero, then the updated $\mathbf{W}$ can be obtained by solving the equation. If there are big-size data or high-dimensional data, the gradient descent method is recommended. Following that, $\mathbf{C}$ and $\mathbf{D}$ can be updated.

When it turns to $\mathbf{M}$, the optimization problem can be transformed from (8) to (10).

$$\min_{\mathbf{M}} \|\mathbf{Z} - \mathbf{B} \otimes \mathbf{M}\|_F^2, \quad (10)$$
$$s.t. \ \mathbf{Z} = \mathbf{X}\mathbf{W} - \mathbf{Y}.$$

According to the definition of F-norm, this problem can be decoupled into $n \times c$ subproblems [30] and represented as

$$\min_{M_{ij}} (Z_{ij} - B_{ij}M_{ij})^2. \quad (11)$$

With $B_{ij}^2 = 1$, (11) is equivalent to (12).

$$\min_{M_{ij}} (Z_{ij}B_{ij} - M_{ij})^2. \quad (12)$$

With the nonnegative constraint, M_{ij} is calculated as

$$M_{ij} = \max(Z_{ij}B_{ij}, 0). \quad (13)$$

Accordingly, $\mathbf{M}$ can be updated as

$$\mathbf{M} = \max(\mathbf{Z} \otimes \mathbf{B}, 0). \quad (14)$$

Up to now, all variables are updated in the iteration and we present the optimization process in Algorithm 1.

After optimization, we obtain weight matrix $\mathbf{W}$ learned across all views and then sort all features according to their importance. The importance is measured by the l_2-norm value of each row vector of $\mathbf{W}$, $\|\mathbf{w}_i\|_2 (i = 1, 2, \ldots, d)$. Feature selection can be completed with features ranked in descending order.

2.4. Convergence

In this subsection, we analyze the convergence of Algorithm 1. We need to guarantee the objective function decreases in each iteration of the optimization algorithm. The following lemma is used to verify its convergence.

Lemma 1. *For any nonzero values $a, b \in \mathbb{R}$, the following inequality holds:*

$$2ab \leq \left(a^2 + b^2\right) \Rightarrow a - \frac{a^2}{2b} \leq b - \frac{b^2}{2b}. \tag{15}$$

Theorem 1. *The objective Function (1) monotonically decreases in the iteration of Algorithm 1.*

Proof. According to Step 6 and Step 7 in Algorithm 1, we have $\mathbf{W}_{t+1}$ and $\mathbf{M}_{t+1}$ as follows:

$$\mathbf{W}_{t+1} \Leftarrow \min_{\mathbf{W}} \|\mathbf{XW}_t - \mathbf{Y} - \mathbf{B} \otimes \mathbf{M}_t\|_F^2 + \alpha Tr(\mathbf{W}_t^T \mathbf{C}_t \mathbf{W}_t) + \beta Tr(\mathbf{W}_t^T \mathbf{D}_t \mathbf{W}_t), \tag{16}$$

$$\mathbf{M}_{t+1} \Leftarrow \min_{\mathbf{M}} \|\mathbf{XW}_{t+1} - \mathbf{Y} - \mathbf{B} \otimes \mathbf{M}_t\|_F^2. \tag{17}$$

Firstly, according to (16) and (17), there is

$$\begin{aligned}&\|\mathbf{XW}_{t+1} - \mathbf{Y} - \mathbf{B} \otimes \mathbf{M}_{t+1}\|_F^2 + \alpha Tr(\mathbf{W}_{t+1}^T \mathbf{C}_t \mathbf{W}_{t+1}) + \beta Tr(\mathbf{W}_{t+1}^T \mathbf{D}_t \mathbf{W}_{t+1}) \\ &\leq \|\mathbf{XW}_{t+1} - \mathbf{Y} - \mathbf{B} \otimes \mathbf{M}_t\|_F^2 + \alpha Tr(\mathbf{W}_{t+1}^T \mathbf{C}_t \mathbf{W}_{t+1}) + \beta Tr(\mathbf{W}_{t+1}^T \mathbf{D}_t \mathbf{W}_{t+1}) \leq \|\mathbf{XW}_t - \mathbf{Y} - \mathbf{B} \otimes \mathbf{M}_t\|_F^2 + \alpha Tr(\mathbf{W}_t^T \mathbf{C}_t \mathbf{W}_t) + \beta Tr(\mathbf{W}_t^T \mathbf{D}_t \mathbf{W}_t).\end{aligned} \tag{18}$$

Thus, according to the definition of $\mathbf{C}$, we have

$$\begin{aligned}\alpha Tr(\mathbf{W}_{t+1}^T \mathbf{C}_t \mathbf{W}_{t+1}) &= \sum_{i=1}^{d} \frac{\|(\mathbf{w}_i)_{t+1}\|_2^2}{2\|(\mathbf{w}_i)_t\|_2} \\ &= \alpha \sum_{i=1}^{d} \|(\mathbf{w}_i)_{t+1}\|_2 - \alpha(\sum_{i=1}^{d} \|(\mathbf{w}_i)_{t+1}\|_2 - \sum_{i=1}^{d} \frac{\|(\mathbf{w}_i)_{t+1}\|_2^2}{2\|(\mathbf{w}_i)_t\|_2}) = \alpha \Phi(\mathbf{W}_{t+1}) - \alpha(\sum_{i=1}^{d} \|(\mathbf{w}_i)_{t+1}\|_2 - \sum_{i=1}^{d} \frac{\|(\mathbf{w}_i)_{t+1}\|_2^2}{2\|(\mathbf{w}_i)_t\|_2})\end{aligned} \tag{19}$$

We also perform the same transformation with $Tr(\mathbf{W}_{t+1}^T \mathbf{D}_t \mathbf{W}_{t+1})$, $Tr(\mathbf{W}_t^T \mathbf{C}_t \mathbf{W}_t)$ and $Tr(\mathbf{W}_t^T \mathbf{D}_t \mathbf{W}_t)$. We can rewrite (18) as

$$\begin{aligned}&\Psi(\mathbf{W}_{t+1}, \mathbf{M}_{t+1}) + \alpha \Phi(\mathbf{W}_{t+1}) + \beta \Omega(\mathbf{W}_{t+1}) - \alpha(\sum_{i=1}^{d} \|(\mathbf{w}_i)_{t+1}\|_2 - \sum_{i=1}^{d} \frac{\|(\mathbf{w}_i)_{t+1}\|_2^2}{2\|(\mathbf{w}_i)_t\|_2}) - \beta(\sum_{v}^{m} \|(\mathbf{W}_v)_{t+1}\|_F - \sum_{v}^{m} \frac{\|(\mathbf{W}_v)_{t+1}\|_F^2}{2\|(\mathbf{W}_v)_t\|_F}) \\ &\leq \Psi(\mathbf{W}_t, \mathbf{M}_t) + \alpha \Phi(\mathbf{W}_t) + \beta \Omega(\mathbf{W}_t) - \alpha(\sum_{i=1}^{d} \|(\mathbf{w}_i)_t\|_2 - \sum_{i=1}^{d} \frac{\|(\mathbf{w}_i)_t\|_2^2}{2\|(\mathbf{w}_i)_t\|_2}) - \beta(\sum_{v}^{m} \|(\mathbf{W}_v)_t\|_F - \sum_{v}^{m} \frac{\|(\mathbf{W}_v)_t\|_F^2}{2\|(\mathbf{W}_v)_t\|_F}).\end{aligned} \tag{20}$$

According to Lemma 1, we arrive at

$$\Psi(\mathbf{W}_{t+1}, \mathbf{M}_{t+1}) + \alpha \Phi(\mathbf{W}_{t+1}) + \beta \Omega(\mathbf{W}_{t+1}) \leq \Psi(\mathbf{W}_t, \mathbf{M}_t) + \alpha \Phi(\mathbf{W}_t) + \beta \Omega(\mathbf{W}_t). \tag{21}$$

Thus, Algorithm 1 decreases the optimization problem in (1) for each iteration so (1) will converge to its global optimum according to its convexity.

Algorithm 1 The optimization algorithm for (8)

Input: The feature matrix across all views, $\mathbf{X} \in \mathbb{R}^{n \times d}$; the label matrix, $\mathbf{Y} \in \{0,1\}^{n \times c}$; the parameters α and β
Output: The weight matrix across all views, $\mathbf{W} \in \mathbb{R}^{d \times c}$
1: Calculate $\mathbf{B}$ from $\mathbf{Y}$ via (2)
2: Initialize $\mathbf{W}_0$ and $\mathbf{M}_0$
3: Initialize $t = 0$
4: **Repeat**
5: Calculate $\mathbf{C}_t$ and $\mathbf{D}_t$ from $\mathbf{W}_t$
6: $\mathbf{W}_{t+1} = (\mathbf{X}^T\mathbf{X} + \alpha\mathbf{C}_t + \beta\mathbf{D}_t)^{-1}(\mathbf{X}^T\mathbf{Y} + \mathbf{X}^T\mathbf{B} \otimes \mathbf{M}_t)$
7: $\mathbf{M}_{t+1} = \max((\mathbf{X}\mathbf{W}_{t+1} - \mathbf{Y}) \otimes \mathbf{B})$
8: $t = t + 1$
9: Calculate residue via (1)
10: **Until** convergence or maximum iteration number achieved

3. Material and Characterization

3.1. Data Collection and Cleaning

In this study, the fault data were collected from an AC transmission network located in a coastal populous city in Guangdong Province, China. These faults occurred between 2016 and 2019, and the voltage levels varied from 110 to 500 kV. Fault signals were recorded by digital fault recorders (DFRs) installed on substations. The DFR equipment involves PMUs and computer systems to synchronize, store and display analog data for voltage and current signals. These signals can be remotely accessed through a communication network and provide offline data stored in common format for transient data exchange (COMTRADE). The sampling rate is 5 kHz in the dataset. Environmental information and other associated monitoring data were obtained through the inner maintenance system. A patrol report of manual inspection was attached to each fault, describing the inspection result and labeling its cause. The original dataset comprised 551 samples, and 288 of them remained after cleansing. The distribution of fault-cause categories is shown in Figure 1. Lightning, external force and object contact are the three dominant causes. External force refers to collision or damage due to human activity. Object contact is usually caused by floating objects in the air. These are typical causes in a densely populated city, causing more than 90% of known faults.

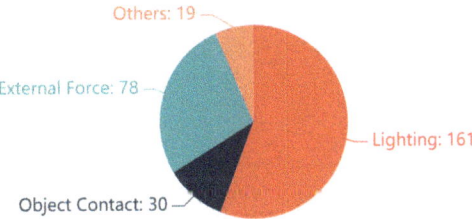

Figure 1. Distribution of transmission line fault cause after cleansing.

3.2. Waveform Characteristics

It is believed that the disturbance variation of electrical quantity after faults occurring contains important transient information for fault diagnosis [33]. The original waveform data are recorded in COMTRADE files with the sampling frequency of 5 kHz. The first step is to acquire fault segments and extract valid waveform segments without disturbance caused by tripping. In this paper, the beginning of valid segments is determined by inspection thresholds based on root mean squared (rms) current magnitude. dI is the difference between consecutive values.

$$dI \geq 0.15 \text{ pu or } I \geq 1.2 \text{ pu.} \tag{22}$$

The start thresholds are determined by inspection to make sure that fault measurements in this study are correctly captured. Since COMTRADE stores not only electrical signals in analog channels but also tripping information in digital channels, one and a half cycles after tripping enabling signal is regarded as the end of the segment. In characterization, we extend previous research work on waveform characterization. The following waveform features are considered and extracted.

1. Maximum Change of Sequence Components: Instantaneous magnitude is calculated relative to prefault amplitude in order to be compatible with measurements from different voltage levels and operation conditions. Karenbauer transformation is used to obtain zero, positive and negative components of three-phase signals, denoted by s, $s = 0, 1, 2$.

$$V_{s(max)} = \frac{\max\left(V_{s(fault)}\right)}{V_{s(pre-fault)}}, s = 0, 2. \tag{23}$$

$$I_{s(max)} = \frac{\max\left(I_{s(fault)}\right)}{I_{s(pre-fault)}}, s = 0, 1, 2. \tag{24}$$

2. Maximum Rate of Change of Sequence Components:

$$\Delta V_{s(max)} = \frac{\max(|\Delta V_s|)}{V_{s(pre-fault)}}, s = 0, 1, 2. \tag{25}$$

$$\Delta I_{s(max)} = \frac{\max(|\Delta I_s|)}{I_{s(pre-fault)}}, s = 0, 1, 2. \tag{26}$$

3. Sequence Component Values at t-cycle: t is set to be 0, 0.5, 1 and 1.5. For instance, $t = 0.5$ means the measuring point is 1/2 cycle from the start.

$$V_{s(t)} = \frac{V_{s(t)}}{V_{s(pre-fault)}}, s = 0, 1, 2. \tag{27}$$

$$I_{s(t)} = \frac{I_{s(t)}}{I_{s(pre-fault)}}, s = 0, 1, 2. \tag{28}$$

4. Custom Time Constant of Sequence Current: Inspired by a linear time-invariant system, time content is introduced to reflect the dynamic response of the network [23]. Time content is the time required to rise from the zero point to $1/e$ of the maximum current. In this study, $1/e$ is replaced with a custom value, m. These features are denoted as $TC\_I_{s(m)}$, $m = 0.1, 0.2, \ldots, 0.9, 1$

5. DC and Harmonic Content: Hilbert–Huang transform is used to conduct spectrum analysis [17]. The harmonic content and DC content are calculated from the ratio of the specific component to the fundamental component. DC and harmonic content are denoted as $Har\_k, k = 0, 3, 5, 7, 9, 11$

6. Wavelet Energy and Energy Entropy: Discrete wavelet transform is applied to decompose fault-phase current signals into three wavelet scales. Wavelet energy E and energy entropy S are calculated for each scale.

$$p_j = \frac{E_j}{\sum_j E_j} = \frac{\sum |C_j|^2}{\sum \sum |C_j|^2}, S_j = -p_j \log_2(p_j). \tag{29}$$

where C_j, E_j, p_j denote wavelet coefficient, wavelet energy and relative energy in scale $j, j = 1, 2, 3$.

7. Maximum DC Current: Equation (30) is used to calculate the maximum DC current on three-phase signals. N_s is the number of data points in one cycle, and $n = 0$ means the triggering point.

$$I_{dc(\max)} = \max(I_{dc\_a}, I_{dc\_b}, I_{dc\_c}), I_{dc} = \frac{\sum_{n=1}^{N_s} i_n - \sum_{n \leq -N_s}^{N_s} i_n}{\max(I_{pre-fault})}. \tag{30}$$

8. Time Domain Factors: Form factor, crest factor, skewness and kurtosis, denoted as t_1–t_4, respectively, are introduced to reflect characteristics of waveform shape and the shock for fault-phase current signals. SD denotes their standard deviation.

$$t_1 = \frac{\sqrt{\frac{1}{N_s}\sum_{n=1}^{N_s}(i_n)^2}}{\frac{1}{N_s}\sum_{n=1}^{N_s}|i_n|}, t_2 = \frac{\max(i_n)}{\frac{1}{N_s}\sum_{n=1}^{N_s}|i_n|}, t_3 = \frac{\sum_{n=1}^{N_s}(i_n - \bar{i})^3}{SD^3 N_s}, t_4 = \frac{\sum_{n=1}^{N_s}(i_n - \bar{i})^4}{SD^4 N_s} \tag{31}$$

9. Approximation Constants δ for Neural Waveform: In order to learn more from the front wave, the waveform of rms neutral voltage/current is approximated by (32), as introduced in [33].

$$f(t, \delta) = 1 - e^{-\delta t}, \tag{32}$$

where t is time step and δ is the approximation constant. Equation (32) estimates the closest value with regard to the actual waveform in per unit value.

10. Fault Inception Phase Angle (FIPA): FIPA is calculated based on the trigger time after the last zero crossing point prior to fault happening.

All waveform features are listed in Table 2. Faulted phase features are included in the next subsection.

Table 2. Feature pools.

Pool Type	Feature	Total Number
Waveform	Maximum sequence voltage/current	5
	Maximum change of three-phase signals and sequence components	6
	Sequence component values	24
	Custom time constant of sequence current	30
	DC and harmonic content	6
	Wavelet energy and energy entropy	6
	Maximum DC current	1
	Form factor, crest factor, skewness and kurtosis	4
	Approximation constants	2
	FIPA	1
Contextual	Time stamp: season, day/night, mouth, hour	4
	Location: landform, zone	2
	Meteorological data: weather, temperature, humidity, rainfall, cloud cover, maximum wind speed, wind scale	7
	Protection data: reclosing, fault phase, fault duration, tripping time, breaker quenching time, reclosing time, number of triggering	7
	Others: voltage level, number of faults	2

3.3. Contextual Characteristics

Most monitoring technologies are developed for specified causes and work independently with interconnected data. In this study, due to data restriction, available nonwaveform data include time stamps, meteorological data, geographical data, protection data and query information. These informative values are preprocessed and integrated into the pool of candidate contextual features, as shown in Table 2. Considering that there is no accurate discretization standard, we only discretize text data roughly if necessary. The time stamp information is discretized twice based on season and day/night as a contrast of months and daytime. As for dynamic records such as meteorological value, the records closest to the fault time are retained. Protection data are feedback information of protection devices after fault, usually obtained from the production management system. Although these collected data are related to fault events, they are not suitable for fault cause identification. These irrelevant features pose a great challenge in feature selection.

4. Experiments and Discussion

4.1. Experiment Setup

To validate the effectiveness and efficiency of HMVFS, we conducted comparison experiments using the mentioned field data previously. Three strategies for utilizing multiview data with feature selection were considered, namely single-view learning, feature concatenation after selection and feature selection after concatenation. The last two are the simplest early fusion methods. Single-view learning is represented via best single view (BSV) method, through which the most informative view achieves the best performance among views. As for the dataset in this paper, contextual features are more representative than hand-crafted waveform features. Feature concatenation after selection (FSFC) employs a feature selection technique separately and concatenates features selected from different views. Feature selection after concatenation (FCFS) concatenates original feature sets of two views and then performs feature selection. Adaptable feature selection methods listed in the next subsection are applied to select discriminative features.

The fault dataset was split into training data and testing data in a stratified fashion according to the ratio of 3:1. All samples were normalized by standard deviation after zero-mean standardization. Then, feature selection methods were used to seek the optimal feature combination using training sets and transform all samples for fault-cause classification. ML classifiers were utilized to finish the classification. In the presence of imbalanced data, criteria such as G-mean and accuracy were used to quantitatively assess classification performance. Since G-mean is a metric within biclass concepts, its microaverage was computed and adopted. The final results of each metric were calculated as the average of the 5 trials.

4.2. Comparison Feature Algorithms

As reviewed in [34], there are many feature selection methods. We conducted comparison experiments between our MVFS and several typical feature selection algorithms, namely Fisher score (F-Score), mutual information (MI), joint mutual information (JMI), joint mutual information maximization (JMIM), ReliefF, Hilbert–Schmidt independence criterion lasso (HSIC Lasso) [35] and recursive feature elimination (RFE). F-Score ranks features through variance similarity calculation, and the same rank can be obtained by analysis of variance (ANOVA). MI ranks features according to values of their mutual information with class labels. JMI and JMIM are developed from MI [36]. RFE ranks and discards features after training a certain kind of classifier. Starting from all features, the elimination process continues until the feature number or output error is settled to a minimum.

The above algorithms are developed for single-view learning and can be used in BSV, FCFS and FSFC directly. Except for RFE, all of them are filter feature selection approaches, as is HMVFS. Besides, the comparison algorithms designed for multiview learning are kernel canonical correlation analysis (KCCA) [24] and discriminant correlation analysis (DCA) [37]. These feature extraction approaches map multiview data into a common feature space so their results are attached to the comparison in FCFS. As for the proposed algorithm, there are two hyperparameters in HMVFS. In the experiments, these hyperparameters α and β were tuned ranging in $\{10^{-2}, 10^{-1}, 1, 10, 10^2, 10^3\}$ through grid search on the training sets. Moreover, experiments without any feature algorithm were conducted using BSV features and all features, tabbed as RAW_BSV and RAW.

4.3. Overall Classification Performance

In this subsection, we compare the mentioned dimension reduction approach on the basis of SVM to verify the effectiveness of multiview learning and HMVFS. Two concatenating rules were applied to FSFC. The first rule tries to keep 1:1 proportion of waveform and contextual features. There is one more contextual feature when the total number is odd. The second rule holds the same proportion of waveform and contextual features as that in HMVFS.

The results in terms of Gmean with different numbers of selected features are shown in Figure 2. By comparing single-view feature selection methods among strategies, we notice that most of them perform best in BSV rather than in FSFC and FCFS. Added fault features from the other view will even degrade their classification, and this indicates that simple concatenation cannot help conventional feature selection methods adapt to multiview classification. A similar conclusion is drawn in [23]. Thus, the introduction of MVL appears vital in particular. HMVFS has comprehensive advantages in the comparison of FSFC and FCFS and achieves the best performance compared with methods in BSV. HMVFS outperforms others in the middle of feature increasing, and its result with 14 selected features is the global or near-global optimum. When features from the other view increase, the performance is degraded to a certain extent, and then it rises to another peak. Most methods in BSV produce a zigzag rise curve and reach their best when almost all view features are selected. They are also inferior to HMVFS in FSFC and FCFS. ReliefF is the best competitor that achieves acceptable performance in different strategies. As for KCCA and DCA, their performance is low. Figure 2 illustrates that HMVFS is more capable of obtaining the best performance combining waveform and contextual features.

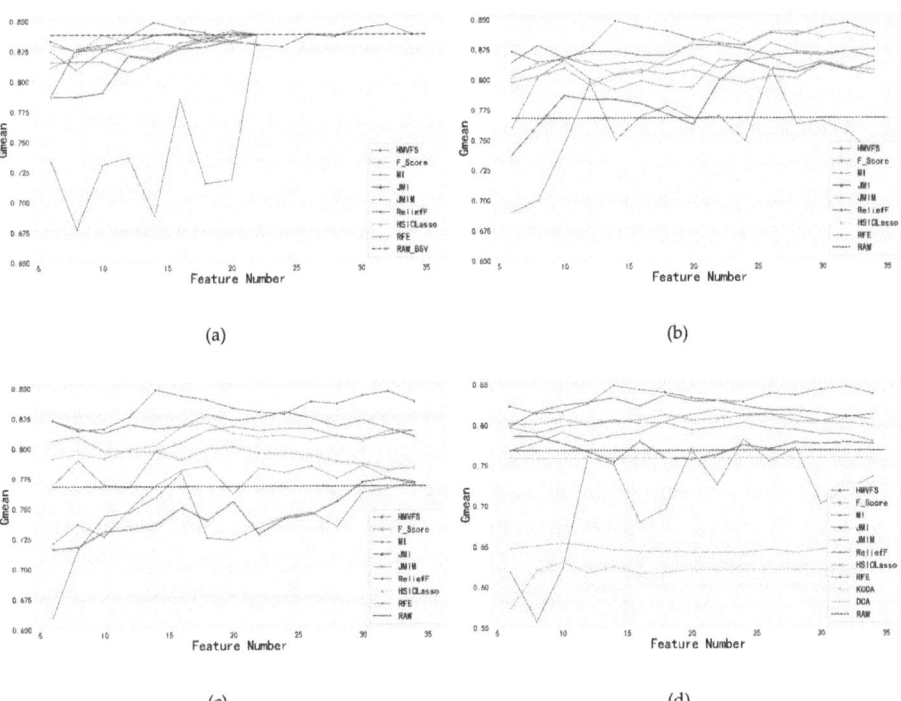

Figure 2. Classification comparison between HMVFS and other feature algorithms in strategies: (**a**) BSV; (**b**) FSFC_rule1; (**c**) FSFC_rule2; (**d**) FCFS.

Due to the limit of yield data condition and fault signature study, irrelevant and redundant features are introduced with increasing feature numbers. This problem is more prominent in the waveform view in both theoretical and experimental studies. The advantage of HMVFS is that it selects features with independent and complementary information of all views, while the single-view methods are easily affected by irrelevant features facing concatenated assembly or meeting the limitation of single-view features. As seen from Figure 2, concatenating and mapping fail to select or transform discriminative features with combined waveform and contextual features. There are two local optimums for HMVFS, and they are better than the performance of competitors, which demonstrates that HMVFS overcomes the negative effect of redundant features in multiview data.

4.4. Parameter Sensitivity

Determination of hyperparameters is an open problem for many algorithms. We conducted parameter sensitivity study by testing different settings of parameters α and β. Since these parameters help HMVFS perform hierarchical feature selection, it is clear that HMVFS will be sensitive to parameter change, and this study may reveal a hierarchical feature relationship. The candidate set was $\{10^{-2}, 10^{-1}, 1, 10, 10^2, 10^3\}$ for each parameter. Classification performance and average running time are recorded and illustrated in Figure 3.

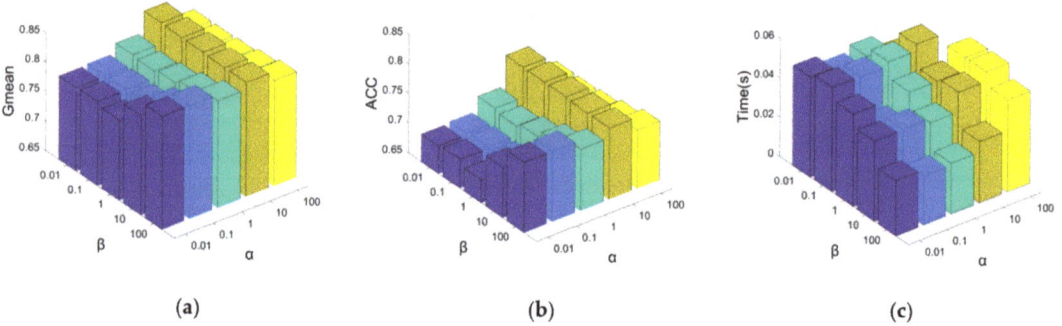

Figure 3. Performance variation of HMVFS with different values for the parameters α and β in terms of (**a**) Gmean; (**b**) ACC; (**c**) time.

It is observed that $\alpha = 10$ is beneficial to final selection and maintains relatively high classification performance, among which lower β has slight advantages. View importance is different in multiview learning. From the perspective of view importance, when only two views exist and one of them is generally better, acceptable performance can be achieved by one view, and additional features are expected for improvement. High-level feature selection is weak because the other view has relatively more redundant features and will be ignored with higher β. Meanwhile, appropriately higher α enhances low-level feature selection to exploit the most representative features from the unimportant view. Moreover, acceptable performance is achieved with $\alpha = 10^{-2}$, $\beta = 10^2$ and $\alpha = 10^{-1}$, $\beta = 10^2$. High-level selection is enhanced, and low-level selection is restrained, which results in limited performance approximating in single-view learning and short convergence time.

4.5. Comparison between ML Classifiers

In order to investigate the effect of classifiers and explore better identification accuracy, we employed different ML learners to complete fault-cause classification with HMVFS. Owing to space limitation and performance stability, F_Score and ReliefF were used for comparison. The typical individual classifiers CN2, LR, KNN, SVM and ANN, which have been proven effective in fault-cause identification studies, were tested, and the results are presented in this subsection. Ensemble models promote fault-cause identification by combining individual learners [22], so we also explored the performance of various ensemble models, including random forest (RF), AdaBoosting, stacking ensemble and dynamic ensemble. META-DES, DES-Clustering and KNORA-U are dynamic ensemble techniques based on metalearning, clustering and k-nearest neighbors, respectively. Classification models were developed using Python machine learning library, scikit-learn and DESlib. Table 3 presents the best performance for each combination of feature selection methods and classifiers. Considering some data may be similar, AUC is introduced as a supplement criterion, which is derived from receiver operating characteristic (ROC) analysis and calculated as the area under the ROC curve.

Table 3. Best performance comparison with different ML classifiers.

Classifier	Feature Selection	Feature Number	Gmean	ACC	AUC
CN2	F_Score	39	0.707	0.581	0.834
	ReliefF	33	0.707	0.580	0.836
	HMVFS	**28**	**0.730**	**0.612**	**0.841**
LR	F_Score	16	0.833	0.756	0.889
	ReliefF	**15**	**0.833**	**0.756**	**0.896**
	HMVFS	33	0.831	0.752	0.896
KNN	F_Score	14	0.838	0.764	0.891
	ReliefF	11	0.835	0.760	0.895
	HMVFS	**7**	**0.848**	**0.778**	**0.909**
SVM	F_Score	18	0.812	0.728	0.908
	ReliefF	18	0.837	0.761	0.906
	HMVFS	**14**	**0.849**	**0.779**	**0.921**
ANN	F_Score	18	0.837	0.761	0.891
	ReliefF	**15**	**0.850**	**0.780**	**0.911**
	HMVFS	36	0.842	0.769	0.915
RF	**F_Score**	**27**	**0.878**	**0.821**	**0.926**
	ReliefF	12	0.876	0.819	0.935
	HMVFS	9	0.875	0.817	0.935
AdaBoost	F_Score	36	0.781	0.684	0.797
	ReliefF	19	0.777	0.679	0.830
	HMVFS	**14**	**0.784**	**0.690**	**0.846**
META-DES	F_Score	19	0.876	0.816	0.930
	ReliefF	11	0.872	0.812	0.928
	HMVFS	**12**	**0.881**	**0.824**	**0.937**
DES-Clustering	F_Score	32	0.872	0.812	0.916
	ReliefF	13	0.875	0.817	0.932
	HMVFS	**10**	**0.882**	**0.827**	**0.945**
KNORA-U	F_Score	15	0.872	0.812	0.926
	ReliefF	14	0.870	0.809	0.932
	HMVFS	**12**	**0.884**	**0.829**	**0.942**
Stacking	F_Score	16	0.880	0.824	0.930
	ReliefF	13	0.874	0.814	0.936
	HMVFS	**11**	**0.886**	**0.831**	**0.939**

As seen from the table, HMVFS outperforms F_Score and ReliefF except with LR and ANN. It is observed that HMVFS always takes fewer features to achieve the best performance in the remaining comparisons. In the group of RF, the best scores of F_Score, ReliefF and HMVFS are very close to each other because RF has the ability of variable selection. Thus the features that function in final classification are similar if selected feature subsets are large enough to contain valuable features. Except for mentioned learners, HMVFS has advantages in both score and feature number.

From the perspective of learners, the classification performance improves with the enhancement of model complexity. CN2 as a rule-based learner cannot cope with multiview features to achieve acceptable performance. Individual learners cannot achieve accuracies greater than 0.8, which are apparently inferior to most ensemble models. Among ensemble models, stacking ensemble realizes the best fault-cause identification in this study. The experimental results of ML classifiers indicate that HMVFS is more suitable for classifiers with high generalization and that ensemble models can bring significant improvement for fault-cause identification.

5. Conclusions

Associated multisource data for transmission line fault-cause diagnosis are divided and extracted as waveform and contextual features in this paper. MVL is introduced to appropriately combine these features for performance improvement. A novel hierarchical multiview feature selection method based on an ε-dragging technique and sparsity regularization is proposed to perform hierarchical feature selection with multiview data. The ε-dragging is applied in the loss function to enlarge sample distance between classes. $l_{2,1}$-norm and F-norm conduct row-wise and view-level selection, respectively, which can be viewed as the low-level and high-level feature selection. We also develop the optimization algorithm and prove its convergence theoretically. The proposed HMVFS is evaluated by comparisons on yield data. The results reveal that HMVFS outperforms conventional feature selection methods in single-view and early fusion strategies. The further experiments concerning ML classifiers also demonstrate the superiority and effectiveness of the proposed method with high generalization learners. This study has shown the combined use of waveform and contextual features with HMVFS can cause significant improvement for fault-cause identification. In future work, more multiview data and further fault signature study are needed to refine the feature pools, and the performance of HMVFS is expected to be further improved.

Author Contributions: Conceptualization, S.J. and X.P.; methodology, H.Y. and C.S.L.; software, S.J.; experiment, validation and analysis, S.J. and H.Y.; investigation, S.J. and X.P.; resources, X.P.; data curation, S.J.; writing—original draft preparation, S.J. and X.P.; writing—review and editing, S.J., H.Y., C.S.L. and L.L.L. All authors have read and agreed to the published version of the manuscript.

Funding: This research was supported by the National Natural Science Foundation of China (61903091) and the Science and Technology Project of China Southern Power Grid Company Limited (031800KK52180074).

Institutional Review Board Statement: Not applicable.

Informed Consent Statement: Not applicable.

Data Availability Statement: Not applicable.

Conflicts of Interest: The authors declare no conflict of interest.

References

1. Ferreira, V.H.; Zanghi, R.; Fortes, M.Z.; Sotelo, G.G.; Silva, R.; Souza, J.; Guimarães, C.; Gomes, S., Jr. A survey on intelligent system application to fault diagnosis in electric power system transmission lines. *Electr. Power Syst. Res.* **2016**, *136*, 135–153.
2. Chen, Y.; Fink, O.; Sansavini, G. Combined fault location and classification for power transmission lines fault diagnosis with integrated feature extraction. *IEEE Trans. Ind. Electron.* **2018**, *65*, 561–569. [CrossRef]
3. Minnaar, U.J.; Gaunt, C.T.; Nicolls, F. Characterisation of power system events on South African transmission power lines. *Electr. Power Syst. Res.* **2012**, *88*, 25–32. [CrossRef]
4. Cai, Y.; Chow, M. Cause-Effect modeling and spatial-temporal simulation of power distribution fault events. *IEEE Trans. Power Syst.* **2011**, *26*, 794–801. [CrossRef]
5. Gui, M.; Pahwa, A.; Das, S. Bayesian network model with Monte Carlo simulations for analysis of animal-related outages in overhead distribution systems. *IEEE Trans. Power Syst.* **2011**, *26*, 1618–1624. [CrossRef]
6. Núñez, V.B.; Meléndez, J.; Kulkarni, S.; Santoso, S. Feature analysis and automatic classification of short-circuit faults resulting from external causes. *Int. Trans. Power Syst.* **2013**, *23*, 510–525. [CrossRef]

7. Liang, Y.; Li, K.; Ma, Z.; Lee, W. Typical fault cause recognition of single-phase-to-ground fault for overhead Lines in nonsolidly earthed distribution networks. *IEEE Trans. Ind. Appl.* **2020**, *56*, 6298–6306. [CrossRef]
8. Xu, L.; Chow, M.; Taylor, L.S. Power distribution fault cause identification with imbalanced data using the data mining-based fuzzy classification *E*-algorithm. *IEEE Trans. Power Syst.* **2007**, *22*, 164–171. [CrossRef]
9. Xu, L.; Chow, M.; Timmis, J.; Taylor, L.S. Power distribution outage cause identification with imbalanced data using artificial immune recognition system (AIRS) algorithm. *IEEE Trans. Power Syst.* **2007**, *22*, 198–204. [CrossRef]
10. Xu, L.; Chow, M. A classification approach for power distribution systems fault cause identification. *IEEE Trans. Power Syst.* **2006**, *21*, 53–60.
11. Cai, Y.; Chow, M.; Lu, W.; Li, L. Statistical feature selection from massive data in distribution fault diagnosis. *IEEE Trans. Power Syst.* **2010**, *25*, 642–648. [CrossRef]
12. Chang, G.W.; Hong, Y.; Li, G. A hybrid intelligent approach for classification of incipient faults in transmission network. *IEEE Trans. Power Deliv.* **2019**, *34*, 1785–1794. [CrossRef]
13. Morales, J.; Orduña, E.A.; Rehtanz, C. Identification of lightning stroke due to shielding failure and backflashover for ultra-high-speed transmission line protection. *IEEE Trans. Power Deliv.* **2014**, *29*, 2008–2017. [CrossRef]
14. Jiang, X.; Stephen, B.; McArthur, S. Automated distribution network fault cause identification with advanced similarity metrics. *IEEE Trans. Power Deliv.* **2021**, *36*, 785–793. [CrossRef]
15. Liang, H.; Liu, Y.; Sheng, G.; Jiang, X. Fault-cause identification method based on adaptive deep belief network and time–frequency characteristics of travelling wave. *IET Gener. Transm. Distrib.* **2019**, *13*, 724–732. [CrossRef]
16. Tse, N.C.F.; Lai, L.L. Wavelet–based algorithm for signal analysis. *EURASIP J. Adv. Signal Process.* **2007**, *2007*, 1–10. [CrossRef]
17. Malik, H.; Sharma, R. Transmission line fault classification using modified fuzzy Q learning. *IET Gener. Transm. Distrib.* **2017**, *11*, 4041–4050. [CrossRef]
18. Tse, N.C.F.; Chan, J.Y.C.; Lau, W.H.; Poon, J.T.Y.; Lai, L.L. Real-time power-quality monitoring with hybrid sinusoidal and lifting wavelet compression algorithm. *IEEE Trans. Power Deliv.* **2012**, *27*, 1718–1726. [CrossRef]
19. Tse, N.C.F.; Chan, J.Y.C.; Lau, W.H.; Lai, L.L. Hybrid wavelet and Hilbert transform with frequency shifting decomposition for power quality analysis. *IEEE Trans. Instrum. Meas.* **2012**, *61*, 3225–3233. [CrossRef]
20. Asman, S.H.; Aziz, N.; Amirulddin, U.; Kadir, M. Decision tree method for fault causes classification based on RMS-DWT analysis in 275 kV transmission lines network. *Appl. Sci.* **2021**, *11*, 4031–4051. [CrossRef]
21. Qin, X.; Wang, P.; Liu, Y.; Guo, L.; Sheng, G.; Jiang, X. Research on distribution network fault recognition method based on time-frequency characteristics of fault waveforms. *IEEE Access* **2018**, *6*, 7291–7300. [CrossRef]
22. Dehbozorgi, M.; Rastegar, M.; Dabbaghjamanesh, M. Decision tree-based classifiers for root-cause detection of equipment-related distribution power system outages. *IET Gener. Transm. Distrib.* **2020**, *14*, 5809–5815. [CrossRef]
23. Minnaar, U.J.; Nicolls, F.; Gaunt, C. Automating transmission-line fault root cause analysis. *IEEE Trans. Power Deliv.* **2016**, *31*, 1692–1700. [CrossRef]
24. Shi, Y.; Ji, H. Kernel canonical correlation analysis for specific radar emitter identification. *Electron. Lett.* **2014**, *50*, 1318–1319. [CrossRef]
25. Ramachandram, D.; Taylor, G. Deep multimodal learning: A survey on recent advances and trends. *IEEE Signal Process. Mag.* **2017**, *34*, 96–108. [CrossRef]
26. Lai, C.S.; Yang, Y.; Pan, K.; Zhang, J.; Yuan, H.L.; Wing, W.; Gao, Y.; Zhao, Z.; Wang, T.; Shahidehpour, M.; et al. Multi-view neural network ensemble for short and mid-term load forecasting. *IEEE Trans. Power Syst.* **2020**. [CrossRef]
27. Lai, C.S. Compression of power system signals with wavelets. In Proceedings of the 2014 International Conference on Wavelet Analysis and Pattern Recognition, Lanzhou, China, 13–16 July 2014. [CrossRef]
28. Lai, C.S. High impedance fault and heavy load under big data context. In Proceedings of the 2015 IEEE International Conference on Systems, Man, and Cybernetics, Hong Kong, China, 9–12 October 2016. [CrossRef]
29. Xu, Z.; Yang, P.; Zhao, Z.; Lai, C.S.; Lai, L.L.; Wang, X. Fault diagnosis approach of main drive chain in wind turbine based on data fusion. *Appl. Sci.* **2021**, *11*, 5804. [CrossRef]
30. Xiang, S.; Nie, F.; Meng, G.; Pan, C.; Zhang, C. Discriminative least squares regression for multiclass classification and feature selection. *IEEE Trans. Neural Netw. Learn. Syst.* **2012**, *23*, 1738–1754. [CrossRef] [PubMed]
31. Zhu, X.; Li, X.; Zhang, S. Block-row sparse multiview multilabel learning for image classification. *IEEE Trans. Cybern.* **2016**, *46*, 450–461. [CrossRef]
32. Zhang, Y.; Wu, J.; Cai, Z.; Yu, P.S. Multiview multilabel learning with sparse feature selection for image annotation. *IEEE Trans. Multimed.* **2020**, *22*, 2844–2857. [CrossRef]
33. Zin, A.; Karim, S. Protection system analysis using fault signatures in Malaysia. *Int. J. Electr. Power Energy Syst.* **2013**, *45*, 194–205. [CrossRef]
34. Lai, C.S.; Zhong, C.; Pan, K.; Ng, W.W.Y.; Lai, L.L. A deep learning based hybrid method for hourly solar radiation forecasting. *Expert. Syst. Appl.* **2021**, *177*, 114941. [CrossRef]
35. Yamada, M.; Jitkrittum, W.; Sigal, L.; Xing, E.; Sugiyama, M. High-dimensional feature selection by feature-wise Kernelized Lasso. *Neural Comput.* **2014**, *26*, 185–207. [CrossRef]

36. Bennasar, M.; Hicks, Y.; Setchi, R. Feature selection using Joint Mutual Information Maximisation. *Expert Syst. Appl.* **2015**, *42*, 8520–8532. [CrossRef]
37. Haghighat, M.; Abdel-Mottaleb, M.; Alhalabi, W. Discriminant correlation analysis: Real-time feature level fusion for multimodal biometric recognition. *IEEE Trans. Inf. Forensics Secur.* **2016**, *11*, 1984–1996. [CrossRef]

Article

Calculation Method for Electricity Price and Rebate Level in Demand Response Programs

Hirotaka Takano [1,*], Naohiro Yoshida [1], Hiroshi Asano [1,2], Aya Hagishima [3] and Nguyen Duc Tuyen [4]

1. Department of Electrical, Electronic and Computer Engineering, Gifu University, Gifu 501-1193, Japan; z4526070@edu.gifu-u.ac.jp (N.Y.); hasano@gifu-u.ac.jp (H.A.)
2. Energy Innovation Center, Central Research Institute of Electric Power Industry, Yokosuka 240-0196, Japan
3. Department of Internationalization and Future Conception, Kyushu University, Fukuoka 815-0811, Japan; ayahagishima@kyudai.jp
4. Power System Department, Hanoi University of Science and Technology, Hanoi 11615, Vietnam; tuyen.nguyenduc@hust.edu.vn
* Correspondence: takano@gifu-u.ac.jp; Tel.: +81-58-293-2720

Abstract: Demand response programs (DRs) can be implemented with less investment costs than those in power plants or facilities and enable us to control power demand. Therefore, they are highly expected as an efficient option for power supply–demand-balancing operations. On the other hand, DRs bring new difficulties on how to evaluate the cooperation of consumers and to decide electricity prices or rebate levels with reflecting its results. This paper presents a theoretical approach that calculates electricity prices and rebate levels in DRs based on the framework of social welfare maximization. In the authors' proposal, the DR-originated changes in the utility functions of power suppliers and consumers are used to set a guide for DR requests. Moreover, optimal electricity prices and rebate levels are defined from the standpoint of minimal burden in DRs. Through numerical simulations and discussion on their results, the validity of the authors' proposal is verified.

Keywords: demand response programs; social welfare maximization; utility function; power supply–demand balance; electricity price; rebate level

1. Introduction

Demand response programs (DRs) are defined as changes in electricity-consuming patterns in response to changes in electricity price or to incentive payment [1]. There are two major categories in DRs: one is the price-based DR, and the other is the incentive-based one. Time of use (TOU), real-time pricing (RTP), and critical-peak pricing (CPP) are well-known as the former. Unit prices of the electric power in these DRs become expensive during the periods of high electricity costs or critical power grid's conditions (peak periods) in comparison with those in off-peak periods. On the other hand, peak time rebate (PTR) and critical peak rebate (CPR) are categorized into the latter. In incentive-based DRs, power suppliers (or power producers, retailers, etc.) reward consumers, who respond to the request of DRs, with money rebates. Since DRs bring controllability in the power demand without huge investment costs in power plants or facilities, they have been attracting attention as one of the most economical and sustainable alternatives to traditional power supply–demand-balancing operations. Therefore, many DR-related studies have been carried out [2–4], as well as demonstrative field tests.

There are various studies contributing to the design of DRs. In [5], states of DR-related activities are analyzed with highlighted deregulation in electricity markets. The authors in [5] defined evaluation indices and discussed effects of DRs in electricity prices. Reference [6] reviews the means by which power suppliers induce their preferable electricity consumption in DRs. Several mathematical problem frameworks and models are summarized, and their solution techniques are introduced. In [7], DRs are classified from the viewpoints of their control mechanisms, motivations offered to changes in electricity consuming patterns, and decision variables. Besides, several models for optimizing control strategies of the DRs are categorized in association with the application targets. The authors in [8] focused on price-based DRs and evaluated their advantages and disadvantages with experimental results. This reference also includes a review of case study results of the DRs in several countries. Reference [9], similarly, summarizes results of case studies, and its authors analyzed them to discuss preferable price settings in price-based DRs.

Although these references indicate a great deal of useful information in the design of DRs, there are still difficulties on settings of electricity prices or rebate levels while ensuring resources of DRs. In fact, electricity prices or rebate levels in field tests have been decided, relying on knowledge, experience, and experimental results [10–16], and thus, it is difficult to discuss on the appropriateness of their settings. For these reasons, the design of efficient DRs becomes a crucial component for advancing technologies of the power grids' management.

This paper presents a theoretical approach that calculates electricity prices and rebate levels in DRs. To set the basis of discussion, the framework of social welfare maximization (SWM), which has often been used to represent models of electricity markets [6,7,17–21], is applied. First, the authors set the utility functions of the power suppliers and the consumers and represent the power supply–demand-balancing operation under the SWM framework. In the process, contributions of the DRs become measurable as an increment/decrement in the utility functions. As a result, we can treat the DR-originated changes in the power demand as an influential factor in the power supply–demand management. Next, acceptable conditions of electricity prices and rebate levels are derived as a guide for the DR request, and then, their optimal values are defined in consideration of burden on both power suppliers and consumers. Finally, the validity of the authors' proposal is verified through numerical simulations with a model constructed by using the actual record of the electricity consumption.

2. Formulation of Power Supply–Demand-Balancing Operation under SWM

SWM is formulated as a problem to maximize the weighted sum of utility functions in a society without regarding to how the profit is distributed in each member of the society [22–24]. Since the members of the society are classified into power suppliers (or power producers, etc.) and consumers, the social welfare function is written as:

$$SW = \sum_{t=1}^{T} SW_t = \sum_{t=1}^{T} \left(\sum_{i=1}^{NS} U_{1,t}(s_{i,t}) + \alpha \sum_{j=1}^{NC} U_{2,t}(d_{j,t}) \right), \tag{1}$$

where t is the time slot ($t = 1, \cdots, T$); i is the number assigned to the power suppliers ($i = 1, \cdots, NS$); $s_{i,t}$ is the electric power fed from power supplier i and an element of vector s_t; $U_{1,t}(\cdot)$ is the utility function of the power suppliers; j is the number assigned to the consumers ($j = 1, \cdots, NC$); $d_{j,t}$ is the power consumption in the consumer j and an element of vector d_t; $U_{2,t}(\cdot)$ is the utility function of the consumers; α is the weighting coefficient.

In the design of DRs, we can regard the power suppliers and the consumers as aggregated ones. Besides, the coefficient α in Equation (1) equals to 1, if we align the units of the utility functions, e.g., into the price.

The suppliers' utility is expressed with the sum of the income by selling electricity and the operational costs in the power supply, while the consumers' utility is described with the sum of the satisfaction obtained in exchange for consuming electricity and the electricity costs [6,25]. Their utility functions are represented as:

$$U_{1,t}(s_t) = p_t s_t - E_t(s_t), \text{ for } \forall t, \tag{2}$$

$$U_{2,t}(d_t) = F_t(d_t) - p_t d_t, \text{ for } \forall t, \tag{3}$$

where p_t is the standard price of electric power; $E_t(\cdot)$ is the operational cost of the power suppliers; $F_t(\cdot)$ is the satisfaction of the power consumers.

Since $p_t s_t$, $p_t d_t$, and $E_t(s_t)$ are expressed with the price, all units in Equation (1) were unified when we evaluated $F_t(d_t)$ in Equation (3) by the price. In this case, the SWM problem can be formulated as:

$$\max_{s,d} SW, \tag{4}$$

$$SW_t = F_t(d_t) - E_t(s_t) + p_t(s_t - d_t), \text{ for } \forall t, \tag{5}$$

$$\text{s.t. } s_t = d_t, \text{ for } \forall t, \tag{6}$$

$$G^{\min} \leq s_t \leq G^{\max}, \text{ for } \forall t, \tag{7}$$

where $G^{\max}$ and $G^{\min}$ are the maximum and the minimum values of the power supply, respectively.

Equation (6) shows the balance of the power supply and demand, and the constraint Equation (7) restricts the controllability of the power supply, depending on specifications of the target power grid, e.g., the maximum and the minimum outputs of power generation units. Hence, the optimal solution of the formulated SWM problem represents the power supply–demand operation that maximizes the social welfare.

If Equation (5) is a convex function, we can apply Lagrange relaxation [6,26,27], and Lagrange multipliers correspond to shadow prices. Lagrangian function and Karush–Kuhn–Tucker conditions are represented as:

$$\mathcal{L}_t(s_t, d_t, \lambda_t, \mu_{1,t}, \mu_{2,t}) = SW_t + \lambda_t(s_t - d_t) + \mu_{1,t}\left(s_t - G^{\min}\right) + \mu_{2,t}(G^{\max} - s_t), \tag{8}$$

$$\begin{cases} -\frac{\partial U_{1,t}(s_t)}{\partial s_t} + \lambda_t + \varepsilon_t - \mu_t = 0 \\ \frac{\partial U_{2,t}(d_t)}{\partial d_t} - \lambda_t = 0 \end{cases}, \tag{9}$$

where λ_t, $\mu_{1,t}$, and $\mu_{2,t}$ are the Lagrange multipliers.

The authors assumed simple utility functions in the numerical simulations of this paper, and thus, we can solve the target problem based on the Newton-Raphson method. Otherwise, any of other techniques, e.g., intelligent optimization algorithms, will be useful for solving SWM problems. For detailed definitions of the assumed functions, refer to Section 4.

3. Calculation Methodology of Electricity Prices and Rebate Levels

The power suppliers bring the electricity consumption closer to the target value, which is preferable in the power supply–demand management by changing the electricity price or rewarding the money rebate to the consumers. The target electricity consumption in DRs is defined as:

$$d'_t = d^*_t + \Delta d_t, \tag{10}$$

where d^*_t is the standard electricity consumption, which is the actual consumption without DRs; Δd_t is the change in electricity consumption by the DR request.

In the actual power grids' operations, d^*_t is replaced with the estimated value, because we cannot know it. This replacement brings an uncertainty to Equation (10), and therefore, can lead to new challenges in the design of DRs. Detailed discussion on the issues remains as a future work of this study.

Under the following assumption, the authors set the acceptable conditions of electricity prices and rebate levels and define their optimal values.

Assumption 1. *Consumers buy electricity to maximize their utility.*

This assumption activates Equation (11), which is derived by Equations (8) and (9):

$$p_t^* = \left.\frac{\partial F_t}{\partial d_t}\right|_{d_t=d_t^*}, \tag{11}$$

where p_t^* is the standard electricity price, which is the actual price without DRs.

3.1. Definition of the Optimal Electricity Price and Its Calculation

The power suppliers decide the electricity price in the price-based DR to control the power demand. The optimal electricity price is defined as:

$$p_t' = p_t^* + \Delta p_t, \tag{12}$$

where Δp_t is the change of electricity price in the DR.

According to Equations (2) and (3), the values of the utility of the power suppliers and the consumers in the DR are calculated as:

$$U_{1,t}(d_t') = p_t'd_t' - E_t(d_t'), \tag{13}$$

$$U_{2,t}(d_t') = F_t(d_t') - p_t'd_t'. \tag{14}$$

In addition, the DR-originated changes in them are calculated as:

$$\Delta U_{1,t} = U_{1,t}(d_t') - U_{1,t}(d_t^*) = (p_t'd_t' - p_t^*d_t^*) - \Delta E_t, \tag{15}$$

$$\Delta U_{2,t} = U_{2,t}(d_t') - U_{2,t}(d_t^*) = \Delta F_t - (p_t'd_t' - p_t^*d_t^*), \tag{16}$$

where ΔE_t is the change in the suppliers' utility by the DR and is written as: $E_t(d_t') - E_t(d_t^*)$; ΔF_t is the change in the consumers' utility by the DR and is described as: $F_t(d_t') - F_t(d_t^*)$.

In the DR, each of the power suppliers' utility and the consumers' one (or each sum of them during the target period) is greater than or equal to zero ($U_{1,t}(d_t') \geq 0$; $U_{2,t}(d_t') \geq 0$). Since the power suppliers request the DR cooperation considering their economic efficiency, the changes in the utility of the power suppliers are also greater than or equal to zero ($\Delta U_{1,t} \geq 0$). By contrast, without any incentive, the changes in the utility of the consumers are negative ($\Delta U_{2,t} < 0$), because their utility function is maximized (approximately maximized in the actual situations) at the standard electricity consumption. By assumption 1, we can represent the electricity price that maximizes the consumers' utility at d_t' as:

$$p_t = \left.\frac{\partial F_t}{\partial d_t}\right|_{d_t=d_t'}. \tag{17}$$

If the power suppliers set higher electricity price p_t^+ than its standard, the consumers' utility decreases according to Equation (16), and thus, the electricity consumption is reduced ($\Delta d_t < 0$). Meanwhile, the electricity consumption is encouraged by setting lower electricity price p_t^- ($\Delta d_t > 0$). The acceptable conditions p_t^+ and p_t^- to achieve the target of the price-based DR are separately derived as:

$$\left.\frac{\partial F_t}{\partial d_t}\right|_{d_t=d_t'} \leq p_t^+ \leq \frac{F_t(d_t')}{d_t'}, \tag{18}$$

$$\frac{p_t^*d_t^* + \Delta E_t}{d_t'} \leq p_t^- \leq \left.\frac{\partial F_t}{\partial d_t}\right|_{d_t=d_t'}. \tag{19}$$

When the electricity price does not satisfy both Equations (18) and (19), the DR request brings negative economic impacts on the utility of the power suppliers, or the consumers' cooperation cannot reach the target of DR request. With a view to minimizing the burden in the price-based DR, the authors defined the optimal electricity price as:

$$p'_t = \left.\frac{\partial F_t}{\partial d_t}\right|_{d_t=d'_t}. \quad (20)$$

3.2. Definition of the Optimal Rebate Level and Its Calculation

If there is no incentive payment, the utility of the consumers decreases, depending on contribution to the DR. This is because the consumers must accept less satisfaction than that in the standard electricity consumption. In the incentive-based DR, values of the utility of the power suppliers and the consumers and their changes are separately calculated as:

$$U_{1,t}(d'_t) = p^*_t d'_t - E_t(d'_t), \quad (21)$$

$$U_{2,t}(d'_t) = F_t(d'_t) - p^*_t d'_t, \quad (22)$$

$$\Delta U_{1,t} = p^*_t \Delta d_t - \Delta E_t, \quad (23)$$

$$\Delta U_{2,t} = \Delta F_t - p^*_t \Delta d_t. \quad (24)$$

These equations are similar to Equations (13)–(16); however, the electricity price is fixed to p^*_t in the incentive-based DR. Although the consumers accept less satisfaction, their utility recovers to its original level by compensating the decrement of Equation (24). Therefore, we can set the acceptable condition for the unit price of money rebate r_t as:

$$\frac{p^*_t \Delta d_t - \Delta F_t}{|\Delta d_t|} \leq r_t \leq \frac{p^*_t \Delta d_t - \Delta E_t}{|\Delta d_t|}. \quad (25)$$

Under this condition, the DR does not bring negative impacts to both the power suppliers and the consumers. The authors defined the optimal rebate level to induce the active cooperation of the consumers as:

$$r'_t = \frac{p^*_t \Delta d_t - \Delta E_t}{|\Delta d_t|}. \quad (26)$$

4. Numerical Simulation Model

To apply the authors' proposal, actual utility functions are needed. Since there are no established utility functions, these functions were made in this paper using a widely used function for the fuel cost of power generation units and a record of smart power meters. Discussion on their appropriateness remains as a future work of this study.

In this paper, the standard electricity prices in the SWM framework were replaced with the annual average price as:

$$p^*_t = p^* (= 23.90), \text{ for } \forall t. \quad (27)$$

In the numerical simulations, p^* was set to 23.90 JPY/kWh, which is the Japanese annual average in 2015. The utility functions are shown below.

4.1. Utility Function for Suppliers

Thermal power generation has taken a large portion in the power supply, e.g., approximately 85% in Japan [28], and its fuel cost, as is well-known, has powerful influence on the operational costs of the power suppliers. The fuel costs of thermal power units are traditionally approximated as quadratic functions by means of generating power [29–34]. For these reasons, the authors added the following assumptions to make the operational cost function.

Assumption 2. *Total operational cost in the power supply is approximated as the quadratic function relying on the fuel costs of thermal power units.*

Assumption 3. *The power suppliers decide the electricity price to maximize their utility function at the annual average of electricity consumption (103.94 kW in this paper).*

The normalized utility function of the power suppliers was defined as:

$$U_1^\dagger(s_t) = -A^\dagger s_t^2 + \left(p_1^\dagger - B^\dagger\right)s_t - C^\dagger, \text{ for } \forall t, \quad (28)$$

where $p_1^\dagger$ is the electricity price in the normalized function; $A^\dagger$, $B^\dagger$, and $C^\dagger$ are the coefficients for the normalized function.

With reference to [35,36], the coefficients of the fuel cost function were set to 2.73×10^{-7} JPY/kWh^2, 2.27 JPY/kWh, and 1.50×10^5 JPY. In addition, the maximum and the minimum outputs of power generation ($G^{\max}$ and $G^{\min}$) were set to 175 kW and zero, respectively, based on the electricity consumption of 500 households in the record. With these values, the coefficients in Equation (28) were assumed to 4.63×10^{-5}/kWh^2, 1.20×10^{-7}/kWh, and 0. Owing to Assumption 3, we can calculate $p_1^\dagger$ as:

$$p_1^\dagger = 2A^\dagger \bar{s} + B^\dagger, \quad (29)$$

where $\bar{s}$ is the annual average of electricity consumption.

By multiplying $\frac{p^*}{p_1^\dagger}$ on both sides in Equation (29), we can update the coefficients as A and B, and as a result, the functions of the power suppliers' utility and the operational cost were written as:

$$U_1(s_t) = -As_t^2 + (p^* - B)s_t, \text{ for } \forall t, \quad (30)$$

$$E_t(s_t) = As_t^2 + Bs_t \, (= E(s_t)), \text{ for } <!-- \forall t, \quad (31)$$

where $A = \frac{p^*}{p_1^\dagger} A^\dagger$; $B = \frac{p^*}{p_1^\dagger} B^\dagger$.

Figure 1 displays the resulting operational cost function of the suppliers, and its coefficients are summarized in Table 1.

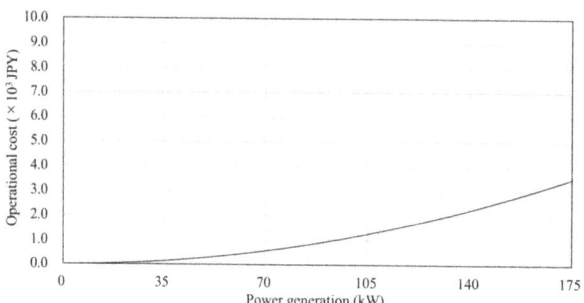

Figure 1. Operational cost function.

Table 1. Coefficients in Figure 1.

A ($\times 10^{-1}$ JPY/kWh^2)	B ($\times 10^{-4}$ JPY/kWh)	C (JPY)
1.15	2.99	0

4.2. Utility Function for Consumers

The consumers' satisfaction is assumed by several functions such as logarithmic or sigmoidal functions [37–42]. In this paper, the functions of hourly satisfaction were made by relying on the record of smart power meters for 500 households. Figure 2 shows the profiles of the hourly total electricity consumption of the 500 households, which are samples including the highest or the lowest electricity consumption for one year. The hourly cumulative frequency distributions of the electricity consumption in each household are displayed in Figures 3 and 4.

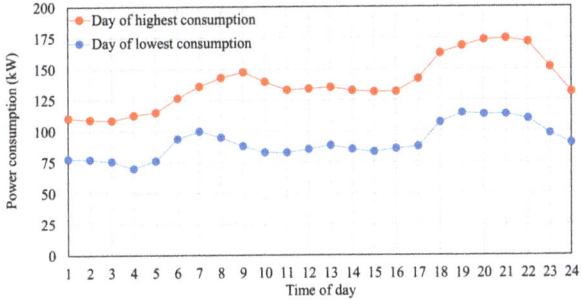

Figure 2. Example of profiles of hourly electricity consumption.

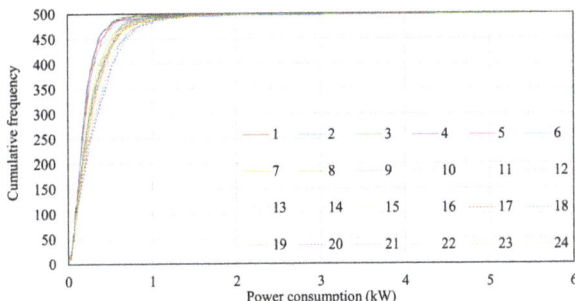

Figure 3. Hourly cumulative frequency distributions in the day of the highest electricity consumption.

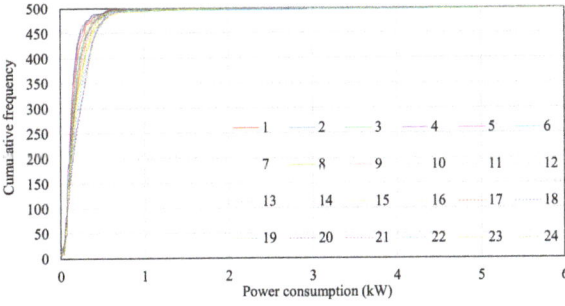

Figure 4. Hourly cumulative frequency distributions in the day of the lowest electricity consumption.

Based on these data, the normalized utility function of the consumers was approximated as:

$$U_{2,t}^{\dagger}(d_t) = X_t^{\dagger} \ln\left(Y_t^{\dagger}\left(d_t + Z_t^{\dagger}\right)\right) - p_{2,t}^{\dagger} d_t, \text{ for } \forall t, \tag{32}$$

where $p_{2,t}^{\dagger}$ is the electricity price in the normalized function; $X_t^{\dagger}$, $Y_t^{\dagger}$, and $Z_t^{\dagger}$ are the coefficients for the hourly normalized function of consumers' satisfaction.

Equation (32) suggests how many consumers are satisfied with the electricity consumption d_t. With assumption 1, $p^\dagger_{2,t}$ can be calculated as:

$$p^\dagger_{2,t} = \frac{X^\dagger_t}{d^*_t + Z^\dagger_t}, \text{ for } \forall t. \tag{33}$$

As with Equations (30) and (31), we can derive the functions of the consumers' hourly utility and satisfaction as:

$$U_{2,t}(d_t) = X_t \ln(Y_t(d_t + Z_t)) - p^* d_t, \text{ for } \forall t, \tag{34}$$

$$F_t(d_t) = X_t \ln(Y_t(d_t + Z_t)), \text{ for } \forall t, \tag{35}$$

where $X_t = \frac{p^*}{p^\dagger_1} X^\dagger_t$; $Y_t = Y^\dagger_t$; $Z_t = Z^\dagger_t$.

Figures 5 and 6 show the hourly satisfaction functions, and their coefficients are summarized in Tables 2 and 3.

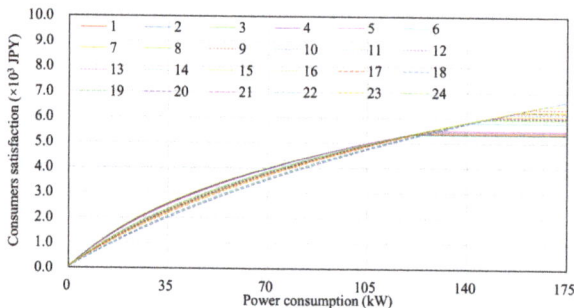

Figure 5. Hourly satisfaction functions in the day of the highest electricity consumption.

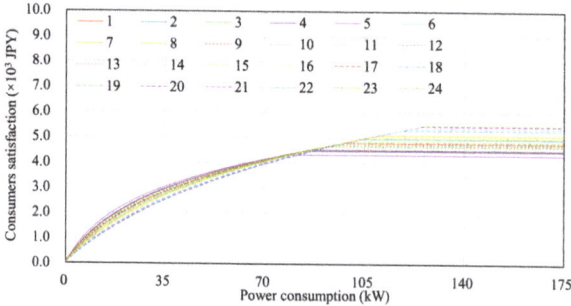

Figure 6. Hourly satisfaction functions in the day of the lowest electricity consumption.

Table 2. Coefficients in Figure 5 (the day of the highest electricity consumption).

t	X_t ($\times 10^{-3}$)	Y_t ($\times 10^2$)	Z_t	t	X_t ($\times 10^{-3}$)	Y_t ($\times 10^2$)	Z_t
1	3.41	3.16	32.03	13	4.55	1.85	54.75
2	3.35	3.25	31.15	14	4.44	1.93	52.78
3	3.32	3.32	30.41	15	4.39	1.97	51.77
4	3.50	2.99	33.77	16	4.39	1.97	51.47
5	3.60	2.84	35.59	17	4.89	1.63	62.21
6	4.11	2.22	45.55	18	5.93	1.18	85.10
7	4.56	1.85	54.56	19	6.25	1.08	92.87
8	4.92	1.61	62.80	20	6.52	1.02	99.01
9	5.15	1.49	67.62	21	6.58	1.00	100.57
10	4.76	1.71	59.44	22	6.43	1.04	97.03
11	4.46	1.91	53.44	23	5.34	1.40	71.73
12	4.49	1.89	53.69	24	4.35	2.01	50.33

Table 3. Coefficients in Figure 6 (the day of the lowest electricity consumption).

t	X_t ($\times 10^{-3}$)	Y_t ($\times 10^2$)	Z_t	t	X_t ($\times 10^{-3}$)	Y_t ($\times 10^2$)	Z_t
1	2.15	8.38	12.13	13	2.56	5.65	18.38
2	2.14	8.47	12.09	14	2.44	6.29	16.42
3	2.08	9.03	11.29	15	2.36	6.72	15.42
4	1.89	11.40	9.03	16	2.45	6.20	16.62
5	2.10	8.90	11.43	17	2.52	5.85	17.59
6	2.74	4.92	20.63	18	3.28	3.40	29.87
7	2.97	4.14	24.45	19	3.58	2.87	35.21
8	2.79	4.72	21.59	20	3.53	2.95	34.27
9	2.53	5.80	17.77	21	3.54	2.93	34.57
10	2.34	6.90	14.97	22	3.38	3.20	31.49
11	2.34	6.87	15.15	23	2.91	4.34	23.33
12	2.44	6.26	16.64	24	2.60	5.47	18.54

5. Numerical Simulation Results

Numerical simulations were carried out with the model constructed in Section 4. The following scenarios were assumed:

Scenario 1. *The power suppliers request to reduce the electricity consumption at any time slot in the day of the highest electricity consumption.*

Scenario 2. *The power suppliers request to encourage the electricity consumption at any time slot in the day of the lowest electricity consumption.*

Under these scenarios, the authors calculated the sets of the optimal electricity price or rebate level and the acceptable condition on each time slot.

5.1. Numerical Simulation Results for Price-Based DRs

By using Equations (18)–(20), the optimal electricity prices and the acceptable conditions in the price-based DR were calculated as:

$$p'_t = \frac{X_t}{d'_t + Z_t}, \tag{36}$$

$$\underline{p_t^+} \leq p_t^+ \leq \overline{p_t^+}, \tag{37}$$

$$\underline{p_t^-} \leq p_t^- \leq p'_t. \tag{38}$$

where $\overline{p_t^+} = \frac{X_t}{d'_t} \ln(Y_t(d'_t + Z_t))$; $\underline{p_t^-} = \frac{A\left(d'^2_t - d^{*2}_t\right) + B(d'_t - d^*_t) + p^*_t d^*_t}{d'_t}$.

Tables 4 and 5 summarize the calculation results in the case that the power suppliers set the DR target as a 1% decrease (Scenario 1) or increase (Scenario 2) of the standard electricity consumption. Figures 7 and 8 illustrate changes in the utility functions of the power suppliers and the consumers and those in the social welfare functions.

Table 4. Numerical simulation results of price-based demand response programs (DRs) under scenario 1.

t	p'_t (JPY/kWh)	$\overline{p^+_t}$ (JPY/kWh)	t	p'_t (JPY/kWh)	$\overline{p^+_t}$ (JPY/kWh)
1	24.09	46.61	13	24.07	42.42
2	24.09	46.88	14	24.07	42.79
3	24.09	47.00	15	24.07	42.93
4	24.09	46.15	16	24.07	42.90
5	24.08	45.71	17	24.07	41.52
6	24.08	43.72	18	24.06	39.29
7	24.07	42.33	19	24.06	38.76
8	24.07	41.43	20	24.05	38.35
9	24.07	40.89	21	24.05	38.26
10	24.07	41.86	22	24.05	38.49
11	24.07	42.74	23	24.06	40.43
12	24.07	42.59	24	24.07	42.96

Table 5. Numerical simulation results of price-based DRs under scenario 2.

t	$\underline{p^-_t}$ (JPY/kWh)	p'_t (JPY/kWh)	t	$\underline{p^-_t}$ (JPY/kWh)	p'_t (JPY/kWh)
1	23.84	23.69	13	23.87	23.70
2	23.84	23.70	14	23.86	23.70
3	23.84	23.69	15	23.85	23.70
4	23.82	23.69	16	23.86	23.70
5	23.84	23.69	17	23.86	23.70
6	23.88	23.71	18	23.91	23.71
7	23.89	23.71	19	23.93	23.72
8	23.88	23.71	20	23.92	23.72
9	23.86	23.70	21	23.92	23.72
10	23.85	23.70	22	23.92	23.72
11	23.85	23.70	23	23.89	23.71
12	23.86	23.70	24	23.87	23.70

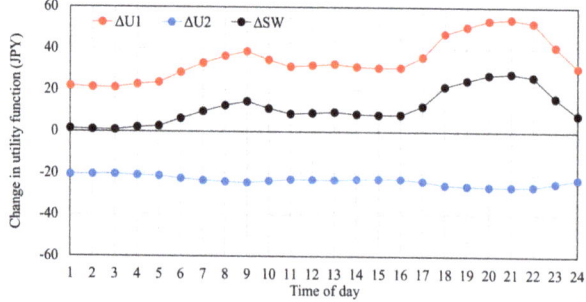

Figure 7. Changes in each function by the 1% DR request in the day of the highest electricity consumption.

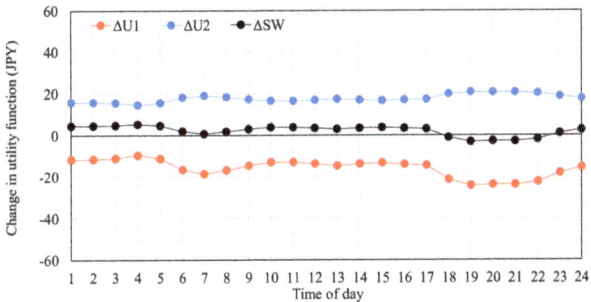

Figure 8. Changes in each function by the 1% DR request in the day of the lowest electricity consumption. In this scenario, the power suppliers could not request the DR cooperation in the target period.

In Table 4, the calculation results in scenario 1 satisfied with their acceptable conditions, and as a result, the optimal electricity prices were slightly increased (less than 1%) as compared to the standard electricity price (23.90 JPY/kWh). By contrast, as shown in Table 5, the power suppliers could not request a 1% increase in the electricity consumption in scenario 2, because the calculation results did not satisfy their acceptable conditions. In particular, the electricity consumption became higher than its annual average (103.94 kW) from 18:00 to 22:00, and therefore, p_t^- exceeded the standard price. In the other periods, the power suppliers can request the $\overline{DR}$ cooperation, until the condition shown in Equation (38) is violated. With reference to Figure 2 we can understand that the calculated electricity prices changed inversely to the profile of electricity consumption in Table 4, while synchronously in Table 5. These results indicated that the consumers can reduce the power demand during the peak periods, and it becomes difficult in the off-peak periods. That is, the calculated prices reflected the controllability in the electricity consumption on each time slot.

In Figures 7 and 8, there were no significant differences in the social welfare by the DR; however, we can confirm that the price-based DRs took burden on the consumers in Figure 7 or the suppliers in Figure 8. As displayed in Figure 9, the optimal electricity price became 25.01 JPY/kWh, and the increment of the price exceeded 1 JPY/kWh when the DR target was set to 7% of the actual electricity consumption. As for the reference, the optimal electricity price in scenario 2 is shown in Figure 10.

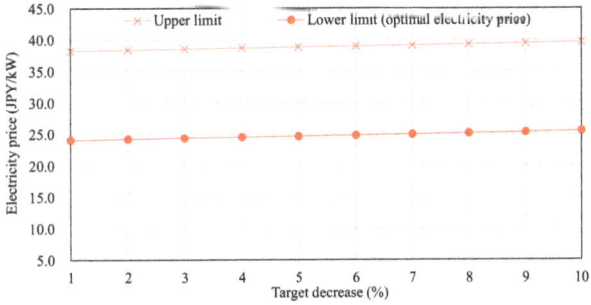

Figure 9. Results on the typical time slots in scenario 1 (at 22:00).

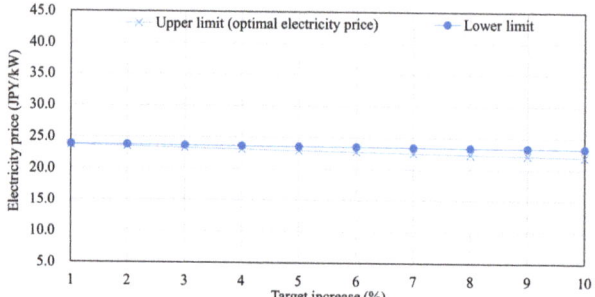

Figure 10. Results on the typical time slots in scenario 2 (at 4:00). In this scenario, the power suppliers could not request the DR cooperation in the target period.

In the numerical simulation model, the authors made the operational cost function of the power suppliers relying on the fuel costs of thermal power units. However, in the actual operations, the other factors such as the surplus power of renewable energy sources have influences on the operational cost. If we reflect them appropriately in the model, the results in scenario 2 can be activated.

From these results, the authors concluded that the authors' proposal functioned properly in the price-based DR.

5.2. Numerical Simulation Results for Incentive-Based DRs

By using Equations (25) and (26), the optimal rebate levels and the acceptable conditions in the incentive-based DR were calculated as:

$$r'_t = \frac{-A(d'^2_t - d^{*2}_t) + (p^* - B)(d'_t - d^*_t)}{|d'_t - d^*_t|}, \tag{39}$$

$$\underline{r_t} \leq r_t \leq r'_t. \tag{40}$$

where $\underline{r_t} = \frac{p^*_t(d'_t - d^*_t)_t - X_t \log \frac{d'_t + Z_t}{d^*_t + Z_t}}{|d'_t - d^*_t|}$.

Tables 6 and 7 summarize the calculation results in the case that the power suppliers set the DR target as a 1% decrease (Scenario 1) or increase (Scenario 2) of the standard electricity consumption. Figures 11 and 12 illustrate changes in the utility functions of the power suppliers and the consumers. Changes in the social welfare functions were omitted in Figures 11 and 12, because they were the same with the changes of the consumers' utility functions in the authors' proposal.

Table 6. Numerical simulation results of incentive-based DRs under scenario 1.

t	$\underline{r_t}$ (×10^{-2} JPY/kWh)	r'_t (JPY/kWh)	t	$\underline{r_t}$ (×10^{-2} JPY/kWh)	r'_t (JPY/kWh)
1	9.31	1.38	13	8.55	7.10
2	9.35	1.08	14	8.59	6.50
3	9.38	0.95	15	8.62	6.26
4	9.24	1.90	16	8.64	6.31
5	9.17	2.42	17	8.36	8.68
6	8.83	5.06	18	7.89	13.40
7	8.57	7.24	19	7.74	14.72
8	8.34	8.85	20	7.64	15.81
9	8.23	9.89	21	7.62	16.07
10	8.42	8.06	22	7.67	15.43
11	8.57	6.57	23	8.15	10.83
12	8.58	6.82	24	8.69	6.23

Table 7. Numerical simulation results of incentive-based DRs under scenario 2.

t	$\underline{r_t}$ (×10^{-2} JPY/kWh)	r'_t (JPY/kWh)	t	$\underline{r_t}$ (×10^{-2} JPY/kWh)	r'_t (JPY/kWh)
1	10.28	5.90	13	9.84	3.41
2	10.28	6.00	14	9.97	4.14
3	10.34	6.37	15	10.03	4.60
4	10.52	7.70	16	9.96	4.02
5	10.33	6.27	17	9.90	3.62
6	9.74	2.22	18	9.30	−0.88
7	9.55	0.80	19	9.10	−2.61
8	9.68	1.94	20	9.13	−2.33
9	9.89	3.55	21	9.12	−2.36
10	10.07	4.75	22	9.24	−1.53
11	10.04	4.77	23	9.61	1.19
12	9.95	4.15	24	9.86	3.01

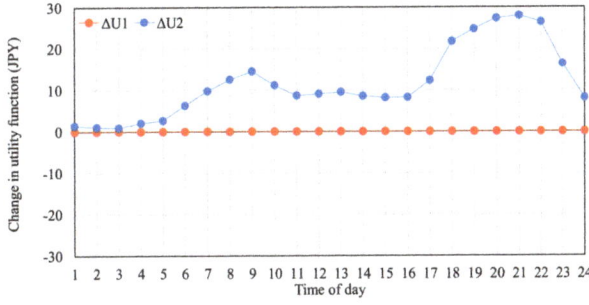

Figure 11. Changes in each function by the 1% DR request in the day of the highest electricity consumption.

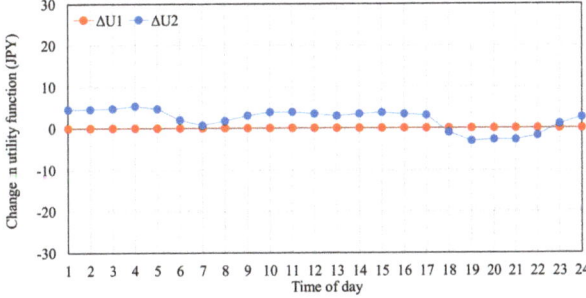

Figure 12. Changes in each function by the 1% DR request in the day of the lowest electricity consumption. In this scenario, the power suppliers could not request the DR cooperation from 18:00 to 22:00.

In Table 6, the calculated rebate levels were in the range of 0.95 JPY/kWh to 16.07 JPY/kWh. Meanwhile, the rebate levels in Table 7 were in the range of −2.61 JPY/kWh to 7.70 JPY/kWh. As opposed to the results of price-based DRs, the time variation in Table 6 and Figure 11 had a similar trend to the electricity consumption. As shown in Table 7 and Figure 12, the optimal rebate levels were lower than $\underline{r_t}$ from 18:00 to 22:00, and thus, the consumers' utility became negative during the periods. It indicated that the power suppliers could not ensure the resources for the DR, and the DR request was impossible during the periods. In Figures 11 and 12, there were no burden on the power suppliers and no decrement in the social welfare excepting the period from 18:00 to 22:00 in scenario 2.

As shown in Figures 13 and 14, the optimal rebate levels reduced in response to the decrease (Figure 13) or increase (Figure 14) of the target values of the DR. In contrast, the lower limits of the rebate levels were gradually raised. This is because the resources for the DR were limited in association with the utility function of the power suppliers.

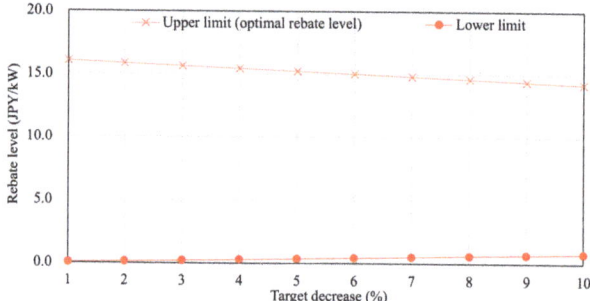

Figure 13. Results on the typical time slots in scenario 1 (at 22:00).

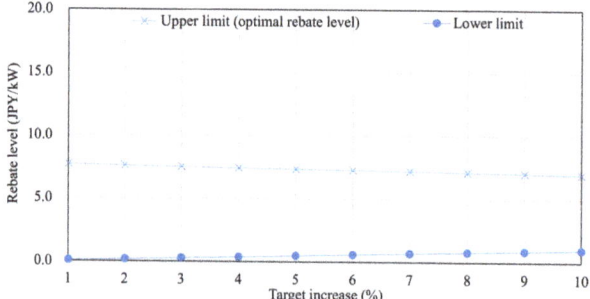

Figure 14. Results on the typical time slots in scenario 2 (at 4:00). In this scenario, the power suppliers could not request the DR cooperation from 18:00 to 22:00.

These results showed that the authors' proposal functioned properly in the incentive-based DR as well.

6. Conclusions

The authors proposed a theoretical approach that calculates electricity prices and rebate levels in DRs, based on the framework of SWM. In the authors' proposal, first, the utility functions of the power suppliers and the consumers were set, and then, the power supply–demand-balancing operation was represented under the SWM framework. Next, the authors derived the acceptable conditions for price-based DRs as Equations (18) and (19) and the acceptable conditions for incentive-based DRs as Equation (25). Besides, the optimal values of electricity prices and rebate levels were defined as Equations (20) and (26), respectively, in consideration of burden on the society. The distinctive feature of the proposed approach was to make influences of the DRs measurable as an increment/decrement in the utility functions. Finally, to verify the validity of the authors' proposal, the numerical simulations were carried out with the model, which was constructed using the approximated fuel cost function of power generation units and the record of smart power meters.

As shown in Section 5, the calculated electricity prices and rebate levels became smaller than the values applied in the demonstrative field tests. This is because the calculation results, as defined in Equations (18)–(20) as well as Equations (25) and (26), strongly depended on the utility functions of the power suppliers and the consumers. In other words, the assumed utility functions have room for discussion on their appropriateness. However, the results of the numerical simulations reflected the controllability in electricity consumption, and we can conclude that the authors' proposal functioned appropriately.

In future works, the appropriateness of the assumed utility functions will be discussed in more detail. Furthermore, the authors will analyze influences of the replacement of the actual electricity consumption, d_t^*, with the estimated one, and the proposed framework will be expanded.

Author Contributions: Conceptualization, H.T. and N.Y.; methodology, H.T. and N.Y.; software, N.Y.; validation, H.T., N.Y., H.A., A.H. and N.D.T.; writing—original draft preparation, H.T., H.A., A.H. and N.D.T.; writing—review and editing, H.T.; supervision, H.T. and H.A.; project administration, H.T. All authors have read and agreed to the published version of the manuscript.

Funding: This research was partly funded by the Japan Society for the Promotion of Science (JSPS; grant number: 19K04325).

Institutional Review Board Statement: Not applicable.

Informed Consent Statement: Not applicable.

Data Availability Statement: Not applicable.

Conflicts of Interest: The authors declare no conflict of interest.

References

1. Palensky, P.; Dietrich, D. Demand Side Management: Demand Response, Intelligent Energy Systems, and Smart Loads. *IEEE Trans. Ind. Inform.* **2011**, *7*, 381–388. [CrossRef]
2. Asano, H.; Nagata, Y. A Survey of Demand Response and Research Activities at CRIEPI. *Rev. Electr. Econ.* **2015**, *62*, 1–7. (In Japanese)
3. Siano, P. Demand Response and Smart Grids—A Survey. *Renew. Sustain. Energy Rev.* **2014**, *30*, 461–478. [CrossRef]
4. Meyabadi, A.F.; Deihimi, M.H. A review of demand-side management: Reconsidering theoretical framework. *Renew. Sustain. Energy Rev.* **2017**, *80*, 367–379. [CrossRef]
5. Albadi, M.; El-Saadany, E. A summary of demand response in electricity markets. *Electr. Power Syst. Res.* **2008**, *78*, 1989–1996. [CrossRef]
6. Deng, R.; Yang, Z.; Chow, M.-Y.; Chen, J. A Survey on Demand Response in Smart Grids: Mathematical Models and Approaches. *IEEE Trans. Ind. Inform.* **2015**, *11*, 570–582. [CrossRef]
7. Vardakas, J.; Zorba, N.; Verikoukis, C.V. A Survey on Demand Response Programs in Smart Grids: Pricing Methods and Optimization Algorithms. *IEEE Commun. Surv. Tutor.* **2014**, *17*, 152–178. [CrossRef]
8. Yan, X.; Ozturk, Y.; Hu, Z.; Song, Y. A review on price-driven residential demand response. *Renew. Sustain. Energy Rev.* **2018**, *96*, 411–419. [CrossRef]
9. Faruqui, A.; Bourbonnais, C. The Tariffs of Tomorrow: Innovations in rate designs. *IEEE Power Energy Mag.* **2020**, *18*, 18–25. [CrossRef]
10. Faruqui, A.; Sergici, S. Household response to dynamic pricing of electricity: A survey of 15 experiments. *J. Regul. Econ.* **2010**, *38*, 193–225. [CrossRef]
11. Heter, K.; Wayland, S. Residential Response to Critical-Peak Pricing of Electricity: California Evidence. *Energy* **2010**, *35*, 1561–1567. [CrossRef]
12. Erucsibm, T. Households' Self-Selection of Dynamic Electricity Tariffs. *Appl. Energy* **2011**, *88*, 2541–2547.
13. Bartusch, C.; Wallin, F.; Odlare, M.; Vassileva, I.; Wester, L. Introducing a demand-based electricity distribution tariff in the residential sector: Demand response and customer perception. *Energy Policy* **2011**, *39*, 5008–5025. [CrossRef]
14. Yousefi, S.; Nighaddam, M.P.; Majd, V.J. Optimal Real Time Pricing in an Agent-based Retail Market Suing a Compre-hensive Demand Response Model. *Energy* **2011**, *36*, 5716–5727. [CrossRef]
15. Yoon, J.H.; Bladick, R.; Novoselac, A. Demand response for residential buildings based on dynamic price of electricity. *Energy Build.* **2014**, *80*, 531–541. [CrossRef]
16. Kawamura, K.; Doki, T.; Oono, Y.; Takano, H.; Murata, J. Analysis of Field Test Results and Proposal of Sustainable De-mand Peak Reduction in Demand Response Programs for Residential Consumers. *IEEJ Trans. Electron. Inf. Syst.* **2017**, *137*, 96–105. (In Japanese)

17. Zou, X. Double-sided auction mechanism design in electricity based on maximizing social welfare. *Energy Policy* **2009**, *37*, 4231–4239. [CrossRef]
18. Yang, P.; Tang, G.; Nehorai, A. A game-theoretic approach for optimal time-of-use electricity pricing. *IEEE Trans. Power Syst.* **2012**, *28*, 884–892. [CrossRef]
19. Li, N.; Chen, L.; Dahleh, M.A. Demand Response Using Linear Supply Function Bidding. *IEEE Trans. Smart Grid* **2015**, *6*, 1827–1838. [CrossRef]
20. Hase, R.; Shinomiya, N. A mathematical modeling technique with network flows for social welfare maximization in deregulated electricity markets. *Oper. Res. Perspect.* **2016**, *3*, 59–66. [CrossRef]
21. Sorin, E.; Bobo, L.; Pinson, P. Consensus-Based Approach to Peer-to-Peer Electricity Markets with Product Differentiation. *IEEE Trans. Power Syst.* **2018**, *34*, 994–1004. [CrossRef]
22. Luenberger, D.G. *Microeconomics Theory*; McGraw-Hill: New York, NY, USA, 1995.
23. Perloff, J.M. *Microeconomics*, 2nd ed.; Addison-Wesley: Boston, MA, USA, 2001.
24. Hausman, D.; McPherson, M.; Sats, D. *Economic Analysis, Moral Philosophy, and Public Policy*, 3rd ed.; Cambridge University Press: Cambridge, UK, 2016.
25. Liu, H.; Shen, Y.; Zabinsky, Z.B.; Liu, C.C.; Courts, A.; Joo, S.K. Social Welfare Maximization in Transmission Enhance-ment Considering Network Congestion. *IEEE Trans. Power Syst.* **2008**, *23*, 1105–1114.
26. Baharlouei, Z.; Hashemi, M.; Narimani, H.; Mohsenian-Rad, H. Achieving Optimality and Fairness in Autonomous De-mand Response: Benchmarks and Billing Mechanisms. *IEEE Trans. Smart Grid* **2013**, *4*, 968–975. [CrossRef]
27. Wang, H.; Gao, Y. Real-time pricing method for smart grids based on complementarity problem. *J. Mod. Power Syst. Clean Energy* **2019**, *7*, 1280–1293. [CrossRef]
28. Agency for Natural Resources and Energy. Japan's Energy White Paper. 2020. Available online: https://www.enecho.meti.go.jp/about/whitepaper/ (accessed on 31 May 2021).
29. Hobbs, B.F.; Rothkopf, M.H.; O'Neill, R.P.; Chao, H.P. *The Next Generation of Electric Power Unit Commitment Models*; Springer: Singapore, 2001; Volume 36.
30. Padhy, N.P. Unit Commitment—A Bibliographical Survey. *IEEE Trans. Power Syst.* **2004**, *19*, 1196–1205. [CrossRef]
31. Saravanan, B.C.; Das, S.; Sikri, S.; Kothari, D.P. A solution to the unit commitment problem—A review. *Front. Energy* **2013**, *7*, 223–236. [CrossRef]
32. Zheng, Q.; Wang, J.; Liu, A.L. Stochastic Optimization for Unit Commitment—A Review. *IEEE Trans. Power Syst.* **2014**, *30*, 1913–1924. [CrossRef]
33. Takano, H.; Asano, H.; Gupta, N. Application Example of Particle Swarm Optimization on Operation Scheduling of Mi-crogrids. In *Frontier Applications of Nature Inspired Computation*; Khosravy, M., Gupta, N., Patel, N., Senju, T., Eds.; Springer: Singapore, 2020; pp. 967–994.
34. Takano, H.; Goto, R.; Hayashi, R.; Asano, H. Optimization Method for Operation Schedule of Microgrids Considering Uncertainty in Available Data. *Energies* **2021**, *14*, 2487. [CrossRef]
35. Investigation R&D Committee on IEEJ. *Electrical Power System Standard Models*; IEEJ: Tokyo, Japan, 1999; Volume 754. (In Japanese)
36. Uchida, N.; Kawata, K.; Egawa, M. Development of test case models for Japanese power systems. In Proceedings of the 2000 Power Engineering Society Summer Meeting, Seattle, WA, USA, 16–20 July 2000. [CrossRef]
37. Wang, L.; Wang, Z.; Yang, R. Intelligent Multiagent Control System for Energy and Comfort Management in Smart and Sustainable Buildings. *IEEE Trans. Smart Grid* **2012**, *3*, 605–617. [CrossRef]
38. Liu, Y.; Zhang, Y.; Chen, K.; Chen, S.Z.; Tang, B. Equivalence of Multi-Time Scale Optimization for Home Energy Man-agement Considering User Discomfort Preference. *IEEE Trans. Smart Grid* **2017**, *8*, 1876–1887. [CrossRef]
39. Li, D.; Chiu, W.-Y.; Sun, H.; Poor, H.V. Multiobjective Optimization for Demand Side Management Program in Smart Grid. *IEEE Trans. Ind. Inform.* **2017**, *14*, 1482–1490. [CrossRef]
40. Takano, H.; Kudo, A.; Taoka, H.; Ohara, A. A basic study on incentive pricing for demand response programs based on social welfare maximization. *J. Int. Counc. Electr. Eng.* **2018**, *8*, 136–144. [CrossRef]
41. Takano, H.; Tanonaka, N.; Kikuda, S.; Ohara, A. A Design Method for Incentive-based Demand Response Programs Based on a Framework of Social Welfare Maximization. *IFAC-PapersOnLine* **2018**, *51*, 374–379.
42. Patnam, B.S.K.; Pindoriya, N.M. Demand response in consumer-Centric electricity market: Mathematical models and optimization problems. *Electr. Power Syst. Res.* **2021**, *193*, 106923. [CrossRef]

Article

Fault Diagnosis Approach of Main Drive Chain in Wind Turbine Based on Data Fusion

Zhen Xu [1,2], Ping Yang [1,*], Zhuoli Zhao [3], Chun Sing Lai [3,4], Loi Lei Lai [3] and Xiaodong Wang [5]

1. Guangdong Key Laboratory of Clean Energy Technology, South China University of Technology, Guangzhou 510640, China; xu.zhen@hgnyjs.com
2. Shenzhen Huagong Energy Technology Co., Ltd., Shenzhen 518000, China
3. Department of Electrical Engineering, School of Automation, Guangdong University of Technology, Guangzhou 510006, China; zhuoli.zhao@gdut.edu.cn (Z.Z.); chunsing.lai@brunel.ac.uk (C.S.L.); l.l.lai@gdut.edu.cn (L.L.L.)
4. Brunel Interdisciplinary Power Systems Research Centre, Department of Electronic and Electrical Engineering, Brunel University London, London UB8 3PH, UK
5. State Power Investment, Nanning 530000, China; bestop2021@163.com
* Correspondence: eppyang@scut.edu.cn

Abstract: The construction and operation of wind turbines have become an important part of the development of smart cities. However, the fault of the main drive chain often causes the outage of wind turbines, which has a serious impact on the normal operation of wind turbines in smart cities. In order to overcome the shortcomings of the commonly used main drive chain fault diagnosis method that only uses a single data source, a fault feature extraction and fault diagnosis approach based on data source fusion is proposed. By fusing two data sources, the supervisory control and data acquisition (SCADA) real-time monitoring system data and the main drive chain vibration monitoring data, the fault features of the main drive chain are jointly extracted, and an intelligent fault diagnosis model for the main drive chain in wind turbine based on data fusion is established. The diagnosis results of actual cases certify that the fault diagnosis model based on the fusion of two data sources is able to locate faults of the main drive chain in the wind turbine accurately and provide solid technical support for the high-efficient operation and maintenance of wind turbines.

Keywords: data fusion; main drive chain; fault diagnosis; wind turbine

1. Introduction

The smart city concept is an advanced trend for the development for cities today and some crucial technologies such as Internet of Things (IoT), renewable energy, and smart grids are integrated to build the intelligent energy system in a smart city [1–5]. To increase the share of renewable energy in electricity generation and avoid the challenges caused by the centralized construction, centralized grid connection, and long-distance transmission of large-scale wind farms far away from the load center, distributed wind turbines are being widely used in development of smart cities [6–8]. The operation and maintenance of wind turbines distributed around the whole smart city are more difficult than the operation and maintenance of wind turbines in a centralized large-scale wind farm. In the wind turbines, electrical components have the highest fault frequency, followed by the main drive chain components. However, the electrical component faults can be located quickly, and the time to recover is short. Compared with the electrical component, the main drive chain component faults have a longer positioning time. Because of their huge size and heavy weight, it is cumbersome to replace them, and they need more time to recover. The outage of wind turbines before the replacement of the main drive chain fault components severely affects the operational reliability. So the wind turbines are in urgent need of economic efficiency of the wind turbines and the accurate and highly efficient remote intelligent operation and maintenance service. Therefore, it is imperative to establish the fault diagnosis system with high precision for the wind turbine main drive chain [9–11].

To improve the accuracy of main drive chain fault diagnosis in wind turbines, scholars worldwide have carried out a lot of research on the fault diagnosis of main drive chain. Among all the researchers, most of them try to alarm the out-of-limit main drive chain based on the real-time monitoring data of the wind farm supervisory control and data acquisition (SCADA) system [12–17] so as to maintain the normal operation state of wind turbines. Based on the real-time monitoring data of the SCADA system and the out-of-limit alarms of the main drive chain, the fault diagnosis framework of the wind turbine is established, and the out-of-limit diagnosis indexes of typical faults are given [18]. The research uses high-frequency SCADA data to extract the core technical indicators to improve the performance of wind turbines [19]. Further, the real-time monitoring data of the SCADA system is used to establish the fault prediction model of wind turbines [20]. More approached-based data analysis of the SCADA are being used today [21,22]. For instance, a data-driven method is proposed to diagnose the pitch fault of wind turbines [23]. Reference [24] authors also use a data-based prognostic system without any additional sensor out of the SCADA. This paper [25] summarizes various types of real-time monitoring systems of wind turbines and states the advantages and disadvantages of various methods for fault monitoring of wind turbines based on the real-time monitoring data of the SCADA system. To improve the accuracy of the main drive chain fault diagnosis in wind turbines, people began to use the professional fault main drive chain diagnosis system for high-frequency vibration data acquisition and fault analysis to its main components [26]. Damage can be detected based on the differences between modified modal displacements in the undamaged and damaged states [27]. Based on the high-frequency vibration signal analysis of the main drive chain fault diagnosis system in reference [28], the gearbox faults under the non-stationary state of speed and load were statistically analyzed. Compared with the out-of-limit alarm signal in the low frequency real-time monitoring system of the SCADA system, the high-frequency signal analysis can locate the gearbox faults more accurately. In references [29–34], the high-frequency resonance vibration signal of bearing is extracted by wavelet analysis. Combined with the classification ability of the support vector machine (SVM) and the dynamic time series processing ability of the hidden Markov model, a new bearing fault diagnosis scheme is proposed so as to improve the accuracy of bearing fault diagnosis. In reference [35], wavelet packet energy entropy is combined with empirical mode decomposition (EMD) to enhance the noise elimination of original vibration data and improve the accuracy of diagnosis. References [36–38] proposed a typical fault diagnosis method of the gearbox based on wavelet decomposition and support vector machine classification, a wind turbine bearing vibration fault diagnosis method based on noise suppression and a fault diagnosis method for the planetary gearbox of main drive chain of wind turbine under non-stationary conditions based on adaptive optimal kernel time-frequency analysis, respectively. References [39–41] show other methods of fault diagnosis based on wavelet transform. They all show that more effective typical fault features are extracted from the high-frequency vibration signals of the professional main drive chain vibration fault diagnosis system. Reference [42] began to introduce the influence of different working conditions on wind turbine fault analysis. Recently, more scholars used advanced algorithms such as neural network and machine learning for fault diagnosis in wind turbines [43–46]. References [47,48] compare the advantages and disadvantages of the neural network model and traditional condition monitoring analysis model in the fault diagnosis of wind turbines. Some other methods proposed fault diagnosis using different technologies such as thermal imaging [49]. They are non-invasive but limited to specific tested objects and show less universality than the commonly used SCADA and vibration fault diagnosis system. For instance, reference [50] uses thermal imaging to evaluate the condition of only angle grinders in wind turbines. Moreover, wind turbines usually operate in wild environments with variable temperatures, and, therefore, evaluating the wind turbine using only thermal imaging is possible to be disturbed in operating states.

Reference [51] uses ultrasonic reflectometry for detecting while the targeted fault to be detected is limited to the lubrication failure of the wind turbine bearing.

The main drive chain vibration fault diagnosis system and wind turbine SCADA system are two of the most popular diagnosis systems for wind turbines. However, their suppliers are independent of each other so the data of these two systems cannot be shared, and, therefore, each of the research projects above is based on one of these two data sources for fault diagnosis.

In this paper, a data interface between the two systems is established through the technical transformation of the two systems by the wind farm owners. An approach of using a data fusion method of two types of data to extract fault features of the wind turbine main drive chain is proposed and paves a new way to improve the accuracy of fault diagnosis of the main drive chain in the wind turbine.

The contributions of this study are as follows:

(1) Proposing a fault diagnosis strategy of the main drive chain in wind turbines based on data fusion, considering both the real-time monitoring data from the SCADA system and the high-frequency vibration data of the main chain.
(2) Proposing the detailed method to classify and extract the fault features based on two types of data and the method for fault diagnosis using the deep autoencoder model.
(3) Conduct case studies in a real wind farm to verify the effectiveness of the proposed strategy, and analyze the experimental results and the benefits to the high-efficient operation and maintenance of wind turbines.

2. Fault Features Extraction of Wind Turbine Main Drive Chain Based on Data Fusion

The entire process of the proposed data-fusion based method is given in Figure 1. The whole process can be divided into two steps: fault features extraction, and fault diagnosis based on data fusion. Details of each step are described separately in Sections 2 and 3.

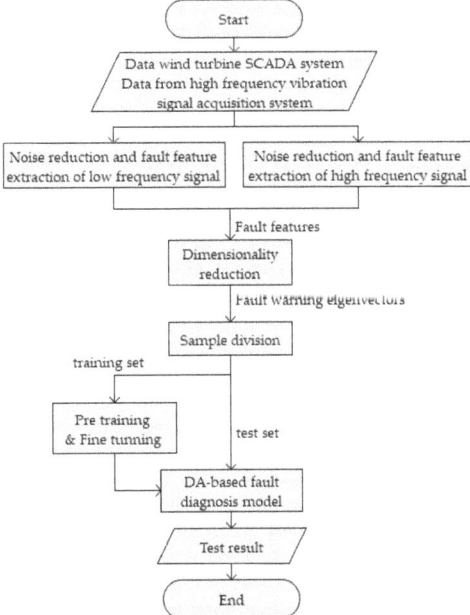

Figure 1. Flow chart of fault diagnosis approach of main drive chain in wind turbine based on data fusion.

This section demonstrates the method of fusing two types of real-time monitoring systems to extract the features of main drive chain faults. This method takes advantage of the globality of the wind turbine SCADA system and the pertinence and depth of the main drive chain vibration fault diagnosis system for the real-time monitoring of the wind turbine main drive chain.

2.1. Process of Fault Features Extraction of the Main Drive Chain in Wind Turbine

The supervisory control and data acquisition (SCADA) system of a wind farm collects real-time operating status data of each wind turbine in the wind farm comprehensively, with the sampling frequency of 1 s. Specifically, it comprehensively monitors the parameters and operating status of each operation control module in a wind turbine, including pitch, yaw, gearbox, generator, hydraulic pump station, nacelle, converter, power grid, safety chain, torque, main shaft, tower base, anemometer, and other modules. With the analysis and judgment for each module's operating status parameters, faults, and trend, normal operation of the turbine is maintained through the approaches of over-limit alarm and over-limit shutdown. However, the fault has already risen to a certain extent when the parameters and trends of modules of the wind turbine exceed the limit, and, therefore, how to trigger early warning becomes a core concern of wind farm owners. In recent years, many wind farm owners have equipped a high-frequency vibration signal acquisition system specifically for the main drive chain in order to improve the accuracy of fault diagnosis to the main drive chain of the wind turbine. They use acceleration sensors and other high-speed sensors to collect the high-frequency vibration signal of the main points of the main drive chain and do time-domain analysis, frequency domain analysis, as well as time-frequency domain analysis to extract more detailed fault features to locate the main drive chain faults more accurately. However, sometimes the added high-frequency vibration signal acquisition system gets noisy result data because it is susceptible to interference from different operating conditions, such as the yaw state of the unit, the rotational speed, and the icing of blades or anemometers. The added high-frequency vibration signal acquisition system is unable to deal with the operating status of the wind turbine or to remove the noise of the vibration data in a well-targeted manner. These deficiencies all make it difficult to reflect the fault state of the main drive chain accurately and comprehensively based on the fault eigenvector extracted from a single data source. That in turn reduces the accuracy of diagnosis results.

Therefore, this paper proposes a method to transform two types of systems technically, the main drive chain vibration fault diagnosis system and the wind turbine SCADA system, to establish a data interface between them and use a method of fusing two kinds of data to extract the fault features of the wind turbine main drive chain. On the one hand, the wind turbine SCADA system has the ability to monitor the overall situation of the wind turbine in real time, which is used to extract low-frequency vibration signals related to drive chain faults, the rotational speed of main shaft and generator, and the operation control mode of the wind turbine. The last two types of signals, the rotational speed of main shaft and generators and the operation control mode of the wind turbine, are highly related to the vibration mode of the main drive chain and provide supplementary knowledge for the denoising of high-frequency vibration signals of the main drive chain. On the other hand, taking advantage of pertinence and depth of the main drive chain vibration fault diagnosis system for real-time monitoring of the wind turbine main drive chain, the high-frequency vibration signals of all added measurement points of the main drive chain can be extracted and use two types of signal, rotational speed of main shaft and generator of the wind turbine and wind turbine operation control mode, to classify the background noises of high-frequency vibration signals. The used sensors and method to equip them are shown in Figures 2 and 3. High-frequency vibration signals are clearly different between different speed ranges of the main shaft and generators and between the power-up and power-down operation intervals of the wind turbine. The effective removal of background noises is beneficial to the extracting of the high-frequency vibration features of the main drive chain itself.

Figure 2. Photos of a vibration sensor and its base used in the main chain of the wind turbine.

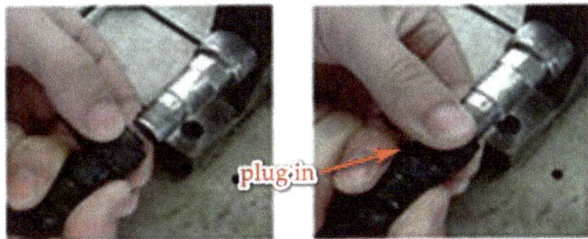

Figure 3. Method to equip the vibrating sensor.

When using the SCADA system to extract low-frequency vibration signals related to drive chain faults, it is also necessary to use two types of signals, the speed of main shaft and generators of the wind turbine and the operation control mode of the wind turbine, to classify its background noise. Between different speed ranges of the main shaft and generators and between the power-up and power-down operation intervals of the wind turbine, background noises are classified and removed to extract the low-frequency vibration characteristics of the main drive chain itself.

2.2. Fault Features Extraction of Wind Turbine Main Drive Chain Based on Data Fusion

Table 1 shows the types and causes of typical faults in the main drive chain of wind turbines. For the typical types of the wind turbine main drive chain faults, the low-frequency vibration signal and high-frequency vibration signal of the main drive chain are denoised, respectively, according to the fault features extraction flowchart given in Figure 4. On the basis of this procedure, it is necessary to extract the low-frequency fault features and high-frequency fault features of each typical fault in the main drive chain, and then eliminate redundant fault features to reduce dimensionality of the fault features to form eigenvectors that characterize the typical faults of the main drive chain.

Among them, the noise reduction of low-frequency and high-frequency vibration signals of the main drive chain, the extraction of low-frequency and high-frequency fault features of typical faults, and the dimensionality reduction of fault features are the core procedures of fault features extraction of the wind turbine main drive chain based on data fusion:

(1) Noise reduction of low-frequency and high-frequency vibration signals of the main drive chain

The nacelle of the wind turbine will still shake and vibrate during normal operation when the main drive chain is not vibrating. The frequency and amplitude of shaking and vibration are closely related to the rotational speed of the main shaft and generators and the operation control mode of the wind turbine. Experimental data indicate that it is a non-linear relationship. To simplify the calculation, this article firstly segments the rotational speed of the main shaft and generators of the wind turbine according to their sizes and classifies the operation control mode of the wind turbine according to the increasing power and decreasing power. Based on the combination of these two approaches, we classify the nacelle shake and vibration during the normal operation of wind turbine and give out the frequency and amplitude of the nacelle shake and vibration background noise in the combination of each rotational speed range of the main shaft and generator as well as the wind turbine power-up or power-down operation.

Table 1. Types and causes of typical faults of main drive chain in wind turbine.

Faulty Module	Main Fault Type	Fault Cause
Gearbox gear	Gear break	Sudden impact overload, bearing damage, shaft bending, continuous contact fatigue, foreign matter mixed in the meshing area, etc.
	Tooth surface wear	Material defects, poor lubrication, foreign matter mixed in the meshing area, etc.
	Tooth surface pitting	Poor lubrication, over-high speed, over-high oil temperature
	Tooth surface bonding	Poor lubrication, over-concentrated local load, over-high oil temperature, over-high speed, etc.
Bearings (gearbox, main shaft, generator, etc.)	Rust and corrosion	Poor sealing, insufficient rust prevention
	Wear	Poor lubrication, foreign matter mixed in, etc.
	Surface peeling	Overload, design or installation defect, foreign matter mixed in, over-small clearances, etc.
	Bonding	Over-small clearance, poor lubrication, overload, rolling body deflection, etc.
	Crack	Impact load, fatigue friction crack, large foreign body stuck in, etc.
Shafting (main shaft, low/high speed shaft in gearbox, etc.)	Shaft misalignment	Design or installation defect, etc.
	Shaft bending	Material and installation defect, stress concentration is not eliminated during the manufacturing process, gearbox damaged, etc.
	Shaft fracture	Material defect, stress concentration is not eliminated during the manufacturing process, gearbox damaged, etc.
Coupling	Misalignment	The gearbox high-speed shaft is misaligned with the generator, bearing air gap is too large, the ball is slightly corroded, etc.
	Grinding disc fracture	Safety cover scratch, the high-speed shaft of the gearbox and the generator are misaligned, etc.
Generator winding fault	Rotor fault	Rotor eccentricity fault, bearing deformation, design defect, poor installation, etc.
	Stator fault	Winding insulation aging

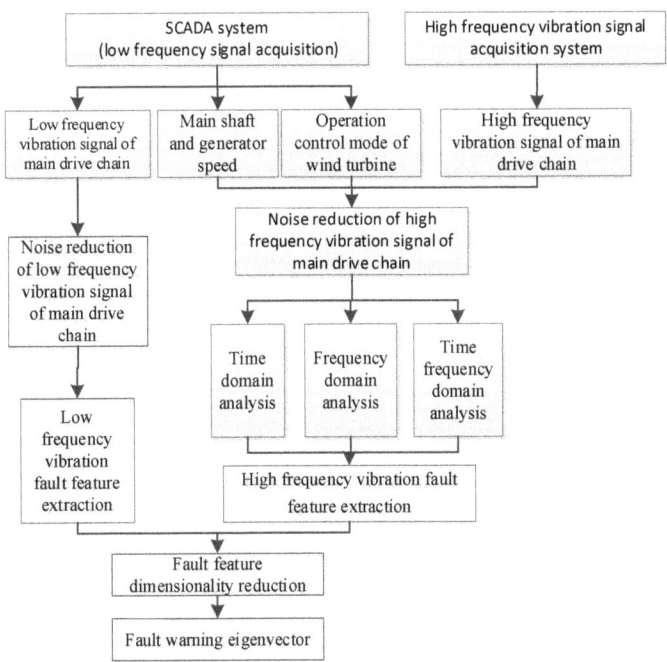

Figure 4. Process of fault feature extraction based on fusion of two types of data.

Assume that the background noise of the nacelle shaking and vibration under the first section of rotational speed ranges of the main shaft and generator and the wind turbine power up operation state are $x1(t)$. Fourier transform of $x1(t)$ is:

$$F1(\omega) = \int_{-\infty}^{+\infty} x1(t)e^{-j\omega t}dt \qquad (1)$$

This will be used as the background noise of the low-frequency and high-frequency vibration signal of the main drive chain in the first section of the main shaft and generator rotational speed ranges and the wind turbine power-up operation. Before extracting the low-frequency and high-frequency fault features, these background noises are eliminated separately.

(2) Extraction of low-frequency and high-frequency fault features of typical faults

After getting the low-frequency vibration signal and high-frequency vibration signal without background noise in the previous section, the frequency domain analysis method is adopted to calculate the following parameters as the low-frequency fault features: the low-frequency radial vibration, the axial vibration amplitude, and the vibration phase difference of the main frequency band. For high-frequency vibration data, it is necessary to calculate dimensional parameters, such as effective value, average amplitude, mean square error, kurtosis, and slope, and non-dimensional parameters, such as kurtosis index, impulse index, and margin index. With the help of the spectrum analysis method, the radial vibration amplitude, axial vibration amplitude, and phase difference of the main frequency band of each frequency band can be extracted as high-frequency fault features.

(3) Dimensionality reduction of fault features

For each type of typical fault, enough fault features should be extracted to determine the type of fault accurately. However, too many redundant fault features will not help increase the accuracy of fault determination, and contradictory samples inside will reduce the accuracy of fault diagnosis. Therefore, it is necessary to reduce the dimensionality of fault features.

In order to judge each type of typical fault, we require not only low-frequency fault and high-frequency fault features data, but also the combination of how the main shaft and generator rotational speed of the wind turbine are segmented according to the sizes and how the operation control mode of the wind turbine is classified according to the power-up and power-down operations. Therefore, it is necessary to take the rotational speed of the main shaft and generators and the operation control mode of the wind turbine as fault features when reducing the dimensionality of fault features. In this paper, the dimensionality reduction algorithm using Principal Component Analysis (PCA) is used to reduce the dimensionality of the fused eigenvectors. Based on the original n-dimensional features, the k-dimensional orthometric eigenvector is extracted as the principal component through centralized processing and calculation of covariance. It uses orthogonal transformation as the mapping matrix, calculates the covariance matrix of data matrix, obtains an eigenvalue and eigenvectors of the covariance matrix, and then selects the eigenvectors corresponding to the k characteristics with the largest eigenvalue (that is, the largest variance) from the matrix. In this way, the data matrix can be transformed into a new space, and dimensionality reduction of data characteristics can be realized. The main processing steps for a high-dimensional space data sample $x \in \mathbb{R}^d$ are: use the orthogonal matrix $A \in R(k \times d)$ to map the sample to a low-dimensional space $Ax \in \mathbb{R}^k$, where $k \ll d$ states that the purpose of dimensionality reduction is to alleviate the curse of dimensionality and classify data better. The specific algorithm is as follows:

Input: n-dimensional sample set $D = \left(x^{(1)}, x^{(2)}, \ldots, x^{(m)}\right)$, the dimension to be reduced to is k.

Output: the sample set D' after dimensionality reduction.

(1) Centralize all samples:

$$x^{(i)} = x^{(i)} - \frac{1}{m}\sum_{j=1}^{m} x_j^{(i)} \tag{2}$$

where m is data volume of sample $x^{(i)}$,

(2) Calculate the covariance matrix XX^T of the sample,
(3) Perform singular value decomposition on the matrix XX^T,
(4) Take out the eigenvectors $w_1, w_2, \cdots, w_k$ corresponding to the largest k singular values, and normalize all the eigenvectors to form an eigenvector matrix W,
(5) For each sample $x^{(i)}$ in the sample set, transform it into a new sample:

$$z^{(i)} = W^T x^{(i)} \tag{3}$$

(6) Obtain the output sample set:

$$D' = z^{(1)}, z^{(2)}, \cdots, z^{(k)} \tag{4}$$

Through the dimensionality reduction processing using the PCA algorithm, the dimension of the fault characteristic vector can be reduced from hundreds to dozens, which can markedly reduce the complexity of the following step of data processing.

3. Fault diagnosis of Wind Turbine Main Drive Chain Based on Fusion of Two Types of Data

Based on low-frequency fault and high-frequency fault features obtained from the fusion of two types of data and fault characteristic variables obtained by combining the main shaft and generator rotation speed of the wind turbine and the operation control mode of the wind turbine, dozens of typical characteristics of fault early warning are generated after dimensionality reduction. However, they are still multi-variable and large-scale data. Moreover, different typical faults have a different number of characteristics of fault early warning. To deal with this kind of multi-variable fault diagnosis problem that input variables need to be adjusted for different typical faults, the Deep Autoencoder (DA) model is a suitable approach for the fault diagnosis model training of different typical fault types. The training of the fault diagnosis model for typical fault types mainly includes the following steps:

(1) Select the low-frequency monitoring data from the SCADA system and the high-frequency vibration monitoring data from the main drive chain vibration fault diagnosis system under the normal state and a typical fault state of the wind turbine main drive chain. Calculate the characteristics of a fault warning and establish a sample data set of this typical fault. Normalize the sample data set and divide it into a training set and test set with a certain proportion.
(2) Determine the number of stacked AEs (Auto Encoders) and establish the DA with multiple hidden layers. The number of input layer neurons is the dimension of the input sample, and the data set is used for pre-training by stacking AEs.
(3) Use the labeled samples in the main drive chain training data set to apply supervised fine-tuning to the entire DA to complete all training processes.
(4) When the entire DA training is completed, establish the DA model for the main drive chain, calculate the reconstruction error R with the test sample set, and integrate the test samples into the DA model for testing.

The fault diagnosis model training process of typical fault types is shown in Figure 5.

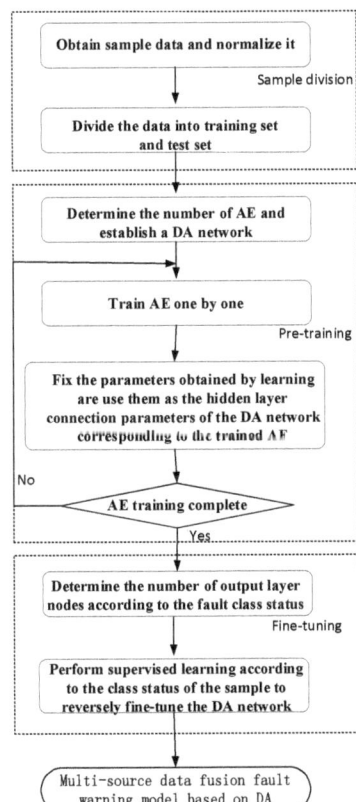

Figure 5. Process of generating DA-based fault diagnosis model.

The DA model training process above is unsupervised learning of the sample data set. The parameters obtained by training can be used as prior information of supervised learning of the DA model. The DA model can be further optimized by using the labeled data set for supervised learning to improve accuracy of fault diagnosis. This fine-tuning process is designed as follows:

Assume that the sample data are:

$$\left\{\left(x^{(1)}, y^{(2)}\right), \cdots, \left(x^{(i)}, y^{(i)}\right), \cdots \left(x^{(m)}, y^{(m)}\right)\right\}_{i=1}^{m} \tag{5}$$

where the category status corresponding to x_i is $y^{(i)} \in \{1, 2, \cdots, k\}$, which is generally is given in the form of label encoder, and k represents the total number of categories. According to the analysis above, the k-dimensional vector output obtained by the classifier represents the conditional probability $h_\theta\left(x^{(i)}\right) = p(y = j|x)$ that the input x is the corresponding category, and the main form is:

$$h_\theta\left(x^{(i)}\right) = \begin{bmatrix} p(y^{(i)} = 1|x^{(i)}; \theta) \\ p(y^{(i)} = 2|x^{(i)}; \theta) \\ \cdots \\ p(y^{(i)} = k|x^{(i)}; \theta) \end{bmatrix} = \frac{1}{\sum_{j=1}^{k} e^{\theta_j^T x(i)}} \begin{bmatrix} e^{\theta_1^T x(i)} \\ e^{\theta_2^T x(i)} \\ \cdots \\ e^{\theta_k^T x(i)} \end{bmatrix} \tag{6}$$

θ is not a column vector but a matrix as

$$\theta = \left(\theta_1^T, \theta_2^T, \cdots, \theta_k^T\right) \tag{7}$$

where each row of the matrix represents the parameter corresponding to a category in the classifier while the count of all categories is k. The supervised global fine-tuning stage aims to do further parameters' adjustment to minimize the value of the objective optimization function. The objective optimization function (or, more exactly, the cost function) is:

$$J(\theta) = -\frac{1}{m}\left[\sum_{i=1}^{m}\sum_{j=1}^{k} 1\{y^{(i)} = j\} \log \frac{e^{\theta_j^T x(i)}}{\sum_{l=1}^{k} e^{\theta_l^T x(i)}}\right] \tag{8}$$

where $1\{\cdots\}$ is the indicator function. The function value is 1 when the value in parentheses is true; otherwise it is 0. We have to minimization the value of $J(\theta)$, and we still use the stochastic gradient descent method to solve it here. The iterative formula is:

$$\theta_j = \theta_j - \alpha \frac{\partial J(\theta)}{\partial \theta_{jl}} \tag{9}$$

$$\frac{\partial J(\theta)}{\partial \theta_{jl}} = -\frac{1}{m}\sum_{i=1}^{m}\left[x^{(i)}(1\{y^{(i)} = j\} - p(y^{(i)} = j|x^{(i)}; \theta))\right] \tag{10}$$

The stochastic gradient descent method is a popular method in the field of machine learning. We repeat the process in Equation (9) until convergence. $\alpha\frac{\partial J(\theta)}{\partial \theta_{jl}}$ in Equation (10) is the partial differential of cost function to θ_j. At the convergence point, the partial differential is 0, and therefore the cost function is minimized.

4. Results

In this section, we analyze the actual data of 66 doubly fed generators of 2MW in a wind farm. This wind farm is fully equipped with a SCADA system and a vibration fault diagnosis system for the main drive chain. Some user interfaces of the software system after upgrading are shown in Figures 6 and 7. The upgrading of the fault diagnosis module for the main chain in the software uses the data-fusion-based fault diagnosis approach we discussed in this article.

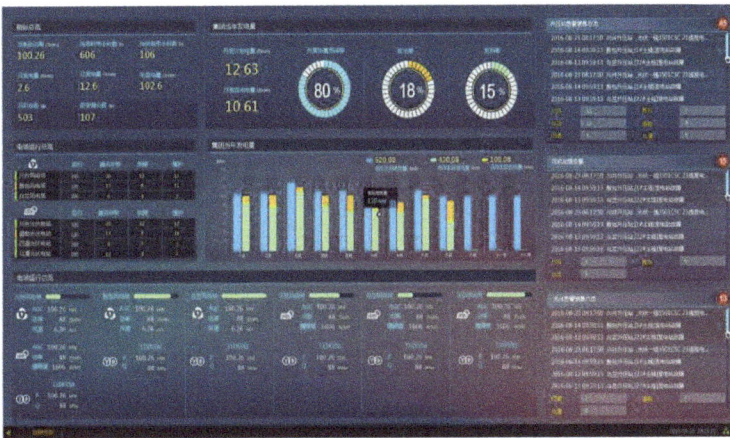

Figure 6. Monitoring platform in operating status.

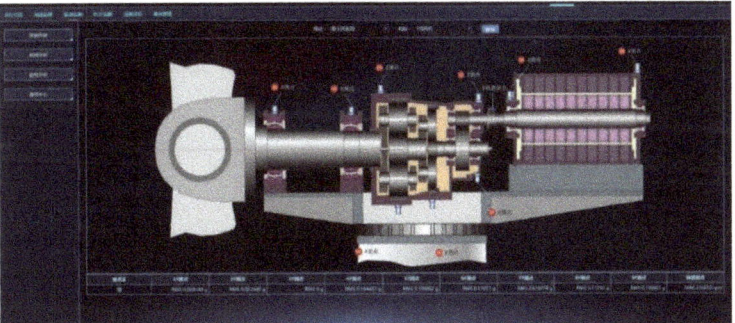

Figure 7. Fault diagnosis module of the monitoring system.

For the typical fault of a gearbox with broken tooth in the main drive chain, the actual data of the wind turbine from 2 May to 29 June 2019 are used to calculate the fault features based on the fusion of the two types of data, and the result is 21 fault warning characteristic variables. The actual data from 2 May to 29 June 2019 are a total of 2000 pairs of fault characteristic data, of which three-quarters are used as the training set of the main drive chain fault diagnosis DA model while the rest are used as the test set. The number of hidden layers of this DA model is set to 17 while the average absolute error of the overall data set is the smallest, and it has the greatest ability to excavate deep features of the input data. The number of hidden nodes in each layer of the DA network are set to 152, 314, and 528, with the consideration to minimize the mean absolute deviation (MAD) and loss indicators.

Because the data sets of the main drive chain under normal operating status are used for the construction of DA model, the reconstruction error under abnormal states is great enough to exceed the monitoring threshold under normal states.

After the training of the fault diagnosis DA model, the test set data are used to test the gearbox broken tooth diagnosis model, and the verified results are shown in Figures 8 and 9.

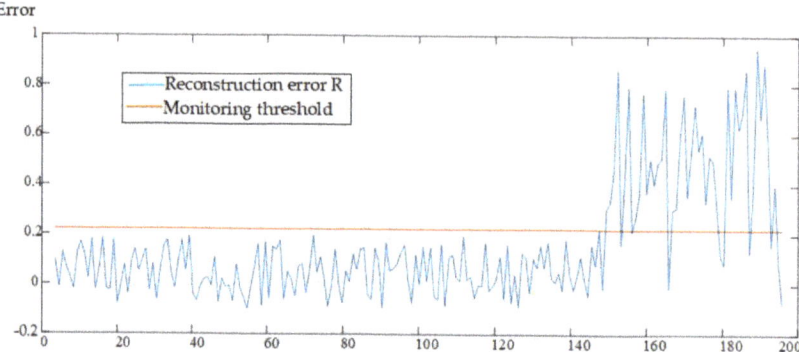

Figure 8. The output of gearbox broken tooth diagnosis model before and after gearbox tooth break of wind turbine A.

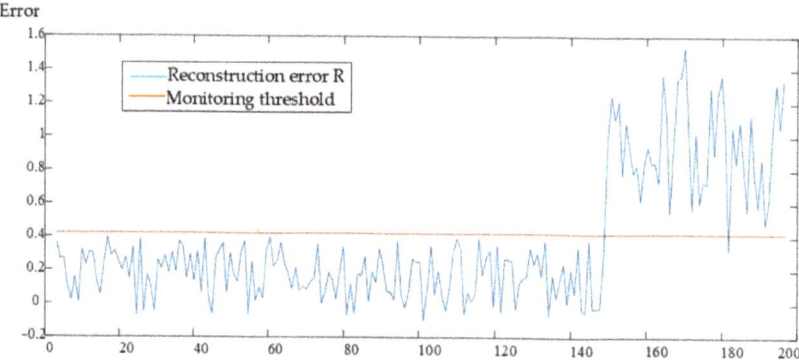

Figure 9. The output of gearbox broken tooth diagnosis model before and after gearbox tooth break of wind turbine B.

As shown in the figures above, before the gearbox teeth of the main drive chain of the wind turbine A and B are broken at a data point of about 150, the model output was within the monitoring threshold range, and the fluctuation range is not large, indicating that the main drive chain gearboxes of the two turbines are operating in normal states. However, the output of the model begins to increase and exceeds the pre-set monitoring threshold after the tooth break. Hence, it is judged that the two wind turbines are malfunctioning, and fault diagnosis as well as early warning are performed.

Consistent with the actual data, the time domain and frequency spectrum diagrams of the vibration signal of the medium-speed shaft of the gearbox of turbine A (9 May 2019) are shown in Figures 10 and 11.

Contrastively, the time domain and frequency spectrum diagrams of the vibration signal of the medium-speed shaft in the gearbox in the early warning state (12 May 2019) are shown in Figures 12 and 13.

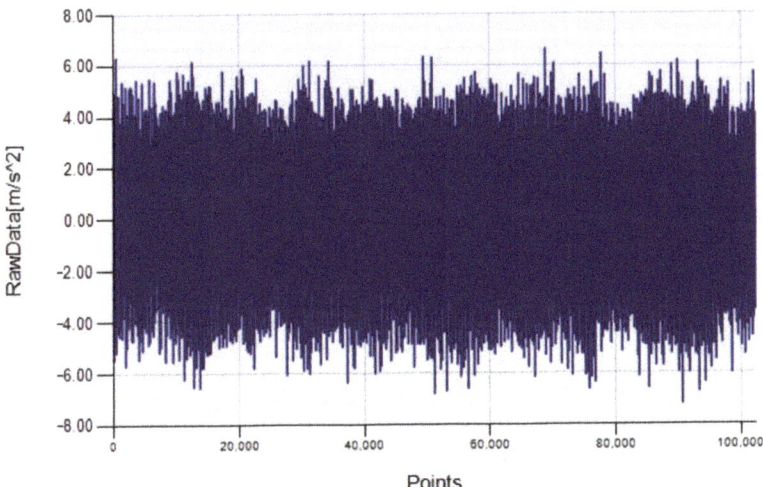

Figure 10. Time domain diagram of vibration signal at the medium-speed shaft position in the gearbox of wind turbine A under normal conditions.

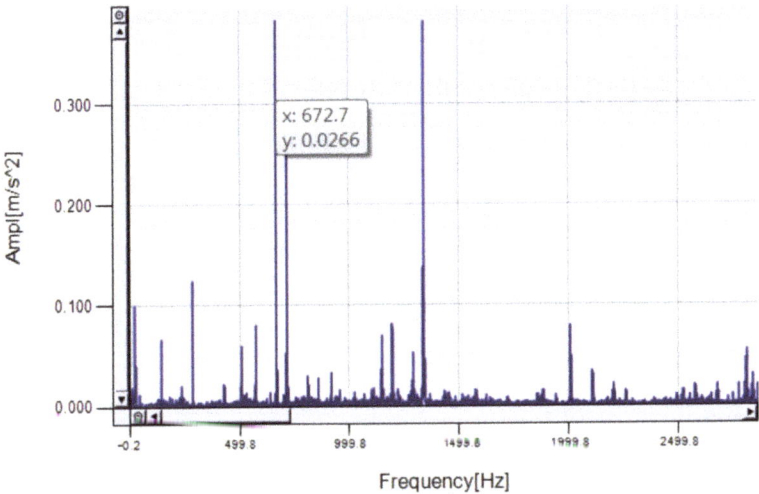

Figure 11. Frequency spectrum of vibration signal at the medium-speed shaft position in the gearbox of wind turbine A under normal conditions.

As shown in Figures 10 and 12, the time domain diagrams under the normal and warning conditions are similar and show little valuable information. However, the hidden differences can be clearly revealed in the diagrams of the frequency spectrum in Figures 11 and 13. The details are described below.

Figure 13 shows a sideband modulation signal of 5.99 HZ (signal in the red box in Figure 13) at a rotational frequency of the medium-speed shaft near the gear mesh frequency (670.833 HZ) from the medium-speed shaft to the high-speed shaft of the gearbox. However, there is no sideband signal such as this under the normal operating state in Figure 12. This difference can be judged as an abnormal condition of the meshing gear of the shaft. The operation and maintenance personnel disassembled the on-site gearbox and found that the fault was the breakage of tooth on the medium-speed gear, as shown in Figure 14. This model correctly warned the early fault of the main drive chain gearbox of turbine A and avoided further expansion of this failure.

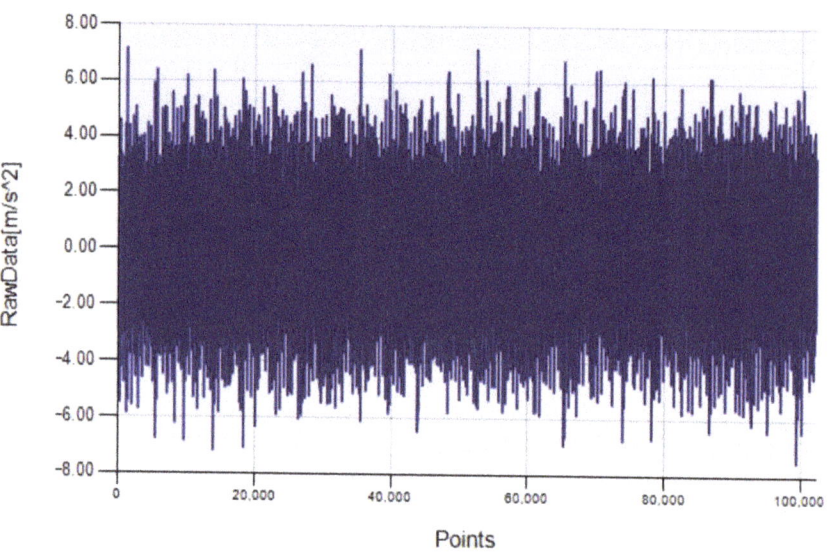

Figure 12. Time domain diagram of vibration signal of medium-speed shaft position in gearbox of wind turbine A under warning conditions.

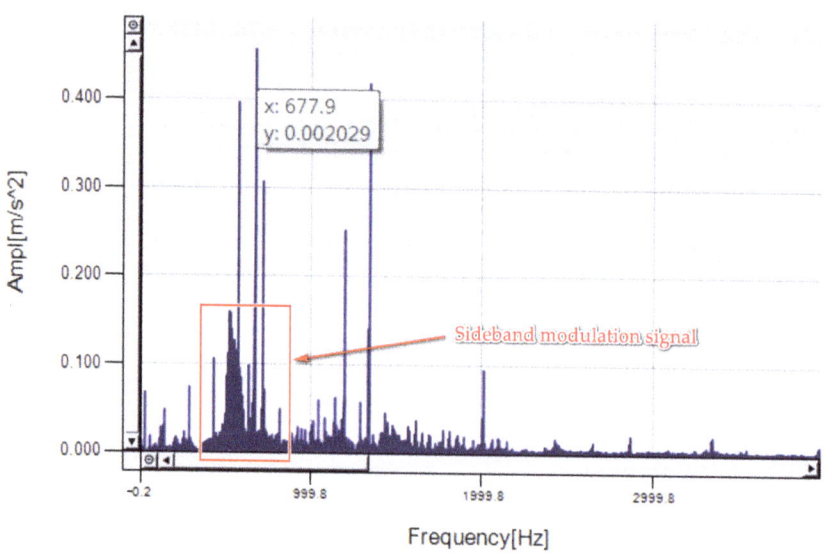

Figure 13. Frequency spectrum of the vibration signal of the medium speed-shaft position in gearbox of wind turbine A under warning conditions.

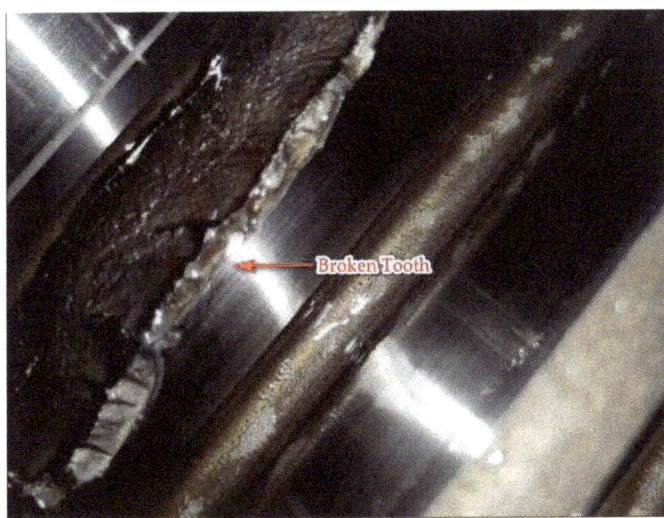

Figure 14. Gearbox med-speed gear broken tooth fault in wind turbine A.

The details of all experiments are not shown completely here. However, a brief table is shown below as Table 2 to demonstrate the results of diagnoses to different typical faults in the main drive chain in wind turbines.

Table 2. Results of diagnoses of to typical faults in wind turbine main drive chain.

Turbine	Fault Type	Fault Location	Abnormal Phenomena	Maintenance after Early Warning	Duration between Early Warning and Alarm from SCADA
#23	Broken tooth	Gear at medium speed shaft	Sideband signal in frequency spectrum	Yes	-
#32	Broken tooth	Minor gear at medium speed shaft	Sideband signal in frequency spectrum	Yes	-
#15	Broken tooth	Minor gear at medium speed shaft	Sideband signal in frequency spectrum	Yes	-
#30	Corrosion Tooth	Planet bearing	Abnormal peak in frequency spectrum near the characteristic frequency of planet bearing	No	14 days
#13	Corrosion	Outer raceway of rear bearing	Exorbitant peak value of vibration signal	No	5 days

As shown in the table above, maintenances are not taken immediately after the early warnings emitted by the diagnosis system in some scenes, and there are considerable interval times before we got alarms from the traditional SCADA system. Additionally, the following maintenances indicate that there is no false alarm. That clearly indicates the efficiency and accuracy of the diagnosis method proposed.

5. Conclusions

This article demonstrates the shortcomings of the commonly used main drive chain fault diagnosis methods that only use a single data source. Then a method of fault features extraction and fault diagnosis based on data source fusion is proposed. The new method makes integrated uses of the globality of the wind turbine SCADA system and the pertinence and depth of the vibration fault diagnosis system for main drive chain in wind turbines to solve the problem that background noises from a single data source are difficult to process.

In the proposed method, fault features of the main drive chain are jointly extracted and a deep self-encoding network fault diagnosis model based on data fusion is established by integrating SCADA real-time monitoring system data with main drive chain vibration monitoring data. The parameters obtained by the unsupervised learning training of the deep auto-encoding network can be used as the prior information of the following supervised learning model. Using labeled data sets for supervised learning further optimizes the deep auto-encoding network model and improves the accuracy of fault diagnosis.

The experimental results show that the diagnosis system using the proposed method accurately located the gearbox broken tooth fault in a wind turbine at a very early phase before the traditional SCADA system raised any alarm. That diagnosis avoided further expansion of this failure followed by greater loss. Obviously, this new approach provides strong technical support for the operation and maintenance of wind turbines with more immediacy and efficiency.

There is a possibility to use the way of fusing data from multiple sources for other problems. However, the scenes and methods proposed in this article are specific and highly concentrated. More complete and pertinent analysis and experiments must be done for another specific problem.

Future related research will be focused on classification and recognition of possible original causes of the detected faults. Various types of faults and the original reasons will be analyzed. It will allow for the pre-analysis and early warning of faults before the manual detection and will improve the efficiency of the operation and maintenance of wind turbines.

Author Contributions: Conceptualization, Z.X. and P.Y.; methodology, Z.X. and P.Y.; software, Z.X.; validation, Z.X., P.Y. and X.W.; formal analysis, Z.X.; investigation, Z.X. and C.S.L.; resources, P.Y, Z.Z. and X.W.; data curation, Z.Z. and X.W.; writing—original draft preparation, Z.X. and X.W.; writing—review and editing, Z.Z., C.S.L. and L.L.L.; visualization, L.L.L.; supervision, P.Y. and L.L.L.; project administration, P.Y. and Z.Z.; funding acquisition, P.Y. All authors have read and agreed to the published version of the manuscript.

Funding: This work was supported by the Research Program of Digital Grid Research Institute, China Southern Power Grid under Grant YTYZW20010.

Institutional Review Board Statement: Not applicable.

Informed Consent Statement: Not applicable.

Data Availability Statement: All data generated or analyzed during this study are included in this article.

Conflicts of Interest: The authors declare no conflict of interest.

References

1. Xing, L.; Jiao, B.; Du, Y.; Tan, X.; Wang, R. Intelligent Energy-Saving Supervision System of Urban Buildings Based on the Internet of Things: A Case Study. *IEEE Syst. J.* **2020**, *14*, 4252–4261. [CrossRef]
2. Zhao, Z.; Yang, P.; Wang, Y.; Xu, Z.; Guerrero, J.M. Dynamic Characteristics Analysis and Stabilization of PV-Based Multiple Microgrid Clusters. *IEEE Trans. Smart Grid* **2019**, *10*, 805–818. [CrossRef]
3. Kirimtat, A.; Krejcar, O.; Kertesz, A.; Tasgetiren, M.F. Future Trends and Current State of Smart City Concepts: A Survey. *IEEE Access* **2020**, *8*, 86448–86467. [CrossRef]
4. Zhao, Z.; Guo, J.; Lai, C.S.; Xiao, H.; Zhou, K.; Lai, L.L. Distributed Model Predictive Control Strategy for Islands Multimicrogrids Based on Noncooperative Game. *IEEE Trans. Ind. Inform.* **2021**, *17*, 3803–3814. [CrossRef]
5. Șerban, A.C.; Lytras, M.D. Artificial Intelligence for Smart Renewable Energy Sector in Europe—Smart Energy Infrastructures for Next Generation Smart Cities. *IEEE Access* **2020**, *8*, 77364–77377. [CrossRef]

6. Lai, C.S.; Jia, Y.; Dong, Z.; Wang, D.; Tao, Y.; Lai, Q.H.; Wong, R.T.; Zobaa, A.F.; Wu, R.; Lai, L.L. A review of technical standards for smart cities. *Clean Technol.* **2020**, *2*, 290–310. [CrossRef]
7. Azevedo Guedes, A.L.; Carvalho Alvarenga, J.; Dos Santos Sgarbi Goulart, M.; Rodriguez y Rodriguez, M.V.; Pereira Soares, C.A. Smart Cities: The Main Drivers for Increasing the Intelligence of Cities. *Sustainability* **2018**, *10*, 3121. [CrossRef]
8. Florescu, A.; Barabas, S.; Dobrescu, T. Research on Increasing the Performance of Wind Power Plants for Sustainable Development. *Sustainability* **2019**, *11*, 1266. [CrossRef]
9. Marti-Puig, P.; Blanco-M, A.; Serra-Serra, M.; Solé-Casals, J. Wind Turbine Prognosis Models Based on SCADA Data and Extreme Learning Machines. *Appl. Sci.* **2021**, *11*, 590. [CrossRef]
10. Rezamand, M.; Kordestani, M.; Carriveau, R.; Ting, D.S.-K.; Orchard, M.E.; Saif, M. Critical Wind Turbine Components Prognostics: A Comprehensive Review. *IEEE Trans. Instrum. Meas.* **2020**, *69*, 9306–9328. [CrossRef]
11. Qin, A.; Hu, Q.; Lv, Y.; Zhang, Q. Concurrent Fault Diagnosis Based on Bayesian Discriminating Analysis and Time Series Analysis with Dimensionless Parameters. *IEEE Sens. J.* **2019**, *19*, 2254–2265. [CrossRef]
12. Liu, Y.; Wu, Z.; Wang, X. Research on Fault Diagnosis of Wind Turbine Based on SCADA Data. *IEEE Access* **2020**, *8*, 185557–185569. [CrossRef]
13. Tautz-Weinert, J.; Watson, S.J. Using SCADA data for wind turbine condition monitoring-a review. *IET Renew. Power Gener.* **2017**, *11*, 382–394. [CrossRef]
14. Wilkinson, M.; Darnell, B.; Van Delft, T.; Harman, K. Comparison of methods for wind turbine condition monitoring with SCADA data. *IET Renew. Power Gener.* **2014**, *8*, 390–397. [CrossRef]
15. Yang, W.; Jiang, J. Wind turbine condition monitoring and reliability analysis by SCADA information. In Proceedings of the 2011 Second International Conference on Mechanic Automation and Control Engineering, IEEE, Inner Mongolia, China, 15–17 July 2011; pp. 1872–1875.
16. Zaher, A.; Mcarthur SD, J.; Infield, D.G.; Patel, Y. Online wind turbine fault detection through automated SCADA data analysis. *Wind Energy* **2009**, *12*, 574–593. [CrossRef]
17. Kermani, M.; Carnì, D.L.; Rotondo, S.; Paolillo, A.; Manzo, F.; Martirano, L. A Nearly Zero-Energy Microgrid Testbed Laboratory: Centralized Control Strategy Based on SCADA System. *Energies* **2020**, *13*, 2106. [CrossRef]
18. Leahy, K.; Gallagher, C.; O'Donovan, P.; Bruton, K.; O'Sullivan, D.T.J. A robust prescriptive framework and performance metric for diagnosing and predicting wind turbine faults based on SCADA and alarms data with case study. *Energies* **2018**, *11*, 1738. [CrossRef]
19. Gonzales, E.; Stephen, B.; Infield, D.; Melero, J.J. On the use of high-frequency SCADA data for improved wind turbine performance monitoring. *J. Phys. Conf. Ser.* **2017**, *926*, 012009. [CrossRef]
20. Kusiak, A.; Li, W. The prediction and diagnosis of wind turbine faults. *Renew. Energy* **2011**, *36*, 16–23. [CrossRef]
21. Yin, H.; Jia, R.; Ma, F.; Wang, D. Wind turbine condition monitoring based on SCADA data analysis. In Proceedings of the 2018 IEEE 3rd Advanced Information Technology, Electronic and Automation Control Conference (IAEAC), Chongqing, China, 12–14 October 2018; pp. 1101–1105. [CrossRef]
22. Pei, Y.; Qian, Z.; Tao, S.; Yu, H. Wind turbine condition monitoring using SCADA data and data mining method. In Proceedings of the 2018 International Conference on Power System Technology (POWERCON), Guangzhou, China, 6–9 November 2018; pp. 3760–3764. [CrossRef]
23. Kusiak, A.; Verma, A. A data-driven approach for monitoring blade pitch faults in wind turbines. *IEEE Trans. Sustain. Energy* **2010**, *2*, 87–96. [CrossRef]
24. Encalada-Dávila, Á.; Puruncajas, B.; Tutivén, C.; Vidal, Y. Wind Turbine Main Bearing Fault Prognosis Based Solely on SCADA Data. *Sensors* **2021**, *21*, 2228. [CrossRef]
25. Crabtree, C.J.; Zappalá, D.; Tavner, P.J. Survey of Commercial Available Condition Monitoring System for Wind Turbines: Supergen Wind Energy Technologies Consortium Report. 2012. Available online: http://www.supergen-wind.org.uk (accessed on 25 May 2014).
26. Tian, S.; Li, Z.; Li, H.; Hu, Y.; Lu, M. Active Control Method for Torsional Vibration of DFIG Drive Chain Under Asymmetric Power Grid Fault. *IEEE Access* **2020**, *8*, 155611–155618. [CrossRef]
27. Duvnjak, I.; Damjanović, D.; Bartolac, M.; Skender, A. Mode Shape-Based Damage Detection Method (MSDI): Experimental Validation. *Appl. Sci.* **2021**, *11*, 4589. [CrossRef]
28. Villa, L.F.; Reñones, A.; Perán, J.R.; de Miguel, L.J. Statistical fault diagnosis based on vibration analysis for gear test-bench under non-stationary conditions of speed and load. *Mech. Syst. Signal Process.* **2012**, *29*, 436–446. [CrossRef]
29. Chen, G. Feature Extraction and Intelligent Diagnosis for Ball Bearing Early Faults. *Acta Aeronaut. Astronaut. Sin.* **2009**, *30*, 362–367.
30. Chen, X.; Chen, Y.; Long, Z.; Zhang, X.; Cheng, Z. Bearing fault diagnosis method based on SVM-HMM. *J. Wuhan Univ. Technol. IAME* **2016**, *38*, 267–270.
31. Wang, B.; Ke, H.; Ma, X.; Yu, B. Fault Diagnosis Method for Engine Control System Based on Probabilistic Neural Network and Support Vector Machine. *Appl. Sci.* **2019**, *9*, 4122. [CrossRef]
32. Zhang, X.; Han, P.; Xu, L.; Zhang, F.; Wang, Y.; Gao, L. Research on Bearing Fault Diagnosis of Wind Turbine Gearbox Based on 1DCNN-PSO-SVM. *IEEE Access* **2020**, *8*, 192248–192258. [CrossRef]

33. Huo, Z.; Zhang, Y.; Shu, L.; Gallimore, M. A New Bearing Fault Diagnosis Method Based on Fine-to-Coarse Multiscale Permutation Entropy, Laplacian Score and SVM. *IEEE Access* **2019**, *7*, 17050–17066. [CrossRef]
34. Wang, Y.; Zhu, Y.; Wang, Q.; Tang, Y.; Duan, F.; Yang, Y. Complex Fault Source Identification Method for High-Voltage Trip-Offs of Wind Farms Based on SU-MRMR and PSO-SVM. *IEEE Access* **2020**, *8*, 130379–130391. [CrossRef]
35. Lv, M.; Su, X.; Chen, C.; Liu, S. Application of Wavelet Packet Energy Entropy and EMD Conjoint Analysis in Fault Diagnosis of Wind Turbine Bearing. *Mach. Electron.* **2018**, *36*, 8–12.
36. Jiang, B.; Cao, H. Fault Diagnosis of GA-SVM Gearbox Based on Wavelet Decomposition and Sample Entropy. *Modul. Mach. Tool Autom. Manuf. Tech.* **2019**, *78*–82. [CrossRef]
37. Yang, D.; Li, H.; Hu, Y.; Zhao, J.; Xiao, H.; Lan, Y. Vibration condition monitoring system for wind turbine bearings based on noise suppression with multi-point data fusion. *Renew. Energy* **2016**, *92*, 104–116. [CrossRef]
38. Feng, Z.; Liang, M. Fault diagnosis of wind turbine planetary gearbox under nonstationary conditions via adaptive optimal kernel time–frequency analysis. *Renew. Energy* **2014**, *66*, 468–477. [CrossRef]
39. Su, N.; Li, X.; Zhang, Q. Fault Diagnosis of Rotating Machinery Based on Wavelet Domain Denoising and Metric Distance. *IEEE Access* **2019**, *7*, 73262–73270. [CrossRef]
40. Zhang, J.; Sun, H.; Sun, Z.; Dong, W.; Dong, Y. Fault Diagnosis of Wind Turbine Power Converter Considering Wavelet Transform, Feature Analysis, Judgment and BP Neural Network. *IEEE Access* **2019**, *7*, 179799–179809. [CrossRef]
41. Rezamand, M.; Kordestani, M.; Carriveau, R.; Ting, D.S.; Saif, M. A New Hybrid Fault Detection Method for Wind Turbine Blades Using Recursive PCA and Wavelet-Based PDF. *IEEE Sens. J.* **2020**, *20*, 2023–2033. [CrossRef]
42. Yan, X. *Research on Fault Early Warning Method for Wind Turbine Based on Condition Identification*; North China Electric Power University: Beijing, China, 2017.
43. Vives, J.; Quiles, E.; García, E. AI techniques applied to diagnosis of vibrations failures in wind turbines. *IEEE Lat. Am. Trans.* **2020**, *18*, 1478–1486. [CrossRef]
44. Lu, L.; He, Y.; Wang, T.; Shi, T.; Ruan, Y. Wind Turbine Planetary Gearbox Fault Diagnosis Based on Self-Powered Wireless Sensor and Deep Learning Approach. *IEEE Access* **2019**, *7*, 119430–119442. [CrossRef]
45. Hsu, J.; Wang, Y.; Lin, K.; Chen, M.; Hsu, J.H. Wind Turbine Fault Diagnosis and Predictive Maintenance Through Statistical Process Control and Machine Learning. *IEEE Access* **2020**, *8*, 23427–23439. [CrossRef]
46. Luo, Z.; Liu, C.; Liu, S. A Novel Fault Prediction Method of Wind Turbine Gearbox Based on Pair-Copula Construction and BP Neural Network. *IEEE Access* **2020**, *8*, 91924–91939. [CrossRef]
47. Schlechtingen, M.; Santos, I.F. Comparative analysis of neural network and regression based condition monitoring approaches for wind turbine fault detection. *Mech. Syst. Signal Process.* **2011**, *25*, 1849–1875. [CrossRef]
48. Ahmad, R.; Kamaruddin, S. An overview of time-based and condition-based maintenance in industrial application. *Comput. Ind. Eng.* **2012**, *63*, 135–149. [CrossRef]
49. Mohammed, A.; Hu, B.; Hu, Z.; Djurovic, S.; Ran, L.; Barnes, M.; Mawby, P.A. Distributed Thermal Monitoring of Wind Turbine Power Electronic Modules Using FBG Sensing Technology. *IEEE Sens. J.* **2020**, *20*, 9886–9894. [CrossRef]
50. Glowacz, A. Ventilation Diagnosis of Angle Grinder Using Thermal Imaging. *Sensors* **2021**, *21*, 2853. [CrossRef]
51. Nicholas, G.; Clarke, B.P.; Dwyer-Joyce, R.S. Detection of Lubrication State in a Field Operational Wind Turbine Gearbox Bearing Using Ultrasonic Reflectometry. *Lubricants* **2021**, *9*, 6. [CrossRef]

Article

Coordinated Operation of Electricity and Natural Gas Networks with Consideration of Congestion and Demand Response

Chun Sing Lai [1,2], Mengxuan Yan [1], Xuecong Li [1,*], Loi Lei Lai [1,*] and Yang Xu [1]

[1] Department of Electrical Engineering, School of Automation, Guangdong University of Technology, Guangzhou 510006, China; chunsing.lai@brunel.ac.uk (C.S.L.); 1122004002@mail2.gdut.edu.cn (M.Y.); young_xuy@163.com (Y.X.)

[2] Brunel Interdisciplinary Power Systems Research Centre, Department of Electronic and Electrical Engineering, Brunel University London, London UB8 3PH, UK

* Correspondence: lixuecong@gdut.edu.cn (X.L.); l.l.lai@gdut.edu.cn (L.L.L.)

Abstract: This work presents a new coordinated operation (CO) framework for electricity and natural gas networks, considering network congestions and demand response. Credit rank (CR) indicator of coupling units is introduced, and gas consumption constraints information of natural gas fired units (NGFUs) is given. Natural gas network operator (GNO) will deliver this information to an electricity network operator (ENO). A major advantage of this operation framework is that no frequent information interaction between GNO and ENO is needed. The entire framework contains two participants and three optimization problems, namely, GNO optimization sub-problem-A, GNO optimization sub-problem-B, and ENO optimization sub-problem. Decision sequence changed from traditional ENO-GNO-ENO to GNO-ENO-GNO in this novel framework. Second-order cone (SOC) relaxation is applied to ENO optimization sub-problem. The original problem is reformulated as a mixed-integer second-order cone programming (MISOCP) problem. For GNO optimization sub-problem, an improved sequential cone programming (SCP) method is applied based on SOC relaxation and the original sub-problem is converted to MISOCP problem. A benchmark 6-node natural gas system and 6-bus electricity system is used to illustrate the effectiveness of the proposed framework. Considering pipeline congestion, CO, with demand response, can reduce the total cost of an electricity network by 1.19%, as compared to −0.48% using traditional decentralized operation with demand response.

Keywords: coordinated operation; natural gas network; electrical network; credit rank indicator

Citation: Lai, C.S.; Yan, M.; Li, X.; Lai, L.L.; Xu, Y. Coordinated Operation of Electricity and Natural Gas Networks with Consideration of Congestion and Demand Response. *Appl. Sci.* **2021**, *11*, 4987. https://doi.org/10.3390/app11114987

Academic Editor: Matti Lehtonen

Received: 11 April 2021
Accepted: 11 May 2021
Published: 28 May 2021

Publisher's Note: MDPI stays neutral with regard to jurisdictional claims in published maps and institutional affiliations.

Copyright: © 2021 by the authors. Licensee MDPI, Basel, Switzerland. This article is an open access article distributed under the terms and conditions of the Creative Commons Attribution (CC BY) license (https://creativecommons.org/licenses/by/4.0/).

1. Introduction

In November 2018, the European Union (EU) presented its long-term vision for a carbon-neutral economy by 2050. While renewable energy sources are adopted in achieving this goal, the EU will require more grid scale storage as the fraction of intermittent renewable energy becomes larger. Without significant grid scale storage, renewable energy sources may have to be taken off the grid or carry out large-scale load shedding to avoid de-stabilizing the grid when supply outstrips demand. Many options are being studied for grid scale storage, but no sustainable, marketable, affordable, and renewable solution is on the table. Power to Gas is a technology that could be useful in the short to medium term, as a component of a comprehensive grid scale storage solution, in support of a power grid supplied by intermittent renewable energy systems (RES) in the long term to promote smart energy for smart cities [1–3]. At the same time, some research deals with the energy, exergy, economic, and exergoenvironmental analyses of hybrid combined system [4].

The use of natural gas to produce electricity is increasing dramatically throughout the world. This trend is being driven by the cost and environmental advantages of gas-fired generation as compared to that of the coal. This increase has also highlighted the need for greater coordination between natural gas and electricity systems. Insufficient coordination limits pipeline operators from providing accurate guidance on gas quantities to supply gas-fired generators each hour [5]. In general, generators overschedule gas deliveries to ensure enough resources to operate. The inefficiency will only increase with the growing use of natural gas for electricity generation. While the two systems have operated for decades, closer coordination could save money, improve efficiencies, and minimize potential disruptions. Improved coordination of these networks will increase energy resiliency and reliability while reducing the cost of natural gas supply for electricity generation.

Natural gas generation can reduce the use of coal-fired power, support integration of renewable resources, and reduce greenhouse gas emission. In power systems with increasing numbers of uncertain renewable sources, gas generation is taking a more important role in supplying short-term flexibility to match unexpected electricity consumption. In the electricity network, the natural gas generation units are rapid response units that overcome generation scarcity caused by sudden variations in the electricity consumption or the renewable generation dispatch. The gas turbines' flexibility relates to the gas network's flexibility [5]. The performance of the gas turbine is affected by some important parameters that consist of the compression ratio, ambient temperature, pressure, humidity, turbine inlet temperature, specific fuel consumption, and air to fuel ratio [6].

Reference [7] examined the renewable energy sources impact, including pumped-storage units and photovoltaic systems on power system security from electricity and natural gas networks perspective. With the large-scale natural gas infrastructure deployment, which leads to changes in capabilities of pipelines, operational procedures, supply, and tariffs, it is important to have coordination between the two networks to enhance future energy system reliability [8].

Natural gas volume is susceptible to price changes and influences the generation and commitment costs of generating units. Naturally, an increasing amount of natural gas usage in the electricity sector has raised difficulties for natural gas network planning and operation. Natural gas pipelines' pressure interruption or loss may contribute to the outage of several natural gas-fired generators and lead to power system security issues. For electricity networks, intermittent renewable energy sources have no fuel costs and only have a fixed operating cost for operation. To achieve robust power system operations, natural gas units may provide flexible dispatch and rapid ramping capability. Turning to variable renewables, it is expected that the intermittency in generation due to, for example, wind will have an impact on other generation units, which will be required to increase and decrease their generation, as the wind generation falls and rises, to meet the shortfall between generation and demand. Naturally, gas turbines will have a greater role to play in generating electricity due to their ability to ramp up/down quickly.

Natural gas fired units (NGFUs) gradually replace coal fired units due to the operation flexibility, great efficiency and little capital costs in many countries [9]. In China, NGFUs occupy 4% of total generating capacity, while in the USA, it reached 39% [10]. Therefore, integrating natural gas systems with power systems is challenging.

A great amount of overall electricity production is produced by gas-fired power plants for several countries. It is projected to rise to 7600 TWh by 2035. With increasing gas demand, gas networks need to expand capacity to provide fuel to additional gas-fired power plants. Reference [11] presents an integrated gas and electricity network expansion planning model. Gas fired generation plants were seen as connections between the two networks. The model concurrently minimizes electricity and gas operational and network development costs.

Various research works were performed on optimal operation strategy and interdependency of the integrated electricity-natural gas system. Reference [7] examined the influence of natural gas infrastructure on the electricity system by focusing on particular constraints of the natural gas network. Reference [12] presented an in-depth integrated model for examining the influence of interdependency between natural gas and electricity networks on power system security, considering constraints of natural gas network in the security-constrained unit commitment (SCUC) problem for power systems. References [13,14] formulated the co-planning process as a mixed-integer nonlinear programming problem to consider recent challenges, including congestions, demand response effect, etc.

The growing reliance of the power system on NG systems has demanded an integrated planning approach for the two systems [12]. Reference [15] proposed a decentralized operation strategy and applied it to coordinate energy flow in a multi-area integrated electricity-natural gas system (IEGS). The complete decentralized operation of interconnected large-scale IEGS is achieved with the iterative alternating direction method of multipliers (ADMM) algorithm. IEGS was represented by several subsystems that were tied by electricity and natural gas networks. An optimal operation strategy based on decentralization was developed for multi-area IEGS.

Reference [10] showed that integrated natural gas and electricity networks operation can be attained in a distributed method with ADMM. Results showed that the ADMM-based approach gives enhanced convergence performances than the classical Lagrange Relaxation (LR), as well as augmented LR-based methods. Reference [16] propose a mixed integer programming model to characterize the electricity portfolio of large consumers. References [17,18] propose electricity and natural gas networks by considering energy hubs. As reported in Reference [19], the changes of natural gas unit's generation dispatch will provide natural gas demand profile changes. The changes may harm the natural gas network's security. Electricity and natural gas infrastructure's coordinated operation can enhance reliability and security for the infrastructure and avoid demand curtailment risks. Reference [20] presented a novel robust operation model for IEGS but without the solution algorithm. With ADMM, reference [21] presented a study on robust optimization for IEGS with distributed structures. Previous works have fixated on the coordination of natural gas and electricity networks at synergistic scheduling level. Electricity network flow rules are summarized in [22] to solve the gas flow problem. Reference [23] presented a methodology for considering the combination of the natural gas network. Coordination with stochastic scheduling were studied in [24]. Scheduling problems, including electricity demand response, were presented in [25,26]. Natural gas prices impacting on the system were examined in [27–29].

Several investigations have fixated on solving steady-state natural gas flow. The challenges are contributed from the non-linear and non-convex Weymouth equation, that is, the natural gas flow is a nonlinear programming (NLP) problem. References [22,23] employed a nonlinear solution or commercial solvers to determine gas flow, including interior point method or Newton-Raphson methods. Approximate piecewise linearization methods were presented in [30] and [31] by transforming to a mixed-integer linear programming (MILP) problem. To eliminate non-convexity, researchers in [10] and [32] employed the special form of the Weymouth equation by transforming the NLP problem to MISOCP via convex relaxation.

Natural gas and electricity networks are owned by various operators and stakeholders. However, both networks are often seen as an integrated system to achieve coordinated operation or establish a third-party coordinator with a lot of information interaction. The main contributions of this work are as follows:

- Proposed a novel CO framework and this changes the traditional decision sequence. With this framework, it is possible to avoid frequent information interaction and without a third coordinator.
- A model is constructed to generate NGFUs' gas consumption constraints. A CR indicator and an update method are introduced, to generate reasonable gas consumption constraints information deliver from gas network operator (GNO) to electricity network operator (ENO).
- Impact of considering classic demand response program [33] and congestion is discussed in a decentralized and coordinated operation. For GNO optimization sub-problem, an improve SCP method based SOC with reasonable initial expansion value is used to determine the steady-state gas flow in a natural gas network.

- Simulations under different scenarios, considering the NG and electricity networks operation, based on a benchmark 6-bus electricity system and 6-node natural gas system were used to verify the proposed framework, considering solution convergence and benefits obtained.
- The impact of the coordinated operation is studied with congestions and demand response. The present method can give a better convergence when compared to a very popular method such as ADMM.

Section 2 presents the formulations of the framework for coordination operation for one IEGS. Section 3 provides a mathematical model for the two sub-optimization problems, namely, GNO optimization and ENO optimization. Section 4 gives the solution methodology for solving the optimization of the two networks based on Weymouth function. Case studies and discussions are given in Section 5. Conclusion and future work are provided in Section 6.

2. Framework for Coordinated Operation

2.1. Coordination of Interdependent Electricity and Natural Gas Systems (IENS): An Overview

Due to the increasing interdependence between power grids and natural gas networks, it may be unreasonable, or physically infeasible, to model and optimize the two energy systems separately in practice. Different coordination strategies are as follows:

- From the perspective of the macro scope of the energy system model, the models can be classified from four aspects: comprehensive evaluation, energy economy, power system planning, and energy system planning [34]. They have different methods, different ranges of use and different fields of application. Energy modeling tools such as LEAP, EnergyPlan, MASSAGE, MARKAL/TIMES have been designated for sustainable energy planning analysis [35]. The main methods of energy system modeling are top-down and bottom-up. The combination of these models leads to a hybrid energy model.
- In IENS, due to the uncertainty of natural gas supply, the balance of supply and demand in power system may be affected for security and economic purposes. Power system researchers have incorporated natural gas transmission constraints into the unit commitment problem of security constraints [12,23,24]. The uncertainty of natural gas supply and the variability of natural gas price are also considered in reference [36] to study the impact of natural gas supply shortage on the optimal dispatch of power system.
- In the natural gas optimization problem, the time-varying gas consumption of gas generating units is simulated to explore the influence of large gas generating units [37–39] on the daily operation efficiency of natural gas network.
- For sequence optimization of power grid and natural gas network [38,39], this model cannot guarantee global optimality of the IENS.
- The collaborative optimization of IENS considers the power grid and natural gas network as a whole to minimize the total cost associated with the two energy systems. It can achieve the best solution for the whole IENS [40–42]. In addition, considering that power grid and natural gas network may belong to different system operators and information exchange may be restricted by policies, researchers explore decentralized algorithms to obtain high-quality coordination solutions of IENS, while maintaining decision independence and information privacy of the two systems [43–45]
- IENS should be considered from operational and long-term planning aspects [46]. Co-planning of power and natural gas networks can be proposed at system level [47] and local level [48]. The integration of natural gas and power sectors usually needs to study the interaction between them and Resource Co-optimization from the perspective of central planners [49]. The mathematical model used in the joint planning of electric power and natural gas shows that some relationships are nonlinear and nonconvex [50]. The most common problem in this paper is MINLP model [51]. In

order to solve the complexity of the model, various techniques of linear and convex reconstruction/relaxation [52] and decomposition method [53] are proposed.

2.2. Coordinated Operation Method Based on ADMM Algorithm

For the synergy between power and gas networks and coordination operators, the synergy among the three decision makers is achieved through consensus based ADMM method. Consensus variables are introduced to reflect the interaction between upper and lower operators. In each iteration, the upper coordination operator checks the convergence and updates the consensus variables, while the lower electric operator and natural gas operator solve the local optimization problem at the same time. The cooperative operation mode of the third party is: Based on the synchronous ADMM decoupling algorithm, a higher-level third-party cooperative operator (CO) is introduced to solve the electricity network subproblem and the natural gas network subproblem simultaneously, and the relevant information is transferred to the third-party scheduling department to realize the cooperative operation.

The structure of ADMM with coordination is shown in Figure 1 composed of five steps:

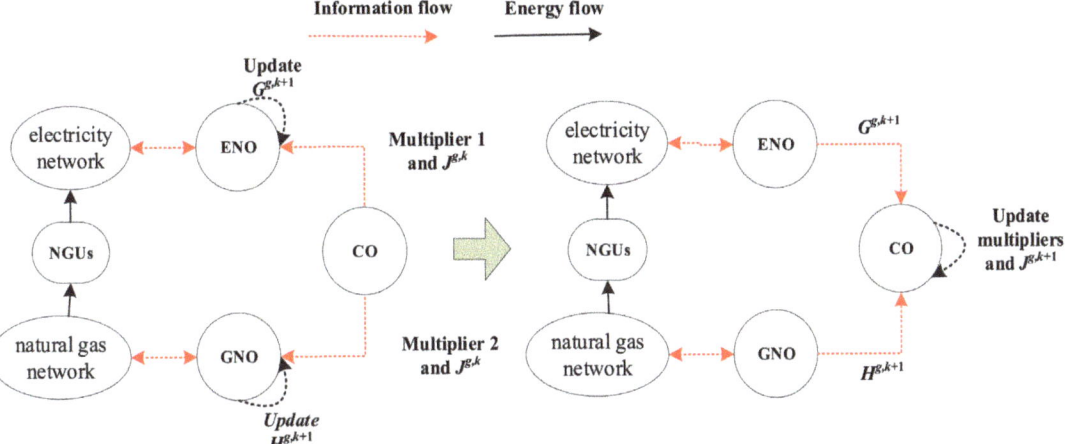

Figure 1. ADMM information flow with coordination operator.

1. Parameter initialization: Initialize the coupling variable values $G_i^{g,1}$ and $H_e^{g,1}$ of the power grid sub-problem and the natural gas network sub-problem respectively. Initialize the coupling variable value $J_e^{g,1}$ of the third-party dispatching department, set the ADMM algorithm step size ρ1 and ρ2. Initialize the Lagrange multiplier $\lambda_{ie,1}^k$ and $\lambda_{ie,2}^k$ of the grid sub-problem and the gas grid sub-problem respectively.
2. Simultaneously solve the electricity network subproblem and the gas network subproblem.
3. Update ADMM multiplier and third-party coupling variable.
4. Convergence criterion: The original residual is the coupling unbalance, and the dual residual is the difference before and after the iteration of the coupling variable.

2.3. Traditional Decentralized Operation

The structure of traditional decentralized operation (DO) is shown in Figure 2 composed of two steps:

Step 1: ENO carries out the optimal local scheduling of electricity network which described as ENO sub-problem, and the corresponding consumption of natural gas information will deliver to the GNO.

Step 2: GNO carries out the optimal scheduling of natural gas network and is described as GNO sub-problem. If GNO cannot supply enough expected gas volume in Step 1 as the premise of resident gas demand has a higher priority, ENO needs to re-dispatch in the electricity network.

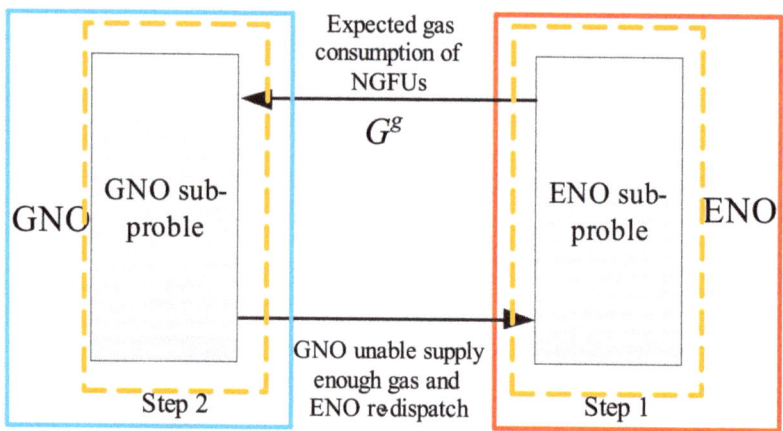

Figure 2. The structure of traditional DO.

The main defect with this traditional DO is that ENO as an advance decision-maker, cannot guarantee the local scheduling result at coupling units will always be a feasible solution in the GNO scheduling feasible domain shown in Figure 3. The decision sequence is ENO-GNO-ENO.

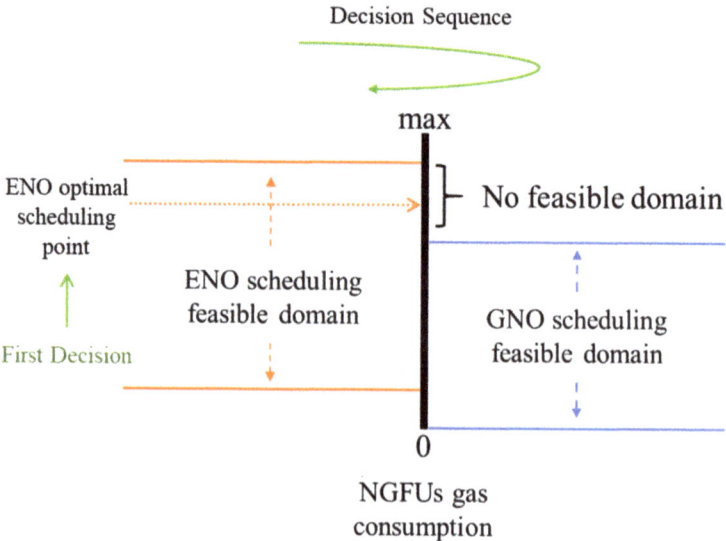

Figure 3. The diagram of traditional DO.

2.4. A New Framework for Coordinated Operation

A new framework is established for CO of electricity and natural gas networks, as presented in Figure 4. It is different from the traditional DO in decision sequence as GNO-ENO-GNO. GNO as advance decision-maker provides NGFUs gas consumption constraints to ENO. When there are multiple coupling units and gas supply capacity is limited, how to generate reasonable constraints for multiple coupling units is extremely important. To solve this problem, to the knowledge of the authors, it is the first time to introduce an indicator, that is, credit rating (CR) for decision making and the entire framework for CO is shown in Figure 5. There are three steps in the procedure.

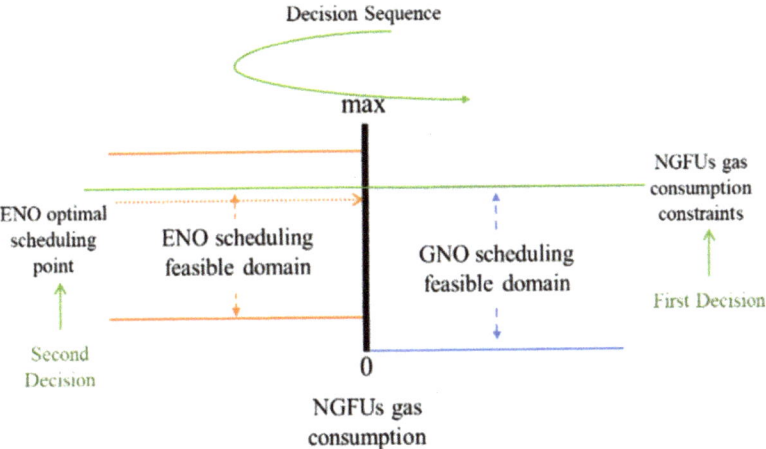

Figure 4. The diagram of the proposed CO framework.

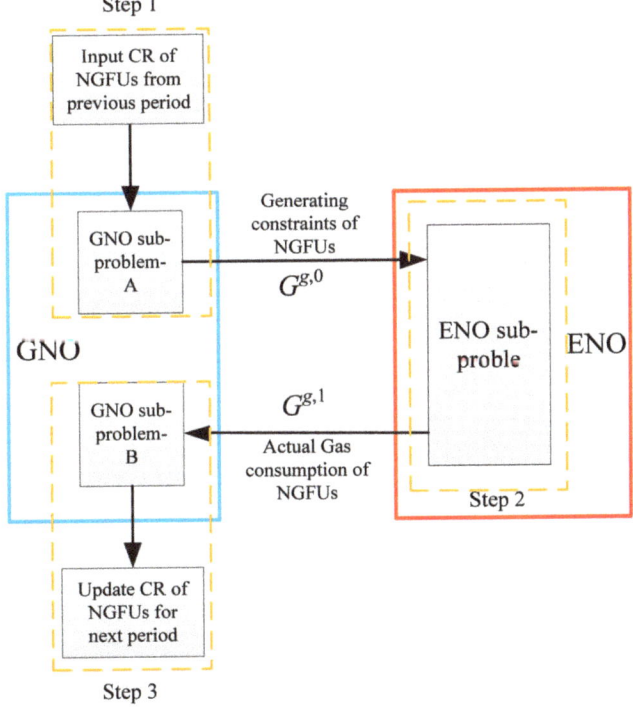

Figure 5. The structure of proposed CO framework.

Step 1: Receive NGFUs' CR from the previous period, GNO calculates the maximum profit with an uncertain gas load of NGFUs, named GNO sub-optimization-A problem. GNO as advance decision-maker provides NGFUs gas consumption constraints to ENO. To generate reasonable gas consumption constraints information that deliver from GNO to ENO. This means that the natural gas network supplies the maximum capacity to NGFUs.
Step 2: ENO set the maximum gas supply of NGFUs as constraints in optimization sub-problem to determine the hourly dispatch, including the dispatching generation and actual gas consumption of NGFUs.
Step 3: After receiving the actual gas load of NGFUs in natural gas network, GNO optimizes the gas network scheduling to achieve the minimal operation cost and updates the CR of NGFUs for the next scheduling period.

Some insights could be summarized as follows:

(1) The GNO and ENO do not have dispatching power and networks information of another party, but the information in coupling units is available for both GNO and ENO.
(2) In a natural gas network, resident natural gas load priority is higher than NGFUs gas load, GNO will cut the gas supply of NGFUs first if there is a natural gas shortage. Thus, in GNO optimization sub-problem, the resident natural gas load is the control parameters and the profit of this part is fixed.
(3) In GNO optimization sub-problem-A, the cost of a gas well is ignored. It assumes that the gas purchase price of NGFUs is greater than the production cost of a gas well, GNO will always benefit from it.
(4) CR is a gas price-related indicator, we assume that the forecast gas contract price of all NGFUs is the same. A higher CR means GNO will provide as much natural gas as possible to this NGFU, so that the gas constraint value deliver to ENO will be greater.

3. Mathematical Model

3.1. GNO Optimization Sub-Problem-A

As shown in (1), the objective of GNO sub-optimization-A problem is to maximize profit. In this sub-problem, the resident natural gas load is fixed, and NGFUs gas load is a variable. The first term of the objective function represents incomes from the resident natural gas load and the second term is expected profits from NGFUs natural gas load.

Natural gas network model is given by Equations (2)–(6) when applied by the steady-state natural gas flow. Constraint (2) represents the pressure of the gas node and gas flow in the pipeline relationship, which is a Weymouth function. The pressure of gas node upper and lower limits is shown in (3). Nodal natural gas balance in the network is given in (4), and the natural gas well production boundary is shown in (5). The gas load of NGFUs is constrained by the unit's maximum gas consumption as given by (6). Constraint (7) represents the relationship between CR, the gas contract price of NGFUs and credit value factor, which is equal to the product of CR and forecast gas contract price.

$$\text{Max } obj\_Ga = \sum_{t=1}^{N_T} \left(\sum_{gl=1}^{N_L} \tau_{gl} L + \sum_{i \in GU} \lambda_i^0 G_{i,t}^g \right) \quad (1)$$

$$F_{mn,t} = \text{sgn}(\omega_{m,t}, \omega_{n,t}) \cdot C_{mn} \sqrt{\left| \omega_{m,t}^2 - \omega_{n,t}^2 \right|} \quad (2)$$

$$\omega_{m,t}^{\min} \leq \omega_{m,t} \leq \omega_m^{\max} \quad (3)$$

$$T^w W - T^g G^g - T^l L = T^f F \quad (4)$$

$$W_s^{\min} \leq W_{s,t} \leq W_s^{\max} \quad (5)$$

$$G_{i,t}^g \leq G_i^{g,\max} \quad (6)$$

$$\lambda_i^0 = \eta_i CR_i^0 \quad (7)$$

After completing the above optimization solution, a series of constraints of NGFUs will be generated, the gas supply to NGFUs of this optimization result is given by (8). This value will deliver to the next step.

$$G_{i,t}^{g,0} = G_{i,t}^{g} = \mathrm{argmax}(obj\_Ga), \forall i \in GU \quad (8)$$

3.2. ENO Optimization Sub-Problem

(1) Objective function

As shown in (9), the objective function of ENO sub-optimization problem is determined by the hourly dispatch. It is to minimize the operating cost over the complete scheduling horizon with security-constraints. The first term of the objective function is the generation costs and startup cost of NGFUs, the second term is the generation costs and startup cost of non-NGFUs. The shutdown cost of units has been converted into the startup cost.

$$\mathrm{Min}\ obj\_E = \sum_{t=1}^{N_T} \left\{ \begin{array}{l} \sum_{i \in GU} \left(G_{i,t}^{g} \eta_i + SU_{i,t} \right) \\ + \sum_{i \in NGU} \left(G_{i,t}^{c} + SU_{i,t} \right) \end{array} \right\} \quad (9)$$

(2) Units and network constraints

Equations (10)–(21) are physical constraints of generating unit. Generation scheduling of each thermal unit is constrained by maximum and minimum output of the unit as presented in (10). The renewable power dispatch at individual hour is constrained by the renewable power forecast given in (11). Thermal units operating ramping up/down constraints are given in (12) and (13). Minimum on/off-limits are enforced by (14) and (15). The relationships between unit states and startup/shutdown indicators are given in (16). The logic between startup indicators and shutdown indicators are given in (17). Constraint (18) represents the generation costs of non-NGFUs. Constraint (19) represents the fuel consumption NGFUs. Constraint (20) shows the startup costs of thermal units.

$$P_i^{\min} u_{i,t} \leq P_{i,t} \leq P_i^{\max} u_{i,t} \quad (10)$$

$$P_{r,t} \leq P_{r,t}^W \quad (11)$$

$$P_{i,t} - P_{i,t-1} \leq UR_i(1 - y_{i,t}) + P_i^{\min} y_{i,t} \quad (12)$$

$$P_{i,t-1} - P_{i,t} \leq DR_i(1 - z_{i,t}) + P_i^{\min} z_{i,t} \quad (13)$$

$$\left[X_{i,t-1}^{on} - T_i^{on} \right] [u_{i,t-1} - u_{i,t}] \geq 0 \quad (14)$$

$$\left[X_{i,t-1}^{off} - T_i^{off} \right] [u_{i,t} - u_{i,t-1}] > 0 \quad (15)$$

$$y_{i,t} - z_{i,t} = u_{i,t} - u_{i,t-1} \quad (16)$$

$$y_{i,t} + z_{i,t} \leq 1 \quad (17)$$

$$G_{i,t}^{c} = \alpha_i^{c} u_{i,t} + \beta_i^{c} P_{i,t} + \gamma_i^{c} P_{i,t}^2, \forall i \in NGU \quad (18)$$

$$G_{i,t}^{g} = \alpha_i^{g} u_{i,t} + \beta_i^{g} P_{i,t} + \gamma_i^{g} P_{i,t}^2, \forall i \in GU \quad (19)$$

$$SU_{i,t} = su_i y_{i,t} \quad (20)$$

Constraint (21) limits the gas consumption of NGFUs, the information of this constraint is from GNO optimization sub-problem-A.

$$G_{i,t}^{g} \leq G_{i,t}^{g,0}, \forall i \in GU \quad (21)$$

For simplicity, the electric power transmission is modeled by (22)–(26) in DC power flow form. Constraint (22) shows the system power balance at each hour. Constraint (23) is the power balance for individual bus. Constraint (24) represents the DC power flow in branch br with first and last buses at j and k. Constraint (25) means line capacity limit, pf^{max} is power flow limits. Constraint (26) is the phase angle for reference bus.

$$\sum_{i \in GU} P_{i,t} + \sum_{i \in NGU} P_{i,t} + \sum_{r=1}^{N_R} P_{r,t} = \sum_{b=1}^{N_B} D_{b,t} \tag{22}$$

$$K_P \cdot P_i + K_W \cdot P_r - K_D \cdot D = K_L \cdot pf \tag{23}$$

$$pf_{br} = \frac{\theta_j - \theta_l}{x_{jl}}, (j, l \in br) \tag{24}$$

$$|pf_{br}| \leq pf_{br}^{max} \tag{25}$$

$$\theta_{ref} = 0 \tag{26}$$

(3) Demand response constraints

The price-elastic load in the electricity network is based on the model reported in References [54,55]. Equations (27)–(30) represent electrical load deviation and electricity price deviation in the demand response program. In this paper, the time-of-use electricity price is adopted to describe the price change in one day. In this demand response strategy, the satisfaction of customers should be considered. At the same time, customers' satisfaction cannot break the limit as shown in (31). Constraint (32) represents the load is shiftable and the sum of electrical load cannot be changed in one day.

$$D^{dev} = E \cdot p^{dev} \tag{27}$$

$$D^{dev} = [\Delta D_1 / D_1^{ini}, \cdots, \Delta D_{N_T} / D_{N_T}^{ini}] \tag{28}$$

$$p^{dev} = [\Delta p_1 / p_1^{ini}, \cdots, \Delta p_{N_T} / p_{N_T}^{ini}] \tag{29}$$

$$D_t = D_t^{ini} + \Delta D_t \tag{30}$$

$$CS = 1 - \frac{\sum_{t=1}^{N_T} |\Delta D_t|}{\sum_{t=1}^{N_T} |D_t^{ini}|} \geq CS^{min} \tag{31}$$

$$\sum_{t=1}^{N_T} D_t^{ini} = \sum_{t=1}^{N_T} D_t \tag{32}$$

After ENO completing optimization sub-problem, the actual gas consumption of NGFUs will be fixed and delivered to GNO for Step 3 as Equation (33) shown.

$$G_{i,t}^{g,1} = G_{i,t}^{g} = \mathrm{argmin}(obj\_E), \forall i \in GU \tag{33}$$

3.3. GNO Optimization Sub-Problem-B

As shown in (34), the objective function of GNO sub-optimization-B problem is to minimize the cost of gas wells. The resident natural gas load and NGFUs gas load are assumed to be fixed. The natural gas network constraints are similar to GNO sub-optimization-A, consisting of (2)–(6).

$$\mathrm{Min}\ obj\_Gb = \sum_{t=1}^{N_T} \sum_{s=1}^{N_S} \tau_s W_{s,t} \tag{34}$$

s.t Constraints (2)–(6).

Meanwhile, GNO updates the CR as shown in Equation (35).

$$CR_i^1 = 0.5 \left(CR_i^0 + \frac{\sum\limits_{t=1}^{N_T} G_{i,t}^{g,1}}{\sum\limits_{t=1}^{N_T} G_{i,t}^{g,0}} \right) \quad (35)$$

4. Solution Methodology

The Solution of Network Sub-Problem Solving

This work adopted an improved SCP method based on SOC with reasonable initial expansion value for natural gas network sub-problem. The method is suitable for both sub-problems A and B. The steps of solving are derived from the special form of the Weymouth function. Weymouth nonlinear steady-state pipeline flow of constraints (2)–(3) can be transformed as follows and shown in (36)–(38).

$$\left(I_{mn,t}^+ - I_{mn,t}^- \right)(\pi_{m,t} - \pi_{n,t}) = (1/C_{mn})^2 F_{mn,t}^2 \quad (36)$$

$$I_{mn,t}^+ + I_{mn,t}^- = 1 \quad (37)$$

$$\pi_m^{\min} \leq \pi_{m,t} \leq \pi_m^{\max} \quad (38)$$

According to [8], (36) is substituted by McCormick envelope (39)–(43) which converts the whole model into the MISOCP problem.

$$\Pi_{mn,t} \geq (1/C_{mn})^2 F_{mn}^2 \quad (39)$$

$$\Pi_{mn,t} \geq \pi_{n,t} - \pi_{m,t} + \left(I_{mn,t}^+ - I_{mn,t}^- + 1 \right)\left(\pi_m^{\min} - \pi_n^{\max} \right) \quad (40)$$

$$\Pi_{mn,t} \geq \pi_{m,t} - \pi_{n,t} + \left(I_{mn,t}^+ - I_{mn,t}^- - 1 \right)\left(\pi_m^{\max} - \pi_n^{\min} \right) \quad (41)$$

$$\Pi_{mn,t} \leq \pi_{m,t} - \pi_{n,t} + \left(I_{mn,t}^+ - I_{mn,t}^- - 1 \right)\left(\pi_m^{\min} - \pi_n^{\max} \right) \quad (42)$$

$$\Pi_{mn,t} \leq \pi_{n,t} - \pi_{m,t} + \left(I_{mn,t}^+ - I_{mn,t}^- + 1 \right)\left(\pi_m^{\max} - \pi_n^{\min} \right) \quad (43)$$

Constraints (39)–(43) will be equal to constraint (36) if (39) is tight. This work adopts a penalty function by combining SCP method to tight the constraint (39).

(1) Penalty function method

To tighten the inequality as much as possible under constraint (39), the objective function includes a penalty as shown in (44), to make the left term of inequality (39) as small as possible. This kind of solution is often not accurate enough but could be used to determine an initial solution. φ represents a pre-set positive number.

$$\text{Min} \sum_{t=1}^{N_T} \sum_{t=1}^{N_S} \tau_s W_{s,t} + \sum_{t=1}^{N_T} \sum_{mn=1}^{N_P} \varphi \Pi_{mn,t} \quad (44)$$

s.t Constraints (4) and (5), (37)–(43)

(2) Sequential cone programming method

In SCP method, extra concave constraint (45) is adopted to let constraints (39)–(43), (45) equivalent to constraint (36).

$$\Pi_{mn,t} \leq \left(\frac{1}{C_{mn}} \right)^2 F_{mn,t}^2 \quad (45)$$

As such, the key problem becomes the processing of Equation (45), and linearization is a common method. According to Taylor series, it can be approximated as (46). When the right-hand side value of the inequality is close to 0, the solution meets the expected requirements. Where S is an auxiliary variable.

$$\left\{ \begin{array}{l} \Pi_{mn,t}^k - \left(\frac{1}{C_{mn}}\right)^2 \left(F_{mn,t}^{k-1}\right)^2 - \\ \left(\frac{1}{C_{mn}}\right)^2 \cdot 2F_{mn,t}^{k-1} \cdot \left(F_{mn,t}^k - F_{mn,t}^{k-1}\right) \end{array} \right\} \leq S_{mn,t}^k \qquad (46)$$

(3) Solution procedure

The Algorithm 1 solution procedure is given as follows:

Algorithm 1: Improve SCP method for natural gas network sub-problem based on SOC

Step 1 Use penalty function method to solve natural gas steady-state flow problem to get the initial solution of gas flow F_{mn}^0. Set initial iteration number $k = 0$.
Min Min $\left\{ \sum_{t=1}^{N_T} \sum_{s=1}^{N_s} \tau_s W_{s,t} + \sum_{t=1}^{N_T} \sum_{mn=1}^{N_P} \varphi \Pi_{mn,t} \right\}$
s.t Constraints (4) and (5), (37)–(43)
Step 2 Parameter settings. Set a punish growth rate v and maximum penalty factor ϕ^{max}, SCP residual tolerance ξ^Z and ξ^S.
Step 3 Solve the following MISOCP problem:
$Z^k = \text{Min } Z^k = \text{Min}\left\{ \sum_{t=1}^{N_T} \sum_{s=1}^{N_s} \tau_s W_{s,t} + \sum_{t=1}^{N_T} \sum_{mn=1}^{N_P} \phi^{k-1} S_{mn,t}^k \right\}$
s.t Constraints (4) and (5), (37)–(43), (46)
Step 4 Check SCP residuals.
$\left| Z^k - Z^{k-1} \right| \leq \xi^Z$
$\max\left(\sum_{mn=1}^{N_P} S_{mn,t}^k \right) \leq \xi^S$
If yes, end the procedure. Otherwise, update the penalty factor and iteration number.
$\phi^h = \min\left(v\phi^{h-1}, \phi^{max} \right)$
Step 5 Set $k = k + 1$ and repeat *Steps 3–4* until the convergence conditions are met.

The gas flow directions will not change after several iterations. The binary variables in the gas network sub-problem (MISOCP) are fixed.

5. Case Studies and Discussions

All the cases are conducted on a Windows 10 64-bit personal computer with Intel Core i5-6500 3.2 GHz CPU and 8 GB of RAM using MATLAB R2014b in YALMIP with Gurobi 6.5 solver. A 6-bus electricity system and 6-node natural gas system is employed to examine the effectiveness of the proposed CO framework. Figure 6 presents the integrated topology. The test networks have one non-NGFU, seven transmission lines, two natural gas wells, five pipelines, three NGFUs, and one renewable energy source. The network considers varying electricity and natural gas loads, the networks test data can be found in http://motor.ece.iit.edu/data/ (accessed on 22 March 2021). Tables 1 and 2 give the gas source characteristics for S1 and S2 and natural gas line characteristics respectively.

Table 1. Natural gas source characteristics.

Source	Connection Node	Minimum Capacity (kcf/h)	Maximum Capacity (kcf/h)	Cost ($/kcf)
S1	4	2000	5000	3.2
S2	5	1500	6000	2.6

Table 2. Natural gas pipeline characteristics.

Pipeline Number	Start Node	End Node	Pipeline Constant (kcf/Psig)
1	2	1	50.6
2	4	2	50.1
3	5	2	37.5
4	5	3	43.5
5	6	5	45.3

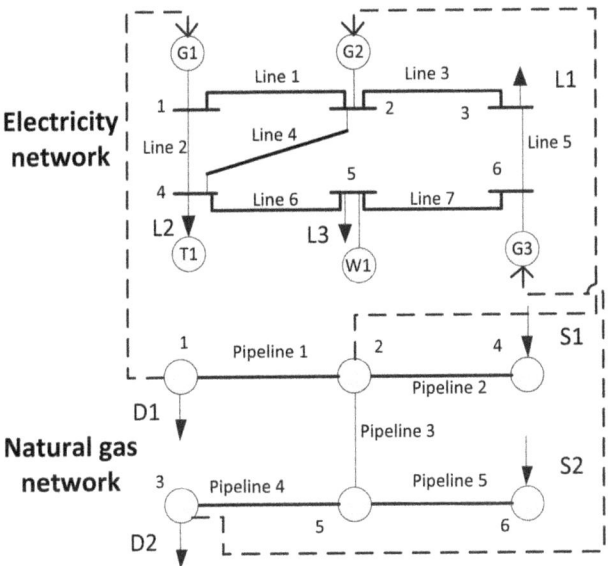

Figure 6. The topology of electricity and natural gas networks.

5.1. Comparison between DO, ADMM and CO

Figure 7 shows the variation of the original residuals and dual residuals with the iteration progress under the synchronous ADMM algorithm. In the 129 iterations, the residuals have a significant downward trend, showing a fluctuating decline. At the beginning of iterations, the magnitude of residual error is about 1e4, which indicates that there is a large gap in the coupling part of gas turbine. In the middle and later stage of the iteration, the residuals do not increase or decrease monotonically, but decrease gradually under small oscillation, and finally reach the convergence requirement.

The main idea of the proposed step-by-step cooperative operation method for a power gas interconnected system is the constructed credit degree and credit value parameters, and the credit value is directly determined by the credit degree. Therefore, this part of the example mainly studies the influence of credit degree on the effectiveness of the proposed method. First, it should be noted that, because the cost, the coefficient of G2 is higher than that of other units. The minimum output is maintained, so the credit degree is set to a fixed value of 1. It mainly analyzes the change and influence of credit degree of G1 and G3. Credit degree indicates the quantitative degree of the relationship between the actual gas consumption of gas units and the constraints generated by gas units. It is suitable for gas network managers to evaluate large gas users. Figure 8 shows the output of gas turbine G1 under different credit combinations under step-by-step coordinated operation, and the credit degree of G1 is 1-1. 9, the interval is 0. 1, G3 credit rating is 1-1. 9, the interval is 0. 1.

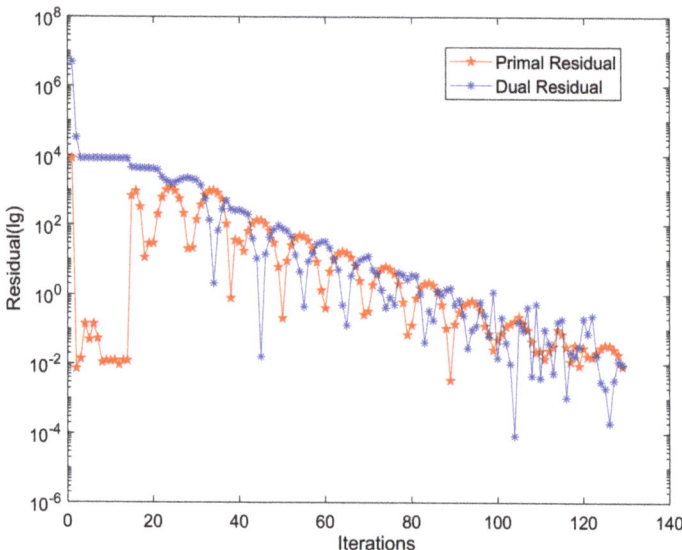

Figure 7. Residual change of synchronous ADMM.

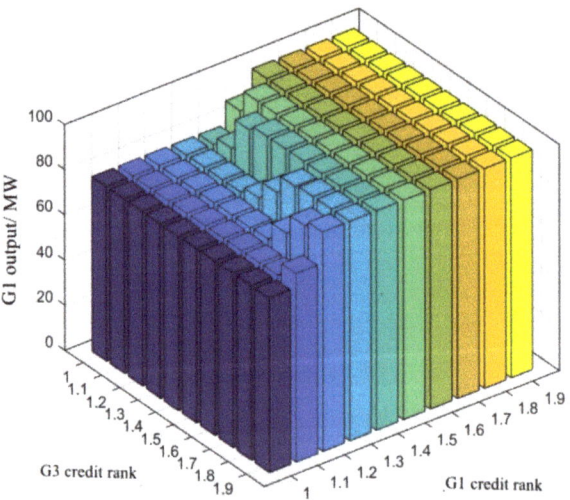

Figure 8. Residual change of synchronous ADMM.

A one-hour operation between electricity and the natural gas network was analyzed. An economic dispatch model was adopted in electricity system and an hourly steady-state gas flow model was adopted in the natural gas system [19]. Concurrently, the gas flow upper limit is similar for the comparison.

This study illustrates the overall advantage obtained from the proposed framework CO, as compared to the other usual methods for coordinated operation of natural gas and electricity networks.

Table 3 shows that performance comparison between DO, alternating direction method of multipliers (ADMM) and CO. It is observed that ADMM requires a longer time to obtain a solution. The convergence time for DO and CO are nearly the same. However, DO cannot produce better solutions under congestions and demand response situations, so CO gives the best overall performance. Due to the information asymmetry of the coupled gas units in the discrete operation, the natural gas required by the gas units in the power grid dispatching has no solution in the gas grid dispatching, that is, the dispatching is unbalanced. The dispatching unbalance of the discrete operation in the table comes from G1, and the specific reasons have been explained in the previous paper. In the operation cost of the two networks, due to the reduction in power generation and gas supply, the operation cost of discrete operation of the two networks is lower than that of other operation modes, but the existence of dispatching imbalance requires the power grid to pay a higher price. In terms of solution time, the optimization time of step-by-step collaborative operation is 1.04s, which is larger than that of discrete operation and much smaller than that of ADMM operation. The reason why the time difference is so large is that the optimization times of each mode are different. The discrete operation mode solves the electric network subproblem and the gas network subproblem once. The step-by-step cooperative operation model solves the electric network subproblem once and the gas network subproblem twice. When the synchronous ADMM runs for 129 iterations, the electric network subproblem and the gas network subproblem are solved for 129 iterations, the asynchronous ADMM only runs for 35 iterations. The solution of the gas network subproblem is the multi-layer iterative sequential cone optimization method, as proposed in this paper.

Table 3. Performance comparison.

Methods	DO	ADMM	Proposed Framework (CO)
CPU time (s)	1.51	241.39	2.29
Gas shortage (kcf)	280.38	0	0
Load shedding (MW)	17.76	0	0
Power generation cost ($)	22,358.3	22,969.2	22,969.5
Load shedding cost ($)	1176	0	0
Total cost of electricity ($)	23,534.3	22,969.5	22,969.5
Gas network income ($)	15,781.9	17,794.9	17,794.9

Under the research background of the collaborative operation of the electricity-gas interconnection system, this case shows the specific modes and advantages and disadvantages of the discrete operation of the electricity-gas interconnection system, the coordinated operation, based on the alternating direction multiplier method, and the proposed step-by-step collaborative operation method. The following observations are obtained:

1. The discrete operation mode of the electricity-gas interconnection system will produce unbalanced coupling dispatch when the gas network is blocked, and the gas generators cannot get enough natural gas supply, which does not meet the optimal economic dispatch decision-making results.
2. Based on the alternating direction multiplier method, the coordinated scheduling of the electricity-gas interconnection system can be realized. It can be divided into asynchronous and synchronous cooperative operation modes with or without the participation of third-party organizations. The above two modes can achieve coordinated scheduling when gas network pipelines are blocked, that is, gas generating units can get enough natural gas supply, and there will be no unbalanced coupling scheduling due to asymmetric information.
3. This case proposes a new idea of realizing step-by-step coordinated operation, based on credit indicators, and establishes a corresponding mathematical model and specific operation framework. This operating mode belongs to the concept of distributed scheduling. Through case analysis, the feasibility and superiority of the proposed method are discussed, and the sensitivity analysis of credit changes is carried out. Compared with other methods, the proposed method has the advantages of simplicity, reliability, and strong applicability.

5.2. Impact Due to Congestion and Demand Response

In this simulation, the impact due to congestion and demand response, with the same fuel price for all units, is adopted for investigation. Load shedding is added to avoid imbalance between load and generation. It is assumed that all NGFUs have an initial credit rank of 0.5, the entire scheduling time is for four days, and an hour is taken as a time step.

The advantages of the proposed framework applied to daily scheduling for a few days are illustrated with the following case studies:

Case 1: Decentralized and coordinated operation without natural gas pipelines congestion.
Case 2: Decentralized and coordinated operation with natural gas pipelines congestion.
Case 3: Decentralized and coordinated operation considering demand response (Based on Case 2).

For Case 1, the natural gas supply capacity of the natural gas network to NGFUs is redundant. Because the gas consumption of NGFUs in ENO optimal scheduling is feasible for GNO, so the cost of ENO in DO is minimum and with no-load shedding.

The hourly cost, total cost of ENO, and hourly generation of G1 are shown in Figure 9. It can be seen that DO hourly cost of the electricity network is less than CO for day 1, the main reason is due to the CR parameter is introduced in CO and has an initial credit rank of 0.5 at first day of optimization. CO has not reached the optimization. In Case 1, without natural gas pipelines congestion, the gas consumption in ENO optimization scheduling is feasible for GNO. Therefore, the cost of ENO in DO is the smallest. After updating the CR, DO total cost of the electricity network is same as that of CO, the scheduling results of the two coordinated operation methods are consistent.

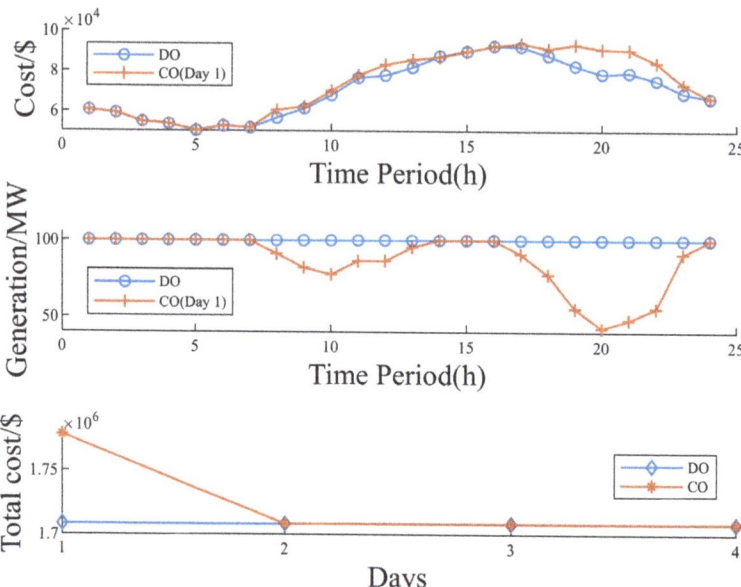

Figure 9. Hourly and total cost of ENO, hourly generation of G1.

Table 4 shows the gap of the iteration process with various expansion points. SCP with initial point achieved by the penalty function is denoted as penalty function (PF) initial. Zero initial means SCP with zero initial points [56]. The gap of the proposed method reduces greatly, and the accuracy will converge 5 iterations later. The zero-point method will converge after 6 iterations. The proposed method has quicker convergence characteristics, while the zero-point method convergence is slower [57].

Table 4. Gap of iteration process with various expansion point.

	Iteration	1	2	3	4	5	6
Gap-S	PF point	1385.308	2.707	0.652	0.366	0.048	N/A
	Zero point	707557.820	471.389	365.477	205.691	0.300	0.038
Gap-Z	PF point	143.076	139.152	0.730	0.552	0.090	N/A
	Zero point	305592.919	70906.237	48.071	15.128	164.559	0.067

Case 2: According to steady-state natural gas flow model, the gas flow in pipelines depends on the first and last node pressure. To simplify the debugging process, we included the gas flow upper limit on pipeline 1 and pipeline 4 in this case.

Pipeline congestion at hours 19–22 is a peak period for both electricity and gas networks in decentralized operation. Pipeline congestion creates the difference between expected generation and actual generation of NGFUs in DO. Figure 10 shows the output of gas-fired units with the optimal economic dispatch of the power grid and the discrete operation of the electricity-gas interconnection system. For the gas unit G1, under the condition of economic dispatch, the output has been kept at full capacity, but when considering the natural gas network constraints, in the discrete operation mode, due to the blockage of pipeline 1 during the period 19–22, G1 cannot get enough gas supply. The output is affected, and there is an imbalance in coupled scheduling. For gas-fired unit G3, in the discrete operation mode, pipeline 3 was blocked during the period 19–21. G3 could not get enough gas supply and there was also unbalanced coupling scheduling. This shows that in the case of natural gas network constraints, once the gas network pipeline is blocked, it will have a negative impact on the grid dispatching. The actual power generation is fewer than the expected one in hour 19–22, and part of the load is not satisfied in DO, so load shedding has to take place. CO total electricity cost is slightly higher than that in DO on the first day, but there is no load shedding. Electricity operation cost will be minimum on the next day in the proposed framework.

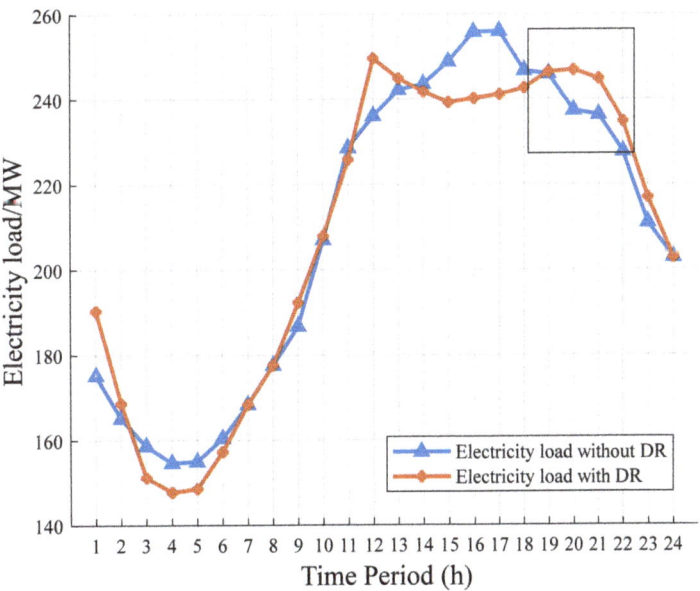

Figure 10. Electricity load profile without and with demand response.

Case 3: The demand response program is considered, and the electricity load is shown in Figure 10. There is increased load in the pipeline congestion period when DR becomes effective. This will cause a more serious gas shortage of NGFUs, and insufficient generation of electricity network as compared to Case 2. Table 5 provides gas shortage and load shedding information with different operation modes. The unbalanced amount of coupled dispatch increased due to the addition of demand response action. The reason is that, due to the consideration of time-of-use electricity prices, electricity price-sensitive loads will be adjusted according to changes in electricity prices, and electricity demand will increase during the period when the gas network pipeline is blocked. This adjustment supply gap is enlarged. As shown in Figure 11, due to the addition of price-based demand response action, the power demand shifts from the peak period (14–19 h) to the valley period (3–6 h), and the load in the valley period increases. The peak load is reduced, so the peak-valley difference is reduced, and the effect of demand response is reflected. However, at 20–22 h, because the demand response action increases the power demand, the load curve is larger than the original curve, so the unbalanced amount of coupled dispatch increases. At 19 h, due to the peak load period, regardless of whether demand response is considered or not, G1 and G3 remain at full output. The impact of the gas network on scheduling does not change due to demand response, so the unbalanced amount of coupled scheduling remains unchanged.

Table 5. Gas shortage and load shedding in DO.

	Case 2		Case 3	
	DO	CO	DO	CO
Gas shortage of NGFUs (kcf)	428.3	0	728.8	0
Load shedding (MW)	29.7	0	46	0

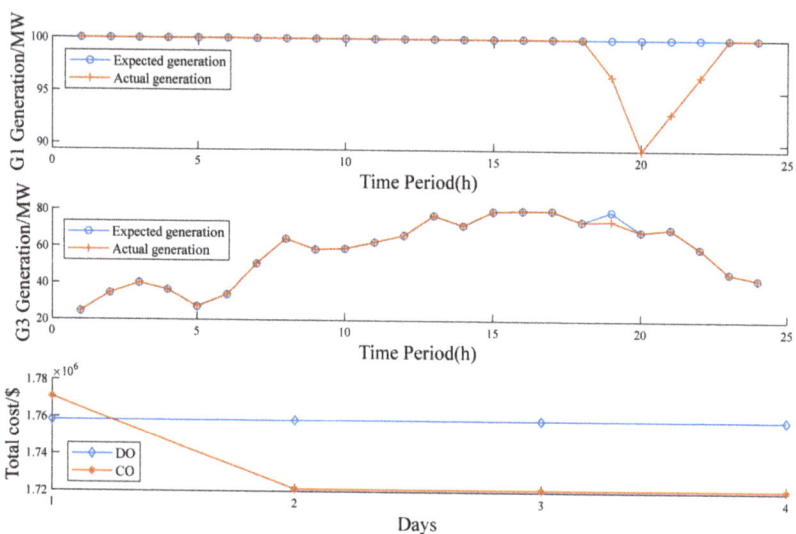

Figure 11. Hourly generation and total cost of ENO.

Table 6 shows the total cost of the electricity network with different operation modes in different cases. In DO pipeline congestion cases, the addition of a demand response program not only promotes the reduction in costs but also increases profit. CO mode can solve this problem with a great extent, not only minimizing the negative impact of DR, but also reducing load shedding.

Table 6. The total cost of the electricity network.

	Decentralized Operation		Coordinated Operation	
	Without DR	With DR	Without DR	With DR
Without pipelines congestion	1708547.2 $	1687651.1 $	1708547.2 $	1687645.0 $
Pipelines congestion	1762052.3 $	1770473.1 $	1721257.2 $	1700798.7 $

The work demonstrates that demand response can achieve benefits, for example, to reduce congestion so as to improve energy supply security and enhance low-carbon economy. The use of gas storage is no double can improve co-ordination with electricity network operation. This gives a direction for future work to study the gas storage investment in detail. A similar approach derived from electrical energy storage (EES) may be considered. At present, EES, such as lithium-ion (Li-ion) batteries, can reduce curtailment of renewables, maximizing renewable utilization by storing surplus electricity. Several techno-economic analyses have been performed on EES, but few have investigated the financial performance. However, [58] presents a state-of-the-art financial model obtaining novel and significant financial and economics results when applied to Li-ion EES. A discounted cash flow model for the Li-ion EES is introduced and applied to examine the financial performance of different EES operating scenarios. It is expected that similar approaches could be investigated for investment purposes for gas storage [59,60]. It is believed that co-ordination of electricity and gas networks could benefit the transition from fossil fuel energy to clean energy with net-zero emission by 2050.

6. Conclusions and Future Work

This paper proposed an innovative coordinated operation framework for natural gas and electricity networks to effectively solve the optimization problem of the operation of integrated gas and electricity systems. The modeling approach developed is applied to demonstrate the benefits of an integrated approach to the operation of interdependent gas and electricity systems. In addition, the novel coordinated operation framework changes the traditional decision sequence. GNO as an advance decision-maker provides NGFUs gas consumption constraints to ENO. With this framework, it is possible to avoid frequent information interaction and without a third coordinator. This framework will not affect the local optimal scheduling for both the electricity network and the natural gas network. The modeling indicates that, in pipeline congestion situations, this framework for CO reduces the impact on the electricity network as compared to that of DO. At the same time, it improves DR program acceptance of electricity networks and prevents massive load shedding. The proposed framework can produce more cost-effective solutions as compared to other methods, such as ADMM. It was demonstrated that CO with demand response can reduce the electricity network total cost by 1.19% when pipeline congestion occurs, as compared to -0.48% using traditional decentralized operation with demand response.

Future work will investigate the application of a coordinated operation framework to a very large-scale, multi-area system with gas and electricity networks managed by different operators. Sensitivity analysis of credit rank will also be studied in detail. The credit rank will be further developed to form in the hope of a guideline or recommended practices. Gas storage in the view of technical, economic, social, and environmental aspects will be looked at.

Author Contributions: Conceptualization, C.S.L., X.L. and L.L.L.; methodology, C.S.L., M.Y., L.L.L. and Y.X.; software, Y.X.; formal analysis, C.S.L., M.Y. and Y.X.; investigation, C.S.L., M.Y. and Y.X.; resources, C.S.L., X.L. and L.L.L.; data curation, C.S.L. and Y.X.; writing—original draft preparation, C.S.L., L.L.L. and Y.X.; writing—review and editing, C.S.L., X.L., and L.L.L.; supervision, L.L.L.;

project administration, X.L.; funding acquisition, X.L. and L.L.L. All authors have read and agreed to the published version of the manuscript.

Funding: The work is sponsored by the Department of Finance and Education of Guangdong Province 2016 [202]: Key Discipline Construction Program, China; and the Education Department of Guangdong Province: New and Integrated Energy System Theory and Technology Research Group [Project Number 2016KCXTD022].

Institutional Review Board Statement: Not applicable.

Informed Consent Statement: Not applicable.

Data Availability Statement: Not applicable.

Conflicts of Interest: The authors declare no conflict of interest.

Nomenclature

Indices and Sets:

t	Index of time periods
i	Index of thermal units
b	Index of buses
r	Index of renewable energy
br	Index of power transmission lines
j, l	Index of buses in the electricity network
m, n	Index of gas nodes
s	Index of gas wells
mn	Index of gas pipelines
gl	Index of resident natural gas load
GU	Set of NGFUs
NGU	Set of non-NGFUs

Parameters:

N_T	Number of time periods
N_B	Number of thermal units
N_R	Number of renewable power generators
N_P	Number of gas pipelines
N_S	Number of gas wells
N_L	Number of resident gas load
P_i^{min}, P_i^{max}	Minimum and maximum output of unit i
UR_i, DR_i	Ramp up/down limits of unit i
T_i^{on}, T_i^{off}	Minimum on/off time of unit i
$\alpha^c, \beta^c, \gamma^c$	Cost coefficient of non-NGFUs
$\alpha^g, \beta^g, \gamma^g$	Fuel coefficient of NGFUs
su_i	Startup cost coefficient of unit i
η	Natural gas contract price of NGFUs.
$P_{r,t}^W$	Forecast of renewable energy r at hour t
pf_{br}^{max}	Maximum power flow of power transmission line br
x_{jk}	Reactance between bus j and k
D^{ini}	Initial electrical load
p^{ini}	Initial electricity price
E	Price-elastic matrix of electrical load
CS^{min}	Minimum customer's satisfaction
C_{mn}	Weymouth constant of pipelines
W_s^{min}, W_s^{max}	Minimum and maximum production of gas well s
τ_s	Cost coefficient of gas well s
τ_{gl}	Gas price of resident natural gas load
L	Resident natural gas load
K_P	Bus-thermal unit incidence matrix
K_W	Bus-renewable unit incidence matrix
K_D	Bus-electrical load incidence matrix

K_L	Bus-branch incidence matrix
T^w	Node-gas well incidence matrix
T^g	Node-NGFUs incidence matrix
T^l	Node-resident natural gas load incidence matrix
T^f	Node-gas pipe incidence matrix
φ	Pre-set constants

Variables:

$G_{i,t}^c$	Cost of non-NGFU i at hour t
$G_{i,t}^g$	Fuel consumption of NGFU i at hour t in electricity network
$P_{i,t}$	Generation dispatch of unit i at hour t
$P_{r,t}$	Generation dispatch of renewable energy r at hour t
$SU_{i,t}$	Startup cost of unit i at hour t
$u_{i,t}$	Status indicator of unit i at hour t
$y_{i,t}, z_{i,t}$	Indicator for startup/shutdown of unit i at hour t
$X_{i,t}^{on}, X_{i,t}^{off}$	On/Off time of unit i at hour t
θ	Bus voltage angle
pf_{br}	Power flow on transmission line br
D	Adjusted electrical load
ΔD_t	Variety in electrical load at hour t
Δp_t	Variety in electricity price at hour t
D^{dev}	Deviation matrix of electrical load
p^{dev}	Deviation matrix of electricity price
CS	Electricity customer's satisfaction
$W_{s,t}$	Production of gas well s at hour t
$\omega_{m,t}$	Gas pressure of gas node m at hour t
$\pi_{m,t}$	Quadratic pressure of gas node m at hour t
$F_{mn,t}$	Gas flow of pipeline mn at hour t
$I_{mn,t}^+, I_{mn,t}^-$	Binary indicators of gas flow direction of pipeline mn at hour t
Π, S	Auxiliary variable
λ^0, λ^1	Initial and updated credit value factor
CR^0, CR^1	Initial and updated credit rank

References

1. Lai, C.S.; Lai, L.L.; Lai, Q.H. *Data Analytics for Solar Energy in Promoting Smart Cities, Smart Grids and Big Data Analytics for Smart Cities*; Springer: Berlin/Heidelberg, Germany, 2020; pp. 173–263.
2. Lai, C.S.; Lai, L.L.; Lai, Q.H. *Blockchain Applications in Microgrid Clusters, Smart Grids and Big Data Analytics for Smart Cities*; Springer: Berlin/Heidelberg, Germany, 2020; pp. 265–305.
3. Lai, C.S.; Lai, L.L.; Lai, Q.H. *Narrowband Internet of Thing-Based Temperature Prediction for Valve-Regulated Lead Acid Battery, Smart Grids and Big Data Analytics for Smart Cities*; Springer: Berlin/Heidelberg, Germany, 2020; pp. 345–363.
4. Esfandi, S.; Baloochzadeh, S.; Asayesh, M.; Ehyaei, M.A.; Ahmadi, A.; Rabanian, A.A.; Das, B.; Costa, V.A.F.; Davarpanah, A. Energy, exergy, economic, and exergoenvironmental analyses of a novel hybrid system to produce electricity, cooling, and syngas. *Energies* **2020**, *13*, 6453. [CrossRef]
5. Clegg, S.; Mancarella, P. Integrated electrical and gas network flexibility assessment in low-carbon multi-energy systems. *IEEE Trans. Sustain. Energy* **2016**, *7*, 718–731. [CrossRef]
6. Alizadeh, S.M.; Ghazanfari, A.; Ehyaei, M.A.; Ahmadi, A.; Jamali, D.H.; Nedaei, N.; Davarpanah, A. Investigation the Integration of Heliostat Solar Receiver to Gas and Combined Cycles by Energy, Exergy, and Economic Point of Views. *Appl. Sci.* **2020**, *10*, 5307. [CrossRef]
7. Shahidehpour, M.; Fu, Y.; Wideman, T. Impact of natural gas infrastructure on electric power systems. *Proc. IEEE* **2005**, *93*, 1042–1056. [CrossRef]
8. Clean Energy Wire. Electricity and Gas Grids Need to Be Better Aligned in New Energy System. Available online: https://www.cleanenergywire.org/news/electricity-and-gas-grids-need-be-better-aligned-new-energy-system (accessed on 22 March 2021).
9. Henderson, M.; Shahidehpour, M. Continuing to grow: Natural gas usage rising in electricity generation [Guest Editorial]. *IEEE Power Energy Mag.* **2014**, *12*, 12–19. [CrossRef]
10. Wen, Y.; Qu, X.; Li, W.; Liu, X.; Ye, X. Synergistic operation of electricity and natural gas networks via ADMM. *IEEE Trans. Smart Grid* **2018**, *9*, 4555–4565. [CrossRef]
11. Chaudry, M.; Jenkins, N.; Qadrdan, M.; Wu, J.Z. Combined gas and electricity network expansion planning. *Appl. Energy* **2014**, *113*, 1171–1187. [CrossRef]

12. Li, T.; Eremia, M.; Shahidehpour, M. Interdependency of natural gas network and power system security. *IEEE Trans. Power Syst.* **2008**, *23*, 1817–1824. [CrossRef]
13. Wang, D.; Qiu, J.; Meng, K.; Gao, X.; Dong, Z.Y. Coordinated expansion co planning of integrated gas and power systems. *J. Mod. Power Syst. Clean Energy* **2017**, *5*, 314–325. [CrossRef]
14. Qiu, J.; Yang, H.; Dong, Z.Y.; Zhao, J.H.; Meng, K.; Luo, F.J.; Wong, K.P. A linear programming approach to expansion co-planning in gas and electricity markets. *IEEE Trans. Power Syst.* **2016**, *31*, 3594–3606. [CrossRef]
15. He, Y.; Yan, M.; Shahidehpour, M.; Li, Z.; Guo, C.; Wu, L.; Ding, Y. Decentralized optimization of multi-area electricity-natural gas flows based on cone reformulation. *IEEE Trans. Power Syst.* **2018**, *33*, 4531–4542. [CrossRef]
16. Canelas, E.; Pinto-Varela, T.; Sawik, B. Electricity portfolio optimization for large consumers: Iberian electricity market case study. *Energies* **2020**, *13*, 2249. [CrossRef]
17. Hosseini, S.E.; Ahmarinejad, A. Stochastic framework for day-ahead scheduling of coordinated electricity and natural gas networks considering multiple downward energy hubs. *J. Energy Storage* **2021**, *33*, 102066. [CrossRef]
18. Hemmati, M.; Abapour, M.; Mohammadi-Ivatloo, B.; Anvari-Moghaddam, A. Optimal operation of integrated electrical and natural gas networks with a focus on distributed energy hub systems. *Sustainability* **2020**, *12*, 8320. [CrossRef]
19. Manshadi, S.D.; Khodayar, M.E. Coordinated operation of electricity and natural gas systems: A convex relaxation approach. *IEEE Trans. Smart Grid* **2019**, *10*, 199–210. [CrossRef]
20. Bai, L.; Li, F.; Jiang, T.; Jia, H. Robust scheduling for wind integrated energy systems considering gas pipeline and power transmission N-1 contingencies. *IEEE Trans. Power Syst.* **2017**, *32*, 1582–1584. [CrossRef]
21. He, C.; Wu, L.; Liu, T.; Shahidehpour, M. Robust co-optimization scheduling of electricity and natural gas systems via ADMM. *IEEE Trans. Sustain. Energy* **2017**, *8*, 658–670.
22. An, S.; Li, Q.; Gedra, T.W. Natural gas and electricity optimal power flow. In Proceedings of the 2003 IEEE PES Transmission and Distribution Conference and Exposition (IEEE Cat. No.03CH37495), Dallas, TX, USA, 7–12 September 2003; pp. 138–143.
23. Liu, C.; Shahidehpour, M.; Fu, Y.; Li, Z. Security-constrained unit commitment with natural gas transmission constraints. *IEEE Trans. Power Syst.* **2009**, *24*, 1523–1536.
24. Alabdulwahab, A.; Abusorrah, A.; Zhang, X.; Shahidehpour, M. Coordination of interdependent natural gas and electricity infrastructures for firming the variability of wind energy in stochastic day-ahead scheduling. *IEEE Trans. Sustain. Energy* **2015**, *6*, 606–615. [CrossRef]
25. Bai, L.; Li, F.; Cui, H.; Jiang, T.; Sun, H.; Zhu, J. Intervaloptimization based operating strategy for gas-electricity integrated energy systems considering demand response and wind uncertainty. *Appl. Energy* **2016**, *167*, 270–279. [CrossRef]
26. Zhang, X.; Shahidehpour, M.; Alabdulwahab, A.; Abusorrah, A. Hourly electricity demand response in the stochastic day-ahead scheduling of coordinated electricity and natural gas networks. *IEEE Trans. Power Syst.* **2016**, *31*, 592–601. [CrossRef]
27. Zhao, B.; Zlotnik, A.; Conejo, A.J.; Sioshansi, R.; Rudkevich, A.M. Shadow price-based co-ordination of natural gas and electric power systems. *IEEE Trans. Power Syst.* **2019**, *34*, 1942–1954. [CrossRef]
28. Morals, M.S.; Lima, J.W.M. Natural gas network pricing and its influence on electricity and gas markets. In Proceedings of the 2003 IEEE Bologna Power Tech Conference, Bologna, Italy, 23–26 June 2003.
29. Wang, C.; Wei, W.; Wang, J.; Liu, F.; Mei, S. Strategicoffering and equilibrium in coupled gas and electricity markets. *IEEE Trans. Power Syst.* **2018**, *33*, 290–306. [CrossRef]
30. Shao, C.; Wang, X.; Shahidehpour, M.; Wang, X.; Wang, B. An MILP-based optimal power flow in multicarrier energy systems. *IEEE Trans. Sustain. Energy* **2017**, *8*, 239–248. [CrossRef]
31. Urbina, M.; Li, Z. A combined model for analyzing the interdependency of electrical and gas systems. In Proceedings of the 2007 39th North American Power Symposium, Las Cruces, NM, USA, 30 September–2 October 2007; pp. 468–472.
32. Andre, J.; Bonnans, F.; Cornibert, L. Optimization of capacity expansion planning for gas transportation networks. *Eur. J. Oper. Res.* **2009**, *197*, 1019–1027. [CrossRef]
33. Xu, F.Y.; Zhang, T.; Lai, L.L.; Zhou, H. Shifting boundary for price-based residential demand response and applications. *Appl. Energy* **2015**, *146*, 353–370. [CrossRef]
34. Subramanian, A.S.R.; Gundersen, T.; Adams, T.A. Modeling and simulation of energy systems: A review. *Processes* **2018**, *6*, 238. [CrossRef]
35. Lai, S.C.; Locatelli, G.; Pimm, A.; Wu, X.; Lai, L.L. A review on long-term electrical power system modeling with energy storage. *J. Clean. Prod.* **2021**, *280*, 124298. [CrossRef]
36. Zhao, B.; Conejo, A.J.; Sioshansi, R. Unit commitment under gas supply uncertainty and gas-price variability. *IEEE Trans. Power Syst.* **2017**, *32*, 2394–2405. [CrossRef]
37. Chertkov, M.; Fisher, M.; Backhaus, S.; Bent, R.; Misra, S. Pressure fluctuations in natural gas networks caused by gas-electric coupling. In Proceedings of the 48th Hawaii International Conference on System Sciences, Hawaii, HI, USA, 5–8 January 2015; pp. 2738–2747. [CrossRef]
38. Hejazi, A.; Mashhadi, H. Effects of natural gas network on optimal operation of gas-fired power plants. In Proceedings of the 6th Conference Thermal Power Plants, Teheran, Iran, 19–20 January 2016; pp. 105–110. [CrossRef]
39. Behrooz, H.; Boozarjomehry, R. Dynamic optimization of natural gas networks under customer demand uncertainties. *Energy* **2017**, *134*, 968–983. [CrossRef]

40. Qadrdan, M.; Wu, J.; Jenkins, N.; Ekanayake, J. Operating strategies for a GB integrated gas and electricity network considering the uncertainty in wind power forecasts. *IEEE Trans. Sustain. Energy* **2014**, *5*, 128–138. [CrossRef]
41. Odetayo, B.; MacCormack, J.; Rosehart, W.; Zareipour, H. A sequential planning approach for distributed generation and natural gas networks. *Energy* **2017**, *127*, 428–437. [CrossRef]
42. Correa-Posada, C.; Sánchez-Martín, P. Integrated power and natural gas model for energy adequacy in short-term operation. *IEEE Trans. Power Syst.* **2015**, *30*, 3347–3355. [CrossRef]
43. Zlotnik, A.; Roald, L.; Backhaus, S.; Chertkov, M.; Andersson, G. Coordinated scheduling for interdependent electric power and natural gas infrastructures. *IEEE Trans. Power Syst.* **2017**, *32*, 600–610. [CrossRef]
44. Li, G.; Zhang, R.; Jiang, T.; Chen, H.; Bai, L.; Li, X. Security-constrained bi-level economic dispatch model for integrated natural gas and electricity systems considering wind power and power-to-gas process. *Appl. Energy* **2016**, *194*, 696–704. [CrossRef]
45. Liu, C.; Shahidehpour, M.; Wang, J. Application of augmented Lagrangian relaxation to coordinated scheduling of interdependent hydrothermal power and natural gas systems. *IET Gener. Transm. Distrib.* **2010**, *4*, 1314–1325. [CrossRef]
46. Alkano, D.; Scherpen, J. Distributed supply coordination for Powerto-Gas facilities embedded in energy grids. *IEEE Trans. Smart Grid.* **2017**, *9*, 1012–1022. [CrossRef]
47. Khaligh, V.; Buygi, M.O.; Moghaddam, A.A.; Guerrero, J.M. Integrated expansion planning of gas-electricity system: A case study in Iran. In Proceedings of the 2018 International Conference on Smart Energy Systems and Technologies (SEST), Sevilla, Spain, 10–12 September 2018; pp. 1–6.
48. International Gas Union. *Underground Storage of Gas*. Report of Working Group Committee 2, International Gas Union, June 2006. Available online: https://bgc.bg/upload_files/file/UGS.pdf (accessed on 1 March 2021).
49. Observ'ER. *Worldwide Electricity Production from Renewable Energy Sources—Stats and Figures Series*; Observ'ER: Paris, France, 2013.
50. Ojeda-Esteybar, D.M.; Rubio-Barros, R.G.; Vargas, A. Integrated operational planning of hydrothermal power and natural gas systems with large scale storages. *J. Mod. Power Syst. Clean Energy* **2017**, *5*, 299–313. [CrossRef]
51. Zeng, Q.; Zhang, B.; Fang, J.; Chen, Z. A bi-level programming for multistage co-expansion planning of the integrated gas and electricity system. *Appl. Energy* **2017**, *200*, 192–203. [CrossRef]
52. Saldarriaga, C.A.; Hincapié, R.A.; Salazar, H. A holistic approach for planning natural gas and electricity distribution networks. *IEEE Trans. Power Syst.* **2013**, *28*, 4052–4063. [CrossRef]
53. Sánchez, C.B.; Bent, R.; Backhaus, S.; Blumsack, S.; Hijazi, H.; Hentenryck, P.V. Convex optimization for joint expansion planning of natural gas and power systems. In Proceedings of the 49th Hawaii International Conference on System Sciences (HICSS), Koloa, HI, USA, 5–8 January 2016; pp. 2536–2545.
54. Saldarriaga-Cortés, C.; Salazar, H.; Moreno, R.; Jiménez-Estévez, G. Stochastic planning of electricity and gas networks: An asynchronous column generation approach. *Appl. Energy* **2019**, *233–234*, 1065–1077. [CrossRef]
55. Kirschen, D.S.; Strbac, G.; Cumperayot, P.; Mendes, D.D. Factoring the elasticity of demand in electricity prices. *IEEE Trans. Power Syst.* **2000**, *15*, 612–617. [CrossRef]
56. Xu, Y.; Zhao, F.; Lai, L.L.; Wang, Y. Integrated electricity and natural gas system for day-ahead scheduling. In Proceedings of the 2019 IEEE International Conference on Systems, Man and Cybernetics (SMC), Bari, Italy, 6–9 October 2019; pp. 2242–2247.
57. Xu, Y.; Lai, L.L.; Zhao, F.; Wang, Y.; Li, X.; Lai, C.S.; Xu, F.Y. Coordinated operation of gas and electricity networks. In Proceedings of the International Conference on Applied Energy 2019, Vasteraås, Sweden, 12–15 August 2019.
58. Lai, C.S.; Locatelli, G.; Pimm, A.; Tao, Y.; Li, X.; Lai, L.L. A financial model for lithium-ion storage in a photovoltaic and biogas energy system'. *Appl. Energy* **2019**, *251*, 113179. [CrossRef]
59. Lai, C.S.; Jia, Y.; Xu, Z.; Lai, L.L.; Li, X.; Cao, J.; McCulloch, M.D. Levelized cost of electricity for photovoltaic/biogas power plant hybrid system with electrical energy storage degradation costs. *Energy Convers. Manag.* **2017**, *153*, 34–47. [CrossRef]
60. Lai, C.S.; McCulloch, M.D. Sizing of stand-alone solar PV and storage system with anaerobic digestion biogas power plants. *IEEE Trans. Ind. Electron.* **2016**, *64*, 2112–2121. [CrossRef]

MDPI
St. Alban-Anlage 66
4052 Basel
Switzerland
Tel. +41 61 683 77 34
Fax +41 61 302 89 18
www.mdpi.com

Applied Sciences Editorial Office
E-mail: applsci@mdpi.com
www.mdpi.com/journal/applsci

www.ingramcontent.com/pod-product-compliance
Lightning Source LLC
LaVergne TN
LVHW070140100526
838202LV00015B/1864